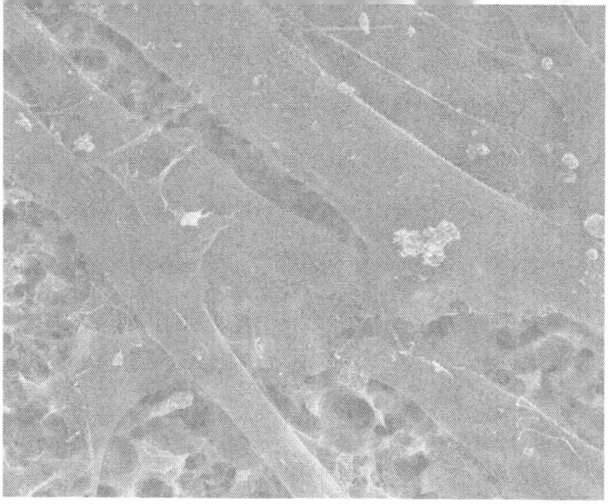

Bionanomaterials for Dental Applications

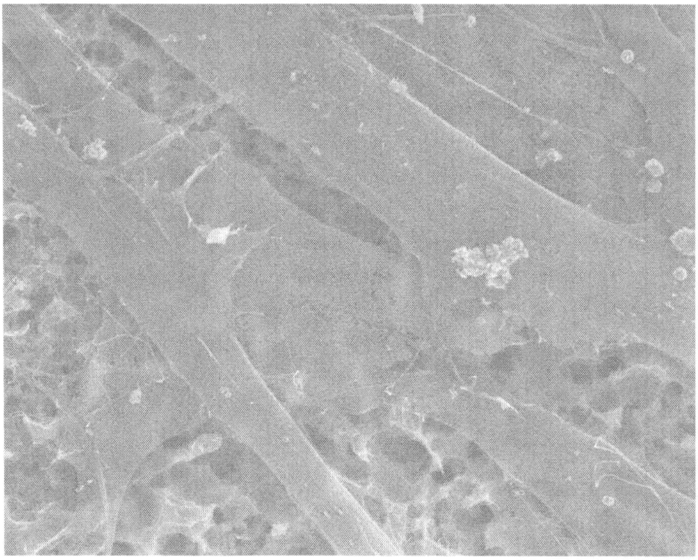

Bionanomaterials for Dental Applications

edited by

Mieczyslaw Jurczyk

PAN STANFORD PUBLISHING

Published by

Pan Stanford Publishing Pte. Ltd.
Penthouse Level, Suntec Tower 3
8 Temasek Boulevard
Singapore 038988

Email: editorial@panstanford.com
Web: www.panstanford.com

British Library Cataloguing-in-Publication Data
A catalogue record for this book is available from the British Library.

Bionanomaterials for Dental Applications

ISBN 978-981-4303-83-5 (Hardcover)
ISBN 978-981-4303-84-2 (eBook)

Printed in the USA

Contents

Preface

Nanotechnology involves the precise manipulation and control of atoms, the building elements of all matter, to create new materials. It is widely accepted that this technology is developing into a major driver for commercial success in the 21st century. Over the last decade, the use of nanostructured metallic and ceramic materials has already changed the approach to materials design in many applications, by seeking structural control at the atomic level and by the tailoring of the mechanical engineering, physico-chemical, and biological properties. Today, it is possible to prepare metal and ceramic nanocrystals with nearly monodispersive size distribution. Nanomaterials demonstrate novel properties compared with conventional (microcrystalline) materials owing to their nanoscale features.

Recently, the mechanical alloying method and the powder metallurgy process for the fabrication of metal/alloy–ceramic nanocomposites with a unique microstructure have been developed. The processes permit the control of microstructural properties such as the size of pore openings, surface properties, and the nature of the base metal/alloy. The availability of large amounts of specifically tailored nanostructured metal/alloy-based powders is crucial for the successful development of new dental implants.

One of the potential applications of nanostructured materials is dentistry. Although Ti is widely used for clinical purposes, some unresolved issues still remain. The clinical failure rate for implant materials occurs in the range from a few to over 10%.

The lack of sufficient bonding of synthetic implants to surrounding body tissues has led to the investigations of novel material formulations. Nanomaterials can be used to synthesize implants with surface roughness similar to that of natural tissues. The mechanical properties are improved, and what is more, the book highlights the enhanced cytocompatibility of the nanomaterials, leading to increased tissue regeneration.

The present research aims to fabricate porous scaffolds to promote bone or tissue ingrowth into pores and provide biological anchorage. Several factors have shown their influence on bone ingrowth into porous implants, such as porous structure (pore size, pore shape, porosity, and interconnecting pore size) of the implant, duration of implantation, biocompatibility, implant stiffness, and micromotion between the implant and the adjacent bone. The architecture of a porous implant has been suggested to have a great effect on implant integration by newly grown bone.

This book is our contribution to this innovative area of bionanomaterials and bionanocomposites for dental applications. Wherever possible, we used our own results to illustrate the discussed subject. The content of this book is classified into 13 chapters. The first chapter emphasizes the motivation for the transformation to the bionanomaterials and synthesis of nanomaterials, aiming at describing the principles and approaches of the synthesis techniques. We provide a comprehensive history of the development of biomaterials, including the existing fabrication methods, with special emphasis on ball milling in high-energy mills. The second chapter focuses on the stomatognathic system. In Chapters 3 and 4, we review the properties of selected biomaterials and the application of nanotechnology in dental materials. Chapter 5 presents a thorough review of the corrosion of metallic biomaterials and implants. The book also describes Ni-free austenitic stainless steel–hydroxyapatite nanocomposites (Chapter 6), Ti-based ceramic nanocomposites (Chapter 7), and shape memory Ni–Ti materials (Chapter 8). Chapters 9 and 10 provide information on the surface treatment of Ti-based nanocrystalline biomaterials and carbon materials. The present state of knowledge related to nanomaterials in preventive dentistry and osteoblast behavior on nanostructured metal implants are presented in Chapters 11 and 12. Chapter 13, the last chapter, focuses on the application of bulk nanostructured materials in dentistry. The objective is to show their unique properties.

Our goal is to provide comprehensive and complete knowledge about bionanomaterials for dental applications to graduate students and researchers, whose background can be in chemistry, physics, chemical engineering, materials science, biomedical science, or even dental science.

I express my appreciation to all of the authors for their contributions.

Mieczyslaw Jurczyk

Chapter 1

Introduction

Mieczyslaw Jurczyk[a,*] and Karolina Jurczyk[b]

[a]*Poznan University of Technology, Institute of Materials Science and Engineering, M. Sklodowska-Curie Sq. 5, 60-695 Poznan, Poland*

[b]*Conservative Dentistry and Periodontology Department, University of Medical Sciences, Bukowska 70 Street, 60-812 Poznan, Poland*

[*]mieczyslaw.jurczyk@put.poznan.pl

1.1 Motivation

In 1959 Richard Feynman, a Physics Nobel laureate, presented his famous idea of nanostructure materials production [17]. He stated: "The principles of physics, as far as I can see, do not speak against the possibility of maneuvering things atom by atom." Feynman proceeded to describe building with atomic precision, and outlined a pathway involving a series of increasingly smaller machines. Today, it is possible to prepare metal, ceramic, and alloy nanocrystals with nearly monodispersive size distribution. Nanostructures represent key building blocks for nanoscale science and technology.

Nanotechnology is a technology that owes its name to the prefix *nano*, a Greek word for *dwarf*, as applied to objects that exhibit billionth (10^{-9}) meter dimensions. Recently, nanotechnology has led to a remarkable convergence of disparate fields including biology,

Bionanomaterials for Dental Applications
Edited by Mieczyslaw Jurczyk

ISBN 978-981-4303-83-5 (Hardcover), 978-981-4303-84-2 (eBook)
www.panstanford.com

applied physics, chemistry, materials science, and computational modeling [3]. Broadly speaking, nanotechnology is the development and use of techniques to construct structures in the physical size range of 1–100 nanometers (nm), as well as the incorporation of these structures into applications. Now, nanotechnology is entering many industry sectors including energy, electronics, aerospace as well as medicine.

Nanoscience and nanotechnologies are not new. Size-dependent properties have been exploited for centuries. For example, Au and Ag nanoparticles have been used as colored pigments in stained glass and ceramics since the 10th century AD. Many chemicals and chemical processes have nanoscale features and, for example, chemists have been making polymers (large molecules made up of nanoscalar subunits) for many decades. But now, due to imaging techniques like the scanning tunneling microscope and the atomic force microscope, the understanding of the nanoworld has improved considerably [14, 16].

During the past few years, interest in the study of bionanostructure materials has been increasing at an accelerating rate, stimulated by recent advances in materials synthesis and characterization techniques and the realization that these materials exhibit many interesting and unexpected properties with a number of potential technological applications [6, 18, 40, 52]. Nanotechnology provides the tools and technology platforms for the investigation and transformation of biological systems, and biology offers inspiration models and bio-assembled components to nanotechnology [21]. For example, the London Centre for Nanotechnology has a wide range of bionanotechnology and health care research programs: bionanoparticles, bionanosensors, biocompatible nanomaterials, advanced medical imaging, technologies for diagnosis, self-assembled biostructures, degenerative disease studies, molecular simulation, lab on a chip and screening, drug screening technologies, and molecular simulation.

Application of new materials such as biomaterials and implants increases steadily. However not all replacement systems have provided trouble-free service. In dental implants the rate of success is 96–98%, which, by millions of implants, gives a significant number of patients in trouble [2]. Therefore, in a failure-free replacement system, no particulate or corrosion debris would be generated and no loosing of the implant components should occur. The source of debris particles is the wear process and grit blast, which includes Al_2O_3, ZrO_2, or SiO_2 particles on the surfaces of especially treated implants.

An appropriate surface modification would prevent their transfer into nearby tissues. The absence of debris particle generation is crucial for the prevention of implant malfunction. The determination of the mechanisms of debris generation and appreciate modification of implant surface bulk structure and properties is one of the main aims of current research projects.

The main purpose of current research is to prevent the failures caused by infection by changing the biomaterial's properties and making them highly friendly for surrounding tissues.

Ti and Ti-based alloys are preferred materials in the production of implants in both medical and dental applications. These biomaterials have relatively poor tribological properties because of their low hardness. One of the methods that allow the change of biological properties of Ti alloys is the modification of its chemical composition. The other way is to produce a composite that will exhibit the favorable mechanical properties of titanium and excellent biocompatibility and bioactivity of ceramic. The most commonly used ceramics employed in medicine are hydroxyapatite (HA), silica, and bioglass. HA shows good biocompatibility because of its similar chemical and crystallographic structure to the apatite of living bone. The ceramic coating on the titanium improves the surface bioactivity but often flakes off as a result of poor ceramic/metal interface bonding, which may cause the surgery to fail. For this reason, composite materials containing titanium and ceramic as a reinforced phase are expected to have broad practical applications.

Since 1996 a research program was initiated at the Institute of Materials Science and Engineering, Poznan University of Technology, in which fine grained, intermetallic compounds were produced by mechanical alloying, high-energy ball milling, hydrogenation-disproportionation-desorption-recombination (HDDR), or mechano-chemical processing (MCP) [22–28, 34–36]. The mechanical synthesis of nanopowders and their subsequent consolidation is an example how this idea can be realized in metals by a so-called bottom-up approach. On the other hand, other methods have been developed, which are based on the concept of the production of nanomaterials from conventional bulk materials via the top-down approach. The investigations by severe plastic deformation (e.g., cyclic extrusion compression method (CEC) or equal channel angular extrusion (ECAE)) [39, 46, 50, 51], show that such a transformation is indeed possible. Currently, at Poznan University of Technology, we facilitate the multidisciplinary interaction of physicists, chemists, materials engineers, biologists, and dentists collaborating on nanoscience, with the goal of integrating nanoscale materials with biological systems. The aim of our research is to develop a new generation of titanium (Ni-free stainless steel)-ceramic bionanocomposites by producing the porous structures with a strictly specified chemical and phase compositions, porosity and surface morphology and, as such, will adhere well to the substrate, show high hardness, high resistance to biological corrosion and good biocompatibility with human tissues.

Nanomaterials can be metals, ceramics, polymers, and composite materials that demonstrate novel properties compared with conventional (microcrystalline) materials due to their nanoscale features. Moreover, researchers have exhibited an increased interest in exploring numerous biomedical applications of nanomaterials and nanocomposites [3, 6, 40]. Till now, it has been shown that implants made from metallic, carbon, or oxide bionanomaterials considerably improved the prosthesis strength and their biocompatibility. These nanocrystalline structures can be produced by non-equilibrium processing techniques such as mechanical alloying [4, 9, 47].

The current projects aim to fabricate Ti-based porous scaffolds to promote bone or tissue ingrowth into pores and provide biological anchorage. Generally, porous metallic scaffolds are fabricated using a variety of processes to provide a high degree of interconnected porosity to allow bone ingrowth. Fabrication technologies include chemical vapor infiltration to deposit tantalum onto vitreous carbon foams, solid freeform fabrication, self-propagating high-temperature synthesis, and powder metallurgy [13, 20, 29, 34, 45, 48]. While these porous metals have been successful at encouraging bone ingrowth both *in vivo* and in clinical trials, the range of materials and microstructures available is still rather limited. It is important to use appropriate surface modification to increase the anti-corrosive and biocompatible properties of Ti implants for long-term clinical applications.

Mechanical alloying, high-energy ball milling, reactive milling, chemical vapor transport, solid–liquid–vapor growth, solvothermal synthesis, solid–gas high-temperature reactions, microwave chemistry, arc furnace techniques, aerosol spray techniques, liquid metals chemistry, and powder metallurgy process for the fabrication of titanium (Ni-free stainless steel)–ceramic nanocomposites with a unique microstructure were developed. Those processes permit the control of microstructural properties such as the size of pore openings, surfaces properties, and the nature of the base metal/alloy. A new type of bulk three-dimensional porous Ti (Ni-free stainless steel)-based nanocomposite biomaterials with desired size of porous and three-dimensional capillary-porous coatings on these nanobiocomposites was developed. Materials with nanoscale grains would offer new structural and functional properties for innovative products in medical/dental applications.

Various methodologies are being used in an effort to improve the interfacial properties between the biological tissues and the existing implants, e.g., Ti and Ti-based alloy. The electrochemical technique, a simpler and faster method, can be used as a potential alternative for producing porous Ti-based metals for medical implants. Good corrosion resistance of the titanium is provided by the passive titanium oxide film on the surface. This layer is important for the good biocompatibility. The native oxide has thickness of a few nanometers. In the case of anodic oxidation, the oxide thickness can be multiplied up to the micrometer range. The structure and thickness of the grown oxide depend on the electrochemical etching conditions, for example: current density, voltage, electrolyte composition. In the electrochemical etching of titanium, electrolytes containing H_3PO_4, CH_3COOH, and H_2SO_4 are used. In Ti anodization, the dissolution is enhanced by HF- or NH_4F-containing electrolytes, which results in pore or nanotube formation. The current density in this case is much higher than in the electrolyte without HF or NH_4F [22]. Fluoride ions form soluble $[TiF_6]^{2-}$ complexes resulting in the dissolution of the titanium oxides. In this way, the dissolution process limits the thickness of the porous layer.

Porous implants layer has lower density than respective bulk, and good mechanical strength is provided by bulk substrate. Hence, the latter is attractive with respect to bulk titanium alloys. The porous layer on the Ti substrate is necessary for osseointegration with bones, which is not normally provided by the native oxide.

On the other hand, Ti and its alloys possess favorable properties, such as relatively low modulus, low density, and high strength. Apart from that, these alloys are generally regarded to have good biocompatibility and high corrosion resistance but cannot directly bond to the bone. In addition, metal implants may loosen and even separate from surrounding tissues during implantation. Titanium- and titanium-based alloys have relatively poor tribological properties because of their low hardness. One of the methods that allow the change of biological properties of Ti alloys is to produce a nanocomposite that will exhibit the favorable mechanical properties of titanium and excellent biocompatibility and bioactivity of ceramic. The most commonly used ceramics in medicine are hydroxyapatite, bioglass, and Al_2O_3 [7, 34].

Current research on the synthesis of nanoscale metallic and composite biomaterials, shows that Ti/(Ni-free stainless steel)–HA nanocomposites posses better mechanical and corrosion properties than microcrystalline titanium/Ni-free stainless steel [49]. In the case of Ti–HA nanocomposities, the Vickers hardness also strongly increases for Ti-20 vol% HA nanocomposites (1030 HV0.2) and is four times higher than that of pure microcrystalline Ti metal (250 HV0.2). The corrosion test results indicated that the microcrystalline titanium possesses lower corrosion resistance and thus higher corrosion current density in Ringer's solutions. The result indicated that there was no significant difference in corrosion resistance among Ti-3 vol% HA ($I_C = 9.06 \times 10^{-8}$ A/cm^2, $E_C = -0.34$ V) and Ti-20 vol% HA ($I_C = 8.5 \times 10^{-8}$, A/cm^2, $E_C = -0.55$ V) although there was a significant difference in porosity. For this reason, they are promising biomaterial for use as heavy load-bearing tissue replacement implants.

The availability of large amounts of specifically tailored nanostructure Ti-based powders is crucial for the successful development of new dental implants. The processing of these nanomaterials and their upscaling to enable industrial use has many challenges. Those new approaches are the gateway for traditional industry to nanotechnology and knowledge-based materials, with positive effects on health issues [1, 29, 45, 53].

1.2 Nanomaterials

One of the first scientific reports is the colloidal gold particles synthesized by Michael Faraday as early as 1857 [18]. By the early 1940s, precipitated and fumed silica nanoparticles were being manufactured and sold in the United States and Germany as substitutes for ultrafine carbon black for rubber reinforcements.

In the 1960s and 1970s, metallic nanopowders for magnetic recording tapes were developed. The first nano-size metallic materials were produced in 1960 by the application of rapid quenching process with the cooling rate of 10^6 K s^{-1} by Pol Duvez and coworkers [30]. Using a quench technique capable of cooling metal/alloy melts to ambient temperatures with such extraordinary cooling rates, the process of nucleation and growth was kinetically bypassed to yield a configuration of frozen liquid or amorphous metal. The nanoalloys with nanometric grains were processed by low-temperature annealing of amorphous alloy [12]. Additionally, in 1972 a new rapid quenching process of melt spinning was used to spin the first nonocrystalline ferrous and ferromagnetic metal ribbons ($Fe_{80}B_{20}$) [10]. The outstanding physical and chemical properties of these materials were the direct consequence of the lack of structural crystalline long-range order and the presence of short-range order. It is important to mention that many, if not all of these amorphous alloys, i.e., metallic glasses, when carefully annealed at low temperatures change to nanostructure alloy phases. Those were the first nanomaterials. They were produced in the form of thin ribbons via rapid solidification processing of melt alloys. They allowed controlled exploration of physical, chemical, mechanical, and other properties as arising from nanostructures. At this time, scientists and engineers became refocused from well-ordered crystalline materials to disordered and nanocrystalline phases. Gleiter observed that nanometer-size crystalline materials being polycrystals with very small crystallite sizes of about 2–10 nm in diameter are composed of randomly oriented high-angle grain boundaries [18].

The first such nanocrystalline phases came from Gleiter research group around 1984 by evaporation of the material in a high-purity inert gas atmosphere followed by condensation and compaction in ultrahigh vacuum [5]. The percentage of metal atoms on the surface of grain increases from a few percent in a 100 nm particle to about 90% in a 1 nm crystallite [19, 43]. As above, these materials should be attractive for the development of engineering materials with an outstanding combination of properties or novel properties. In the meantime, among materials that became studied were nanophases produced by mechanical alloying [4, 7, 47].

Nanomaterials continue to attract a great deal of attention because of their potential impact on an incredibly wide range of industries and markets. Consequently, the technology is evolving rapidly and will develop faster over the coming years. Nanomaterials cross the boundary between nanoscience and nanotechnology and link the two areas together [37]. Nanoscience is the study of phenomena and manipulation of materials at atomic, molecular, and macromolecular scales, where properties differ significantly from those at a larger scale. Nanotechnologies are the design, characterization, production, and application of structures, devices, and systems by controlling shape and size on the nanoscale. Nanoparticles can come in a wide range of morphologies, from spheres, through flakes and platelets, to dendritic structures, tubes, and rods.

It is recognized that the size range that provides the greatest potential and, hence, the greatest interest is that below 100 nm; however, there are still many applications for which larger particles can provide properties of great interest.

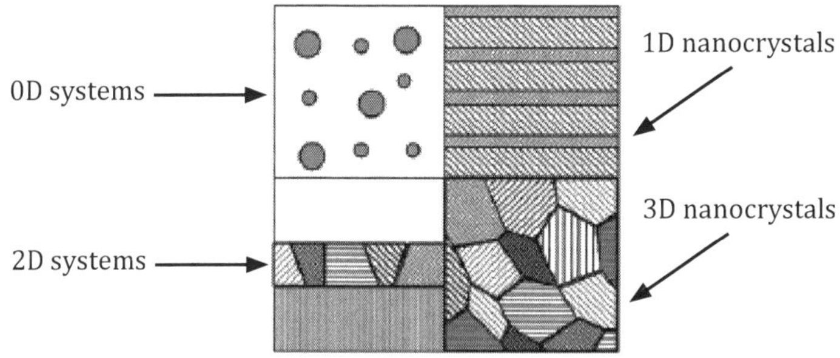

Figure 1.1 Siegel's classification of nanomaterials [43].

According to Siegel, nanostructure materials are classified as zero-dimensional, one-dimensional, two-dimensional, and three-dimensional nanostructures (Fig. 1.1) [43].

The first industrial production of nanomaterials occurred in the early 20th century with the production of carbon black and subsequently, in the 1940s, fumed silica. The real burst in the commercialization of nanoparticle production has occurred over the last 10 years or so. One of the main drivers for this has been the extraordinary growth in the electronics and optoelectronics as well as in power/energy, healthcare/medical, engineering, consumer goods, environmental and electronics industries. Table 1.1 identifies the key applications in healthcare/medical industries.

Table 1.1 A selection of current and future applications using nanoparticles in healthcare/medical industries [2]

Under development	Being introduced	Well established
Nanocrystalline drugs for easier absorption	Molecular tagging using CdSe quantum dots	Ag-based antibacterial wound dressings, ZnO fungicide
Inhalable insulin	Drug carriers for drugs with low water solubility	Au for biolabeling and detection
Nanospheres for inhaling drugs currently injected using biocompatible Si	Coatings for implants such as hydroxyapatite	Magnetic resonance imaging contrast agents using superparamagnetic Fe_2O_3
Bone growth promoters	Marker particles for use in assays	Sunscreens using ZnO and TiO_2
Virus detection using quantum dots		
Anticancer treatments		
Magnetic particles for the repair of the human body with prosthetics or artificial replacement parts		
Antioxidant drugs based on fullerenes		

1.3 Nanobiotechnology

Nanobiotechnology is defined as a field that applies the nanoscale principles and techniques to understand and transform living or non-living biosystems and that uses biological principles and materials to create new devices and systems integrated from the nanoscale [11, 41, 42]. Recent research on biosystems at the nanoscale has created one of the most dynamic science and technology domains at the confluence of physical sciences, molecular engineering, biology, biotechnology, and medicine [11, 45, 53]. This domain includes better understanding of living and thinking systems, revolutionary biotechnology processes, the synthesis of new drugs and their targeted delivery, regenerative medicine, neuromorphic engineering and developing a sustainable environment [42]. Smalley [44] classified nanotechnology into two categories: "wet" nanotechnology (including living biosystems) and "dry" nanotechnology. Research on dry nanostructures is now seeking systematic approaches to engineer man-made objects at the nanoscale and to integrate nanoscale structures into large-scale structures, as nature does.

The relationship between biomaterials and nanotechnology was discussed recently by Williams [54]. Three aspects were considered:

(i) In many medical applications, the form of the biomaterial is specifically designed such that it is presented to the human body as a unit with nanoscale dimensions. It is important that such material has distinctly different properties from the bulk material as a consequence of its occurrence as discrete entities for it to be considered a nanomaterial.

(ii) Biocompatibility phenomena are controlled by nanoscale topographical features. It is known that interactions between cells and biomaterial surfaces are mediated by molecules and structures on cell membranes and within the cytoplasm. Therefore, the behavior of such cells at the interface with a biomaterial may be profoundly influenced by the nanotopography.

(iii) Nanotechnologies have positively influenced the quality of life through advances in medicine and biotechnology; however, it is also possible that they may be associated with health and environmental risks [15, 32, 33]. The rapid progress in the industry causes, for example, huge quantities of nanoparticles in the air produced by combustion. Small size alone is not the critical factor in the toxicity of nanoparticles, but the overall number and thus the total surface area (essentially the dose) are also important. Therefore, we have to note the increasing risk due to the development of new materials and wider applications based on nanotechnology.

A large number of potential applications of nanotechnologies are now opening up. Some reports give an overview of current and potential future developments in nanoscience and nanotechnologies and their health, safety, environmental, social, and ethical implications. Much of nanoscience and many nanotechnologies are concerned with producing new or enhanced materials. Some nano-products are already on the market and enjoying commercial success. For example, nanoparticles are used in some sunscreens to reflect and absorb ultraviolet (UV) light; the football-shaped buckminsterfullerene (C_{60}) and its analogs show great promise as lubricants and, thanks to their cage structures, as drug delivery systems [8, 31].

Applications of nanotechnologies in medicine are especially promising in the long term. These can be expected to enable drug delivery targeted at specific sites in the body so that, for example, chemotherapy is less invasive [8]. Nanotechnology is expected to lead to stronger, longer-lasting implants; sensors that can be used to monitor aspects of human health; and improved artificial cochleae and retinas. However, many of these applications will not be realized for at least 10 years, partly because of the rigorous testing and evaluation that will be required. Antimicrobial wound dressings are already on the market in the United States. These dressings use nanocrystalline Ag to provide a steady dose of ionic Ag to protect against secondary infections and are claimed to be effective against 150 different pathogens.

Little research has been carried out on the toxicity of manufactured nanoparticles, but we can learn from studies on the effects of exposure to mineral dusts in some workplaces and to the nanoparticles in air pollution. Considerable evidence from industrial exposure to mineral dusts demonstrates that the toxic hazard is related to the surface area of the inhaled particles and to their surface activity. Epidemiological studies of urban air pollution support the conclusion that finer particles cause more harm than coarser ones — diesel PM_{10} pollution is implicated in heart and lung disease and asthma, particularly in susceptible people.

There is practically no information on the environmental impacts of nanoparticles. More research on their properties and effects is necessary. It is time to look at the toxicity, epidemiology, persistence, and bioaccumulation of manufactured nanoparticles, including their exposure pathways.

References

1. Agheli, H., Malmström, J., Hanarp, P., and Sutherland, D.S. (2006). Nanostructured biointerfaces, *Mater. Sci. Eng., C*, **26**, pp. 911–917.

2. Arys, A., Philippart, C., Dourov, N., He, Y., Le, Q.T., and Pireaux, J.J. (1998). Analysis of titanium dental implants after failure of osseointegration: combined histological, electron microscopy and X-ray photoelectron spectroscopy approach, *J. Biomed. Mater. Res.*, **43**, pp. 300–312.
3. Baker Jr., J.R. (2005). What is nanomedicine?, *Nanomed. Nanotechnol. Biol. Med.*, **1**, p. 243.
4. Benjamin, J.S. (1976). Mechanical alloying, *Sci. Am.*, **234**, pp. 40–57.
5. Birringer, R., Gleiter, H., Klein, H.P., and Marquardt, P. (1984). Nanocrystalline materials — an approach to a novel solid structure with gas-like disorder?, *Phys. Lett., A*, **102**, p. 365–369.
6. Bugunia-Kubik, K., and Susisaga, M. (2002). From molecular biology to nanotechnology and nanomedicine, *J. Biosystems.*, **65**, pp. 123–138.
7. Cao, W.P., and Hench, L. (1996). Bioactive materials, *Ceramics Int.*, **22**, pp. 493–507.
8. Ciavarella, S., Milano, A., Dammacco, F., and Silvestris, F. (2010). Targeted therapies in cancer, *BioDrugs.*, **24**, pp. 77–88.
9. Chawla, V., Prakash, S., and Sidhu, B.S. (2007). State of the art: applications of mechanically alloyed nanomaterials — a review, *Mater. Manuf. Proc.*, **22**, pp. 469–473.
10. Chen, H.S., and Polk, D.E. (1974). US Patent 3,856,513, Allied Chemical Corporation, USA.
11. Chung, D. (2003). Nanoparticles have health benefits too, *New Sci.*, **179**, pp. 2410–2416.
12. Cohen, M.H., and Turnbull, D. (1961). Composition requirements for glass formation in metallic and ionic systems, *Nature*, **189**, pp. 131–132.
13. Costa, V.C., Costa, H.S., Vasconcelos, W.L., Pereira, M.M., Oréfice, R.L., and Mansur, H.S. (2007). Preparation of hybrid biomaterials for bone tissue engineering, *Mat. Res.*, **10**, pp. 21–26.
14. Dowling, A.P. (2004). Development of nanotechnologies, *Mater. Today*, **7**, pp. 30–35.
15. European Commission, Scientific Committee on Emerging and Newly Identified Health Risks (SCENIHR) (2006). *Opinion on 'the appropriateness of existing methodologies to assess the potential risks associated with engineered and adventitious products of nanotechnologies'*, adopted by the SCENIHR during the 10th plenary meeting, Brussels.
16. Erhardt, D. (2003). Materials conservation: Not-so-new technology, *Nat. Mater.*, **2**, pp. 509–510.

17. Feynman, R.P. (1960). There's plenty of room at the bottom, *Eng. Sci.*, **23**, pp. 22–36.

18. Gleiter, H. (1995). Nanostructured materials: State of the art and perspectives, *Nanostruct. Mat.*, **6**, pp. 3–14.

19. Granqvist, C.J., and Buhrman, R.A. (1976). Ultrafine metal particles, *J. App. Phys.*, **47**, pp. 2200–2219.

20. Guan, L., and Davies, J.E. (2004). Preparation and characterization of a highly macroporous biodegradable composite tissue engineering scaffold, *J. Biomed. Mater. Res.*, A**71**, pp. 480–487.

21. Horton, M.A., and Khan, A. (2006). Medical nanotechnology in the UK: a perspective from the London Centre for Nanotechnology, *Nanomed. Nanotechnol. Biol. Med.*, **2**, pp. 42–48.

22. Jakubowicz, J. (2008). Formation of porous TiO_x biomaterials in H_3PO_4 electrolytes, *Electrochem. Commun.*, **10**, pp. 735–739.

23. Jakubowicz, J., Jurczyk, K., Niespodziana, K. and Jurczyk, M. (2009). Mechanoelectrochemical synthesis of porous Ti-based nanocomposite biomaterial, *Electrochem. Commun.*, **11**, pp. 461–465.

24. Jurczyk, K., Niespodziana, K., and Jurczyk, M. (2006). Preparation and characterization of nanocomposite Ti-hydroxyapatite materials, *Eur. J. Med. Res.*, **11**, Suppl. II, pp. 133.

25. Jurczyk, K., Niespodziana, K., Stopa, J., and Jurczyk, M. (2007). Composite Ti-hydroxyapatite bionanomaterials for application in modern dentistry. *Polish J. Environ. Studies*, **16**, pp. 323–327.

26. Jurczyk, M. (1995). Anisotropic Nd–Fe–Co–Zr–B Powders Prepared by the HDDR Process, *J. Alloys Comp.*, **228**, pp. 172–176.

27. Jurczyk, M., Smardz, L., Smardz, K., Nowak, M., and Jankowska, E. (2003). Nanocrystalline $LaNi_5$-type electrode materials for Ni-MH_x batteries, *J. Solid State Chem.*, **171**, pp. 30–37.

28. Jurczyk, M., and Nowak, M. (2008). Nanomaterials for hydrogen storage synthesized by mechanical alloying, in *Nanostructured Materials in Electrochemistry* (ed. Eftekhari, A.), Wiley-VCH, USA, pp. 349–386.

29. Khang, G., Kim, M.S., and Lee, H.B. (2007). *Manual for Biomaterials/ Scaffold Fabrication Technology*, World Scientific Publishing Co Pte Ltd., Singapore.

30. Klement, J.W., Willens, R.H., and Duwez, P. (1960). Non-crystalline structure in solidified gold-silicon alloys, *Nature*, **187**, pp. 869–870.

31. Kratschmer, W., and Lamb, L.D. (1990). Solid C_{60}: a new form of carbon, *Nature*, **347**, pp. 354.

32. The Royal Academy of Engineering, The Royal Society (2004). Nanoscience and nanotechnologies: Opportunities and uncertainties, The Royal Society & The Royal Academy of Engineering, London.
33. Health and Consumer Protection Directorate General of the European Commission (2004). Nanotechnologies: a preliminary risk analysis, based on a workshop on 1–2 March 2004, Health and Consumer Protection Directorate General of the European Commission, Brussels. http://ec.europa.eu/health/ph_risk/document/ev_20040301_en.pdf.
34. Niespodziana, K., Jurczyk, K., and Jurczyk, M. (2006). The manufacturing of Ti-hydroxyapatite nanocomposites for bone implant applications, *Nanopages*, **1**, pp. 219–229.
35. Niespodziana, K., Jurczyk, K., and Jurczyk, M. (2007). Titanium-ceramic nanocomposite materials, *Adv. Mat. Science*, **7**, pp. 103–107.
36. Okonska, I., Iwasieczko, W., Jarzebski, M., Nowak, M., and Jurczyk, M. (2007). Hydrogenation properties of amorphous 2Mg + Fe/*x* wt% Ni materials prepared by mechanical alloying ($x = 0, 100, 200$), *Int. J. Hydrogen Energy*, **32**, pp. 4166–4190.
37. Pitkethly, M. J. (2004). Nanomaterials — the driving force, *Mater. Today*, 38, **7**, pp. 20–29.
39. Richert, J., and Richert, M. (1986). A new method for unlimited. Deformation of metals and alloys, *Aluminium*, **62**, pp. 604– 607.
40. Roco, M.C. (2001). International strategy for nanotechnology research and development, *J. Nanoparticle Res.*, **3**, pp. 353–360.
41. Roco, M.C. (2003). Nanotechnology: convergence with modern biology and medicine, *Curr. Opin. Biotech.*, **14**, pp. 337–346.
42. Roco, M.C., Williams, R.S., and Alivisatos, P. (eds) (2000). Applications: Biological, medical and health applications, in *Nanotechnology Research Directions*, Kluwer Academic Publishers, Boston, pp. 153–172.
43. Siegel, R.W. (1991). Cluster assembled nanophase materials, *Annu. Rev. Mater. Sci.*, **21**, pp. 559–578.
44. Smalley, R. (2002). Nanotechnology: the wet/dry frontier, in *Proceedings of the Small Wonders Workshop*, Washington DC.
45. Stylios, G.K., Giannoudis, P., and Wan, T. (2005). Applications of Nanotechnologies in Medical Practice, *Injury*, **36**, pp. S6–S13.
46. Sun, P.I., Kao, P.W., and Chang, C.P. (2000). Characteristics of submicron grained structure formed in aluminum by equal channel angular extrusion, *Mater. Sci. Eng., A*, **283**, pp. 82–85.

47. Suryanarayna, C. (2001). Mechanical alloying, *Progr. Mater. Sci.*, **46**, pp. 1–184.

48. Tang, F., Fudouzi, H., Uchikoshi, T., and Sakka, Y. (2004). Preparation of porous materials with controlled pore size and porosity, *J. Eur. Ceram. Soc.*, **24**, pp. 341–344.

49. Tulinski, M., and Jurczyk, M. (2009). Mechanical and corrosion properties of Ni-free austenitic stainless steel/hydroxyapatite nanocomposities, *Solid State Phenomena*, **151**, pp. 213–216.

50. Valiev, R.Z., Korrznikov, A.V., and Mulyukov, R.R. (1993). Structure and properties of ultrafine-grained materials produced by severe plastic deformation, *Mater. Sci. Eng., A*, **168**, pp. 141–148.

51. Valiev, R.Z., Zehetbauer, M.J., Estrin, Y., Höppel, H.W., Ivanisenko, Y., Hahn, H., Wilde, G., Roven, H.J., Sauvage, X., and Langdon, T.G. (2007). The innovation potential of bulk nanostructured materials, *Adv. Eng. Mater.*, **9**, pp. 527–533.

52. Wang, Z.L., Liu, Y., and Zhang, Z., eds. (2003). *Handbook of nanophase and nanostructured materials*, Kluwer Academic/Plenum Publishers, New York.

53. Ward, B.C., and Webster, T.J. (2007). Increased functions of osteoblasts on nanophase metals, *Mat. Sc. Eng., C*, **27**, pp. 575–578.

54. Williams, D. (2008). The relationship between biomaterials and nanotechnology, *Biomaterials*, **29**, pp. 1737–1738.

Chapter 2

Stomatognathic System

Karolina Jurczyk
Conservative Dentistry and Periodontology Department, University of Medical Sciences, Bukowska 70 Street, 60-812 Poznan, Poland
karolajur@gmail.com

2.1 Elements of the Stomatognathic System

Stomatognathic system is a part of the human body influenced by external factors. Initially considered a relationship between occlusion and temporomandibular joint, it turned out to be a more complex morphological and functional unit, consisting of tissues and organs functionally combined with one another [8]. Apart from the temporomandibular joint, we have to stress the importance of the bones of the head, alveolar process, teeth, periodontium, neuromuscular system, blood and lymph vessels, oral mucosa and salivary glands, in the functioning of the stomatognathic system.

Bionanomaterials for Dental Applications
Edited by Mieczyslaw Jurczyk

ISBN 978-981-4303-83-5 (Hardcover), 978-981-4303-84-2 (eBook)
www.panstanford.com

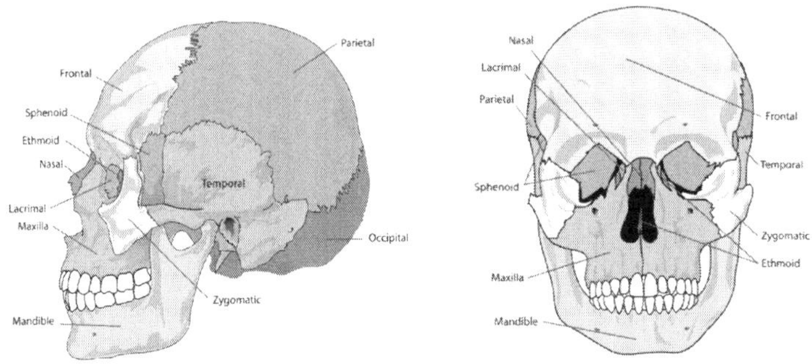

Figure 2.1 Lateral and front view of human skull.

Precise anatomy of the bones of the head, as a part of the skeletal system, has been described by many authors [6, 7]. The human skull is made up of 22 bones, which are, except for the mandible, connected together by sutures, a rigid connection enabling very little movement (Fig. 2.1). Eight bones form the neurocranium, including one frontal, two parietals, one occipital bone, one sphenoid, two temporals, and one ethmoid, which is a protective vault surrounding the brain. However, the remaining 14 bones form the splanchnocranium, the support for the face.

The mandible consists of the horizontal part, from which the body of the mandible perpendicularly extends on both sides of the rami, ending with two processes. Condyle is the superior and posterior projection of the ramus and together with the temporal bone forms the temporomandibular articulation. The other projection is the coronoid process, lying more anterior to the latter, and it provides attachment to the temporalis muscle.

The temporomandibular joint, as a paired articulation, enables remarkable freedom of motion of the mandible, but still controls and limits the motions. The anatomy is complex and not the typical compared with other joints of the body. The articular surface of the condyle is small compared with the glenoid fossa and articular eminence on the temporal bone. The capsule of the joint is not as well organized as the capsules of other joints. Its lateral side is reinforced by temporomandibular ligament consisting of collagen fibers.

The movement and position of the mandible is influenced by many muscles, among them the mastication muscles (Fig. 2.2). This group consists of eight muscles symmetrically positioned on both sides of the skull. The masseter, temporalis, and medial pterygoid serve to elevate the mandible when they constrict (Fig. 2.3). As goes for the lateral pterygoid, when shortening in conjunction with other muscles, as well as the opposite lateral pterygoid, serves to depress, protrude, or shift the mandible laterally.

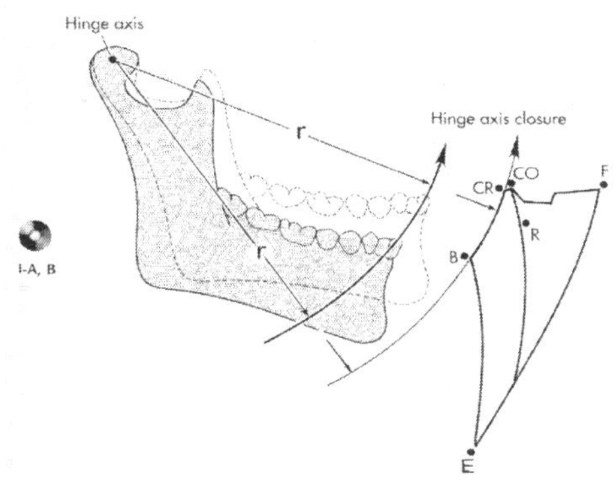

Figure 2.2 Mandibular movement projected on saggital plane.

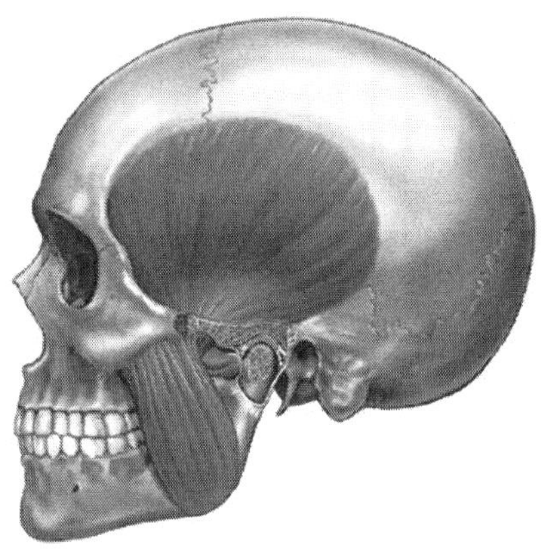

Figure 2.3 Human skull showing masseter and temporalis muscles.

Moreover, the suprahyoid muscles extending from the skull and mandible to the hyoid bone serve to depress the mandible, strain floor of the mouth and cooperate in deglutination. The muscles of facial expression are too weak to be involved in the movement of mandible; however, they play a vital role in the esthetics.

The teeth lie in the sockets of the maxillary alveolar process and the alveolar portion of the body of the mandible, forming two arches [10, 14]. According to the shape of the teeth, their anatomy, function, and position, they are classified as incisors, canines, premolars, and molars. Their position in the arch is shown in Fig. 2.4. Most teeth have identifiable features that distinguish them from others. To simplify the identification of teeth, they are classified by a numbering system (Fig. 2.5).

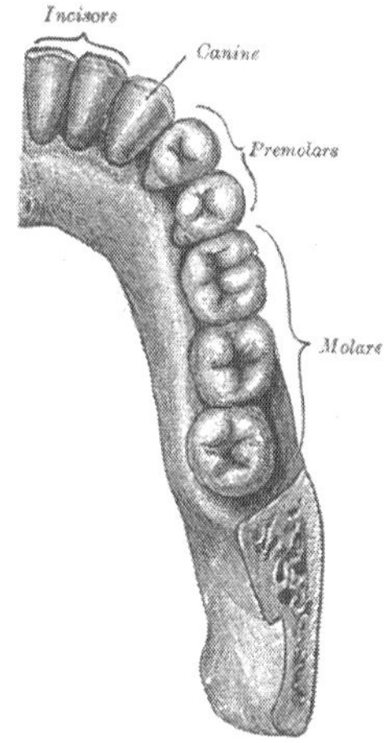

Figure 2.4 View of the lower arch.

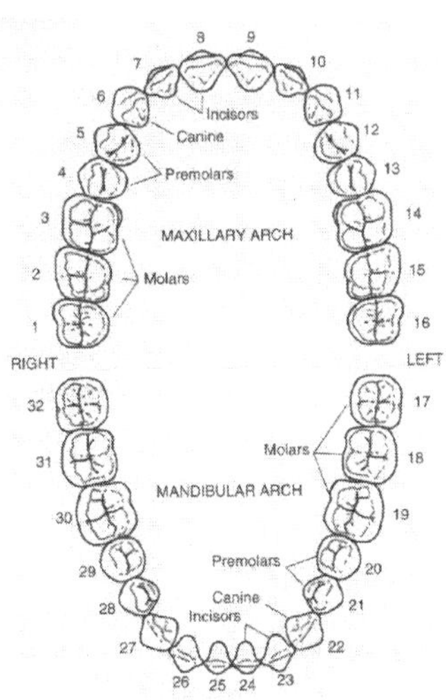

Figure 2.5 Numbering system.

The knowledge of the anatomical shape of the crowns of the teeth is the basis of the choice selection of prosthetic reconstructions of the missing teeth in patients. The position of the teeth in the arch is in respect to their function, meaning that they are supported in the alveolar sockets through a connective tissue called the periodontal ligament to play a distinctive role in the mastication process. The main function of the periodontium is to attach the teeth to the bone and to maintain the integrity of the surface of the masticatory mucosa of the oral cavity [5]. Also called "attachment apparatus," it undergoes many changes with age and also due to functional and oral environment alterations.

Maxillary teeth axes lie oblique to the vertical axis of the cranium. Roots in the dental arch in the maxilla are more closely spaced than are the crowns of the teeth, which appear to slightly tilt outward. As goes for the axes of the mandibular teeth, they are inclined inward relative to the vertical axis of the cranium, such that their crowns on the opposite side of the jaw lie closer than the roots.

The anatomical crown of the tooth is covered by enamel, a highly mineralized tissue, up to the cervical region of the tooth (Fig. 2.6). Underneath lies the dentin, which forms the main core of the tooth. This structure contains 70% of inorganic substance, making it less brittle from the enamel. In the middle of the tooth is the pulp chamber containing vessels and nerves.

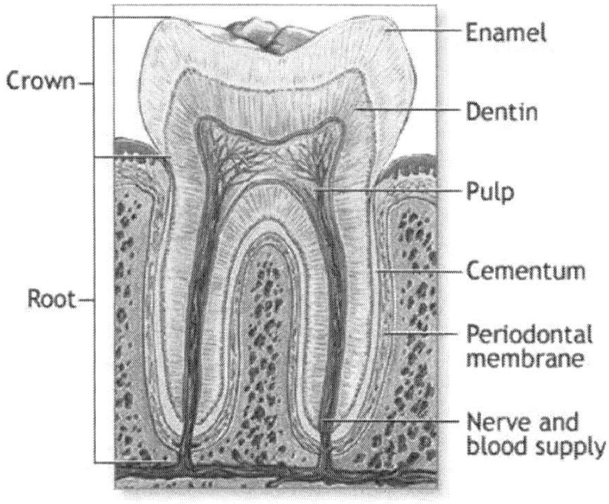

Figure 2.6 The anatomy of the tooth.

2.2 Implantology Issues in the Stomatognathic System

The stomatognathic system, as a morphological and functional unit, composed of many tissues, such as the teeth, jaws, muscles of mastication, epithelium, and temporomandibular joints and nerves that control these structures, takes part in many physiological processes.

The combination of all the structures enables proper speech and the reception, mastication, deglutition of food, and respiration. The basis of proper functioning of the system is the well controlled central nervous system. All the specific elements of the system remain in mutual synchronization, synergy, and cohesion, enabling the system to work properly. A disease or trauma occurring in any of the elements of the system causes the malfunctioning of the whole system, as well as in the biological aspect and also has its psychological influence on the patient. Apart from the health benefits of a proper bite and speech, are the esthetic benefits, skin and muscle tone of the lips and cheeks, restoration of symmetry of the lower facial 1/3, and a radiant smile.

The stomatognathic system is without any doubt a very complex structure. Its proper functioning is controlled by regulation mechanisms, in which a mutual relationship exists between morphology and function and different compensation abilities.

The role of occlusion as a contributing factor in various signs and symptoms in the stomatognathic system has been discussed in literature [12]. In simplified understanding, occlusion refers to the arrangement of maxillary and mandibular teeth and to the way the teeth contact. However, the biological and psychosocial role of this interference is seen the most in case of missing teeth. A variety of impairments, such as functional, aesthetic, and psychological, occur in case of edentulous patients, which have occlusion-free jaws.

Those occlusion deviations lead to further changes in the structures of the stomatognathic system [3, 13]. A hyperactivity of the mastication muscles has been notated, which, in consequence, leads to fatigue and muscle spasm. Moreover, the forces induced by mastication are also forwarded to the temporomandibular joint, which usually adapts to the functional demands. Adaptation may occur through the change in the osseous components of the joint. However, when the adaptive capacity is exceeded, dysfunction or generation of the structure may take place, which is signaled by the patient in the form of pain.

The whole adaptation process is controlled by the central nervous system due to the well innervated periodontium, muscles, oral mucosa, articulations, and tooth pulp. All the information from the peripheral nerves located in the tissues is forwarded later on to the central nervous system. A vital role in the function of proprioception encompasses the muscle spindles and Golgi tendon organs. In a narrow connotation, it can be said that these organs play an intermediate role, by passing information on the instantaneous length of the muscle in which they are located to the CNS. They help in keeping the proportion between muscles tension and stretch and have a part in the control of muscular contraction. In other words, the proprioreceptors may change the activity of the masticatory muscles and adapt the force of the mastication to the physiological limits of the tissues. This mechanism in physiological circumstances protects from the negative factors.

However, when the loss or a change in one of the tissues occurs, it may exceed the adaptive limits of other structures, leading to further disability and collapse of function of the whole unit. At some point, the process speeds up and rapid breakdown ensues.

To discuss the therapeutic or reconstructive issues of the stomatognathic system needs a good knowledge of the morphology and function of all elements. For this purpose we usually use materials, which would be capable of reconstructing the disable tissue, due to aging process, trauma or disease, and taking other its function. Biomedical engineering has rapidly developed over the past years, creating analogs of the tissues but is still far from perfection [9]. Improving the quality of life is a rewarding goal for scientists and engineers. One of the fields of science is based on understanding the relationship between the human system and the environment. Indeed, those fields of research and new technologies in the field of rehabilitation improve the quality of life for disabled as well as abled patients.

Most of the global population, sooner or later, needs a repair or reconstruction within the craniofacial region [4]. It may be caused by a minor pathological process, such as a dental caries occurring in a tooth, or an oncological process leading to a resection of the tissues. Most of the human tissues are hybrid, such as the tooth composed of enamel, dentine, cementum, and pulp tissue. Therefore material engineering is focused on developing single tissues, as well as hybrid organs and interfaces. Understanding the interaction between molecules of the extracellular matrix and attached cells to the materials is essential for proper biomaterial design. Applications of biomaterials need experimental verifications in different laboratory models to prove their functional effectiveness [1]. In addition to the physical, chemical, mechanical, and biological performance, the clinical performance of the material is also reviewed through different trials.

Obtaining the same chemical structures as those of the body tissues, however, does not solve the problem that the artificially obtained material is non-vital. For example, new structures of materials incorporating sensory systems and signal processing are proposed for the next generation of prostheses. They will help better tailor the response in stiffness and damping equipment or provide spectral properties closer to those of natural materials and living.

In the rehabilitation of the oral tissues, the fundamental role is played by osseointegration, due to its excellent results and great diversity of planning in the dental clinic. Osseointegration means a direct connection between bone and implant without interposed soft tissues layers [5]. However, 100% connection of the bone to the implant material does not occur. Therefore, a more suitable approach to understanding this term would be based on stability and not histological criteria: "A process whereby clinically asymptomatic rigid fixation of alloplastic materials is achieved, and maintained, in bone during functional loading" [15].

Although modern implants have improved substantially over the last 50 years, the general concept has remained unchanged: replacing a missing tooth with an inert non-biological material. It can be metal, ceramic, or combination of both. The rate of technological improvements in implants has reached a plateau and new developments will require major changes to the basic approach. The recent researches focused on developing new materials, designs, and the introduction of new surfaces, although not all those tendencies were without pitfalls. For example, the ceramic-coated metal implants turned out to be not so successful in the clinical trial due to the brittle coating, which tends to flake away.

A great number of investigations are focused on implantology, but they should be considered with great caution although basic research is needed to search for new materials and designs. *In vitro* studies bring into our knowledge many interesting findings;

however, we must remember that those trials are conducted on glass disk with artificial environment without the influence of hormones, nervous system and blood flow. It so happens that *in vivo* studies not always confirm the *in vitro* results. Therefore, it is essential to finish all the *in vivo* models to initiate the clinical trials.

2.3 Characteristics of the Stomatognathic System

Within the stomatognathic system, we can distinguish two environments. One of them is composed of hard and soft tissues as well as body fluids. From the biophysical point of view, tissue environment has the characteristics of electrochemical membrane. Moreover, it possesses membranes and membrane potentials, which are responsible for ionoselective effect and metabolism. Body fluids, on the contrary, are characterized by ion conductivity. Into this environment, due to a loss of tissue, biomaterials are introduced to repair and to bring back the function of the system. Mostly, these are metal implants or screws or bone materials used in maxillofacial surgery, implantology, and prosthetics. The osseointegrated interface depends on the physical and biophysical characteristics of the surrounding environment and the physicochemical properties of the surface of the implant.

The second environment is composed of the oral cavity, to which materials are introduced by preventive dentistry, prosthetics, and orthodontics. Those implants have hybrid composite structure, where the core is composed of metal, an electric conductor. However, the surface of the implant is coated by polymer or bioceramics, which has the characteristics of an insulator or a semiconductor. Moreover, the whole structure of the biomaterial forms a complex system, which is an essential issue for the development of corrosion or biodegradation of implants. Some of the prosthetic reconstructions are localized in both environments.

Referring to the complexity of the system, biochemical, biomechanical, and bioelectronic factors should be considered, which are undoubtedly connected to the ongoing metabolic, bacteriological, and immunological processes within the system.

Microflora of the oral cavity is specific: containing unique molecular, structural, and microbial characteristics, each of the body sites harboring a normal microflora, therefore forming an ecosystem [11]. It consists of species which are responsible for protecting the surrounding tissues against the pathogenic bacteria. In a physiological state, all bacteria remain in equilibrium. The change in a pathological state occurs when the influence of the negative factors, such as antibiotic administration, fermentation of food, and a dysfunction of the immunological system, is stronger than the protective ability of the system. The physical and chemical nature of the oral cavity is not uniform and changes depending on the localization. For example, anaerobes can be found in gingival sulcus, whereas aerobes exist on supragingival surfaces, due to presence of oxygen. Also, the presence of food changes the oxygen concentration, pH, the metabolites, which in consequence also has its impact on the bacteria species existing in the oral cavity.

The biocompatibility of biomaterials depends on many factors, mainly on composition, location, and interaction with surrounding tissues within the oral cavity [2]. Due to differences in composition of materials, we encounter different biological responses. However, the reaction of the host body depends on whether the implanted material releases any of its components, which may be toxic, mutagenic, or immunogenic to the tissues. Not without any influence is also the location of the implant. Some materials that may be toxic in contact with oral mucosa are biocompatible with the hard tissues and vice versa. We have to take into consideration the fact that the stomatognathic system is variable, taking for instance the pH, the effects of body fluids, which also have their effect on the biocompatibility of materials. The morphology of the biomaterial surface should be designed so, or to elicit the growth of cells, or to prevent the attachment of other cells or retention of plaque.

References

1. Bayne, S.C. (2007). Dental restorations for oral rehabilitation — testing of laboratory properties versus clinical performance for clinical decision making, *J. Oral Rehab.*, **34**, pp. 921–932.
2. Craig, R.G., and Ward, M.L. (1997). *Restorative Dental Materials*, 10th ed., Mosby, St. Louis.
3. Farantalz, G.J., Beckler, I.M., Gremilion, H., and Pink, F. (1998). The effectiveness of equilibration in the improvement of signs and symptoms in the stomatognathic system, *Intern. J. Periodontics Restorative Dent.*, **18**, pp. 595–599.

4. Ferreira, C.F., Magini, R.S., and Sharpe, P.T. (2007). Biological tooth replacement and repair, *J. Oral Rehab.*, **34**, pp. 933–939.
5. Lindhe, J., Karring, T., and Lang, N.P., (2003). *Clinical Periodontology and Implant Dentistry*, Blackwell Munksgaard, Oxford.
6. Lippert, H. (2004). *Lehrbuch Anatomie*, Urban & Schwarzenberg, München.
7. Moore, K.L. (1985). *Clinically Oriented Anatomy*, 3rd ed., William & Wilkins, Baltimore.
8. Panek, H. (2004). Holistic concept of stomatognathic system, *Dent. Med. Probl.*, **41**, pp. 277–280.
9. Scheller, E.L., Krebsbach, P.H., and Kohn, D.H. (2009). Tissue engineering: state of the art in oral rehabilitation, *J. Oral Rehabil.*, **36**, pp. 368–389.
10. Schroeder, A., Sutler, F., Buser, D., and Krekeler, G. (1996). *Oral Implantology*, Thieme Medical Publishers Inc., New York.
11. Tannock, G.W. (1995). *Normal Microflora: An Introduction to Microbes Inhabiting the Human Body*, Chapman & Hall, London.
12. Turp, J.C., Greene, C.S., and Strub, J.B. (2008). Dental occlusion: a critical reflection on past, present and future concepts, *J. Oral Rehabil.*, **35**, pp. 446–453.
13. Tylman, S.D., Malone W.F. (1978). *Tylman's Theory and Practice of Fixed Prosthodontics*, Mosby Company, St. Louis.
14. Woelfel, J.B., and Scheid, R.C. (1997). *Dental Anatomy, Its relevant to Dentistry*, 5th ed., Williams & Wilkins, USA.
15. Zarb, G.A., and Albrektsson, T. (1991). Osseointegration: A requiem for the periodontal ligament? *Int. J. Periodontics Restorative Dent.*, **11**, pp. 88–91.

Table 3.1 Specific requirements for metallic biomaterials

Material properties	Important for application
Mechanical properties	
Ultimate tensile strength, tensile yield strength elongation at fracture, reduction in area fracture	Endosseous implants like orthopedic implants, dental implants, nails, plates, screws
Toughness	
Young's modulus	
Fatigue strength	
Wear resistance	
Physical properties	
Density	Ultrasonic examinations
Acoustic properties	Pacemaker electrodes
Electrical resistance heart	NMR-examinations
Magnetism	Composite materials
Thermal expansion	
Chemical Properties	
Oxidation	Prerequisite for all biomaterials
Corrosion, degradation	
Fretting corrosion	
Biological Properties	
Bioadhesion (osseoconductivity, osseointegration)	Prerequisite for all biomaterials
Immune reaction (allergic, toxic, mutogenic, carcinogenic)	

The range of applications for biomaterials is large (see Table 3.3). The number of different biomaterials is also significant. However, in general, metallic biomaterials are used for load bearing applications and must have sufficient fatigue strength; ceramic biomaterials are generally used for their hardness and wear resistance for applications such as articulating surfaces in joints and in teeth as well as bone bonding surfaces in implants; polymeric materials are usually used for their flexibility and stability, but have also been used for low friction articulating surfaces.

Table 3.2 Some accepted biomaterials

Metals	Ceramics	Polymers
316L stainless steel	Alumina	Ultra-high molecular weight polyethylene
Co–Cr Alloys	Zirconia	Polyurethane
Titanium	Carbon	
Ti–6Al–4V Noble metal alloys	Hydroxyapatite	

Table 3.3 Dental applications of some biomaterials

Metals	Ceramics	Polymers
Implants	Tooth implants	Orthodontic devices (plates, dentures)
Parts of orthodontic devices	Dental porcelains	
Pins for anchoring tooth	Hydroxyapatite (coatings on metallic pins) (fill large bone voids)	

Interestingly, the separation between the three traditional material classes (i.e., metals, ceramics, and polymers) is gradually being replaced by keywords such as *scaffolds* and *composite biomaterials*. This might be explained by the realization that a single material class does not reflect the complexity of highly structured human tissues, and necessitates the use of advanced biomimetic processing techniques to create intelligent biomaterials of similar functionality [29, 119].

Recently, increasing interest has been shown in ceramic–polymer composites as potential fillers of bone defects [65, 143]. Three of the most commonly used composites — calcium phosphate ceramics, tricalcium phosphate, and hydroxyapatite — have demonstrated adequate biocompatibility and suitable osteoconduction and osseointegration [8]. Bioceramic glasses such as 45S5 Bioglass® have also exhibited the capacity to induce bone-bonding, and even vascularization. However, these ceramics are considered too stiff and brittle to be used alone. The addition of a ceramic to a polymer scaffold has several advantages including combining the osteoconductivity and bone-bonding potential of the inorganic phase with the porosity and interconnectivity of the three-dimensional construct. The most prominent natural polymer used to fabricate matrices in composites is collagen type I, probably due to its prevalence in bone's extracellular matrix and its ability to promote mineral deposition and provide binding sites for osteogenic proteins [115, 135]. Although collagen itself is an inadequate bone graft, when combined with ceramics and growth factors, it becomes a powerful inducer of bone regeneration [137].

Scaffolds comprised of synthetic polymers offer many advantages over natural polymers including reproducibility, unlimited supply, relative lack of immunologic concerns, and tailorable properties such as degradation rates and mechanical strength [8, 63]. Synthetic polymers used for bone regeneration include poly(lactic acid) (PLA), poly(glycolic acid) (PGA), poly(lactic-coglycolic) acid (PLGA), polypropylene fumarate (PPF), and the polyhydroxyalkanoates (PHAs) [73]. Combining polymers with ceramics creates bioactive scaffolds that enhance tissue formation with greater initial strength [40].

Recently, titanium–hydroxyapatite [94] and Ni-free austenitic stainless steel/hydroxyapatite [128] nanocomposite biomaterials were synthesized. These composites could be medically beneficial in applications such as better biomedical implants with improved mechanical properties, corrosion resistance, and biocompatibility.

3.2 Metallic Biomaterials

Metallic materials are used in biomedical devices for various parts of the human body (Table 3.4, Fig. 3.2). Metals and its alloys such as stainless steel, Co–Cr–Mo alloy, commercially pure titanium, and titanium alloys are widely used as biomedical materials and are important in medicine and they cannot be replaced with ceramics or polymers at present mainly because of their high strength and toughness (Table 3.5).

Table 3.4 Biometallic devices and metallic biomaterials used for them

Biometallic devices	Metallic biomaterials
Artificial join	Ti–6Al–4V alloy, Co–Cr alloy, 316 stainless steel
Bone plate	Ti–6Al–4V alloy, 316 stainless steel, Ti–6Al–4V alloy
Clip	Ti–6Al–4V alloy, Co–Cr alloy
Crown, bridge	Au–Cu–Ag alloy, Au–Cu–Ag–Pt–Pd alloy, Ti, Ti–6Al–4V alloy
Dental implant	Ti, Ti–6Al–4V alloy, Ti–6Al–7Nb alloy
Pacemaker	electrode: Pt–Ir alloy, Ti; case: Ti, Ti–6Al–4V alloy
Stent	316 stainless steel, Ni–Ti alloy

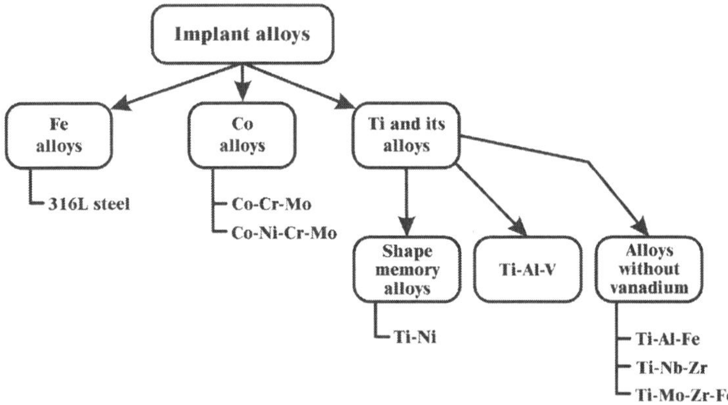

Figure 3.2 Implant alloys.

Table 3.5 Mechanical properties of metallic biomaterials [15, 16]

Metals	Ultimate tensile strength [MPa]	Fatigue strength* [MPa]	Elongation at fracture [%]	BF [$*10^{-3}$]	E [GPa]
CrNi-steels	490–690	200–250	>40	1–12	210
CoCr-alloys	800–1200	550–650	8–40	1.5–2.3	200
CP–Ti	390–450	150–200	22–30	1.4–1.9	100
Ti-6Al-4V	930–1140	350–650	8–15	3.0–5.6	115

* Rotating bending fatigue.

Unfortunately, these materials have exhibited tendencies to fail after long-term use due to various reasons such as high modulus compared to that of bone, low wear and corrosion resistance, and lack of biocompatibility. The various causes for revision surgery are depicted in Fig. 3.3 [46].

From the biomechanical point of view, it is desirable to have a Young's modulus of metallic biomaterials comparable to that of the cortical bone in order to achieve a good load transfer from the implant into the bone, leading to a continuous stimulation of new bone formation (isoelastic material). In order to provide an E modulus comparable to that of compact bone (10–15 GPa), porous sintered implants are required. The reduction of Young's modulus as a function of the porosity can be calculated by means of equation (3.2) [5]:

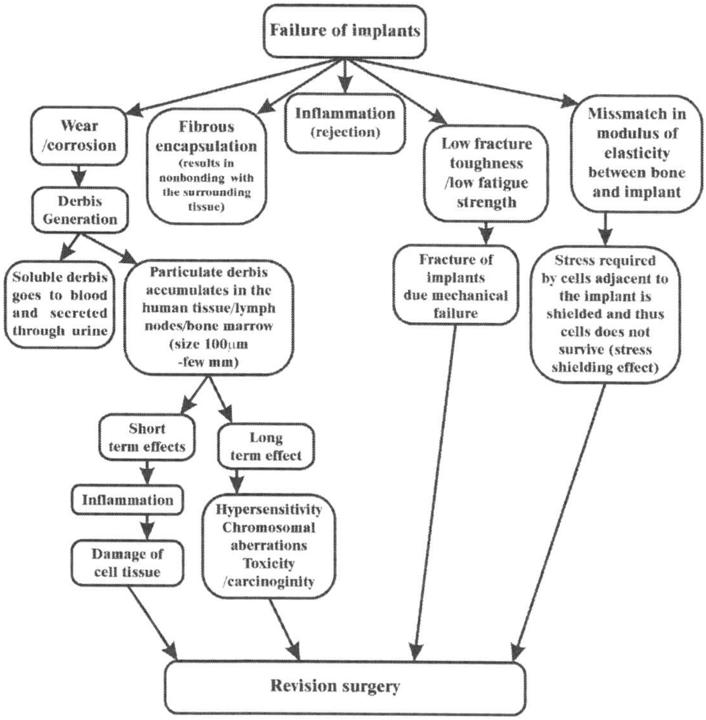

Figure 3.3 Various causes for failure of implants that leads to revision surgery (adopted from [46]).

$$E_p = E_0\,(1 - (1.21 \cdot p^{2/3})) \tag{3.2}$$

where E_p is the Young's modulus of the porous sintered material, E_0 the modulus of the bulk material and p the porosity.

3.2.1 Stainless Steel

In 1913, English metallurgist Harry Brearly discovered that adding chromium to low carbon steel gives it stain resistance. In addition to iron, carbon, and chromium, modern stainless steel may also contain other elements, such as nickel, niobium, molybdenum, and titanium. These elements enhance the corrosion resistance of stainless steel. It is the addition of a minimum of 12% chromium to the steel that makes it resist rust, or stain "less" than other types of steel. The chromium in the steel combines with oxygen in the atmosphere to form a thin, invisible layer of chrome-containing oxide, called the passive film.

Stainless steel is usually divided into five types: ferritic, austenitic, martensitic, precipitation-hardened, and duplex. Only austenitic and martensitic stainless steels are used in medical applications. Austenitic, or 300 series, stainless steels make up over 70% of total stainless steel production. They contain a maximum of 0.15% carbon, a minimum of 16% chromium and sufficient nickel and/or manganese to retain an austenitic structure at all temperatures from the cryogenic region to the melting point of the alloy. Superaustenitic stainless steels, such as alloy AL-6XN and 254SMO, exhibit great resistance to chloride pitting and crevice corrosion due to high molybdenum content (>6%) and nitrogen additions, and the higher nickel content ensures better resistance to stress-corrosion cracking versus the 300 series. The higher alloy content of superaustenitic steels makes them more expensive. They are nominally non-magnetic but usually exhibit some magnetic response depending on the composition and the work hardening of the steel.

On the other hand, martensitic steels are similar to ferritic steels in being based on chromium but have higher carbon levels up as high as 1%. This allows them to be hardened and tempered much like carbon and low-alloy steels. They are used where high strength and moderate corrosion resistance is required. They have generally low weldability and formability and are magnetic.

These types of steel are well-suited for making surgical instruments: they are easy to clean and sterilize, strong and corrosion-resistant. The nickel/chrome/molybdenum alloys are also used for orthopedic implants as aids in bone repair, as a structural part of artificial heart valves, and other implants. Immune system reaction to nickel is a potential complication. Most surgical equipment is made out of martensite steel. It is much harder than austenitic steel, and it is easier to keep it sharp.

Implants are made out of austenitic steel, often 316L and 316LVM, because it is less brittle (see Tables 3.6 and 3.7). Grade 316 is the standard molybdenum-bearing grade, second in importance to 304 amongst the austenitic stainless steels. The molybdenum gives 316 better overall corrosion resistant properties than Grade 304, particularly higher resistance to pit and crevice corrosion in chloride environments. Grade 316L, the low carbon version of 316 and is immune from sensitization (grain boundary carbide precipitation).

The austenitic structure also gives these grades excellent toughness, even down to cryogenic temperatures. Additionally, grade 316 surgical steel is used in the manufacture and handling of food and pharmaceutical products where it is often required in order to minimize metallic contamination.

Table 3.6 Comparison of standardized some 300-type stainless steels

EN-standard steel no. k.h.s DIN	EN-standard steel name	SAE grade	UNS
1.4408	G-X 6 CrNizMo 18-10	316	
1.4307	X2CrNi18-9	304L	S30403
1.4306	X2CrNi19-11	304L	S30403
1.4311	X2CrNiN18-10	304LN	S30453
1.4301	X5CrNi18-10	304	S30400
1.4404	X2CrNiMo17-12-2	316L	S31603
1.4401	X5CrNiMo17-12-2	316	S31600
1.4406	X2CrNiMoN17-12-2	316LN	S31653
1.4432	X2CrNiMo17-12-3	316L	S31603
1.4435	X2CrNiMo18-14-3	316L	S31603
1.4436	X3CrNiMo17-13-3	316	S31600
1.4571	X6CrNiMoTi17-12-2	316Ti	S31635
1.4429	X2CrNiMoN17-13-3	316LN	S31653

UNS, unified numbering system.

Table 3.7 Characteristics of some 300 grade stainless steels (austenitic chromium–nickel alloys)

Type	Characteristics
304	the most common grade; the classic 18/8 stainless steel
304L	same as the 304 grade but contains less carbon to increase weldability; is slightly weaker than 304
304LN	same as 304L, but also nitrogen is added to obtain a much higher yield and tensile strength than 304L
316	the second most common grade (after 304); for surgical stainless steel uses; alloy addition of molybdenum prevents specific forms of corrosion
316L	extra low carbon grade of 316
316LN	same as 316L, but also nitrogen is added to obtain a much higher yield and tensile strength than 316L

Grade 316LVM austenitic stainless steel is initially electric-arc melted and as a refinement to the purity and homogeneity of the metal, 316LVM is Vacuum Arc Remelted (VAR). This process yields a more uniform chemistry with minimal voids and contaminants. A reducing atmosphere is preferred for thermal treatment, but inert gas can be used. 316LVM will fully anneal at 1010–1121°C in just a few minutes. The precipitation of carbides that decreases corrosion resistance in other 300 series alloys is controlled by reduced carbon content in 316LVM. This steel has been used for permanent implants for many years. The corrosion resistance in the annealed condition is good. This stainless steel has good ductility in the cold worked condition. Applications include: suture wire, orthopedic cables, skin closure staples, catheters, stylets, bone pins, and many small machined parts.

Nowadays, stainless steel is a frequently used biomaterial for internal fixation devices because of a favorable combination of mechanical properties, corrosion resistance, and cost effectiveness when compared to other metallic implant materials (Tables 3.8 and 3.9). The biocompatibility of stainless steel has been proven by successful human implantation for decades. Metallurgical requirements are stringent to ensure sufficient corrosion resistance, non-magnetic response and satisfactory mechanical properties [64].

Table 3.8 Typical physical properties for 316L grade stainless steel

Density (kg/m³)	Elastic modulus (GPa)	Mean coefficient of thermal expansion (µm/m/°C)		Thermal conductivity (W/m·K)		Specific heat 0–100°C (J/kg·K)	Electric resistivity (nΩ·m)
		0–100°C	0–315°C	at 100°C	at 500°C		
8000	193	15.9	16.2	16.3	21,5	500	740

Table 3.9 Mechanical properties of 316L stainless steel

Tensile str. (MPa) min	Yield str. 0.2% proof (MPa) min	Elong. (% in 50 mm) min	Hardness rockwell B (HR B) max	Hardness brinell (HB) max
485	170	40	95	217

In orthopedic and trauma surgery, stainless steel is mainly used when stiffness is required. The ductility of stainless steel is higher than that of cp titanium, because of its hexagonal crystal structure. This makes contouring of stainless steel plates easier, compared to titanium plates. Therefore, reconstruction plates made of stainless steel are favored in acetabular and pelvic surgery and at other anatomical locations.

Due to their high nickel content (10–14%), stainless steel implants can cause negative tissue reactions and dermatitis (Table 3.10) [24, 76]. Moreover, the nickel ions released during corrosion of the device are reported to cause allergies and even cancer [76]. The World Health Organization (WHO) estimated that a content of less than 0.2% of nickel is congruous with medical requirements [24]. Therefore the development of materials with improved corrosion resistance and without nickel is absolutely imperative. One of the most promising austenitizing elements to replace nickel is nitrogen [103]. Nitrogen increases the stability of austenite, enhances corrosion resistance, and prevents the formation of sigma phase [124]. By this method, small devices with the ferritic structure can be precisely machined and subjected to nitrogenization of their surfaces in nitrogen gas at temperature of about 1200°C. They are then nickel-free austenitic stainless steels with improved mechanical properties and corrosion resistance [128]. It is expected, that these new stainless steel alloys may have a similar biocompatibility as cp titanium implants.

Table 3.10 Chemical composition of some 300 grade austenitic stainless steels

SAE designation	% Cr	% Ni	% C	% Mn	% Si	% P	% S	% N	Other
304	18–20	8–10.50	0.08	2	0.75	0.045	0.03	0.1	—
304L	18–20	8–12	0.03	2	0.75	0.045	0.03	0.1	—
304N	18–20	8–10.50	0.08	2	0.75	0.045	0.03	0.10–0.16	—
316	16–18	10–14	0.08	2	0.75	0.045	0.03	0.10	2.0–3.0 Mo
316L	16–18	10–14	0.03	2	0.75	0.045	0.03	0.10	2.0–3.0 Mo
316N	16–18	10–14	0.08	2	0.75	0.045	0.03	0.10–0.16	2.0–3.0 Mo
316LVM	17.57	14.68	0,023	1.84	0.37	0.014	0.01	0.03	2.79 Mo 0.03 Cu

3.2.2 Cobalt Alloys

Co–Cr–Mo alloys are well used for biomedical applications such as dental and orthopedic implants owing to their excellent mechanical properties, wear resistance, and biocompatibility [83]. Their biocompatibility is closely linked to the high resistance toward corrosion thanks to the spontaneous formation of a passive oxide film with elevated chemical and mechanical stability [57]. It is indicated that the cast alloys (ASTM-75) show low ductility and fatigue strength compared to those of the forged alloys (ASTM-799). Cast alloys have a coarse grain, casting defects like microporosity, and a cored matrix. For long-term requirements in arthroplasty, a high corrosion resistance combined with a good biocompatibility is needed. The fabrication process with the use of Co, Cr, and Mo allows achieving extremely hard implants with a high corrosion resistance in body fluids [83]. Cobalt-base alloys (e.g., CoCrW, CoCrMo) have a Young's modulus of about 250 GPa combined with a high wear and heat resistance.

Many properties of the alloy originate from the crystallographic nature of cobalt, the solid-solution-strengthening effect of chromium and molybdenum, the formation of extremely hard carbides, and the corrosion resistance imparted by chromium. Various *in vitro* and *in vivo* tests have shown that the alloys are bio-compatible and suitable for use as surgical implants.

It is important to note, what the brittle σ phase formed along the interdendritic region deteriorate mechanical properties [74]. In addition, to increase the ductility of Co–Cr–Mo alloys, the γ phase (fcc) is more proper than the ε phase (hcp) considering the number of the slip plan. Therefore, it is very important for Co–Cr–Mo alloys to suppress the σ phase formation and to keep a matrix as the γ phase. In cast alloys, alloy design is the foremost valuables improving mechanical properties and microstructures. As for alloying elements, Cr and Mo are substantial elements, and C and N are well known interstitial elements. Among the main substantial elements, Cr is known for increasing strength of the cobalt alloys and also known for a ε phase stabilizer and have a tendency of forming an σ phase. In Fe–Cr systems, it is well known that the N addition stabilizes austenite and inhibits an σ phase formation, and increasing Cr content can increase N solubility [74]. This result brings about a significant improvement in the mechanical properties of the Ni-free Co–Cr–Mo alloy and even under the as-cast condition. Thus, the Co–Cr–Mo alloys with enriched Cr content up to 34 wt%, modified by N addition, are suitable for artificial hip and knee joints, and dental implants that are produced on the basis of the casting process such as investment casting.

Available are several popular Co-based medical alloys. They include: vitallium, protasul-2, protasul-10, protasul-21 WF (Table 3.11), endocast, MP35N (Table 3.12). Mechanical properties and corrosion resistance of cobalt-base alloys are function of chemical compositions of Co-base materials and their microstructures (Tables 3.13 and 3.14).

Table 3.11 Chemical composition of protasul-21 WF [wt %]

Cr	Mo	Co	Ni	Fe	C	Si	Mn	W	N
26–30	5–7	bal	<1	<0,75	<0,35	<1	<1	<0,2	<0,25

Table 3.12 Chemical composition of MP35N [wt %]

Cr	Mo	Co	Ni	Fe	C	Si	Mn	Ti	B, S	P
19–21	9–10	bal	33–37	1.0	0.03	0.01	0.15	1.0	0.01	0.015

Alloying theory as used in alloy development or design relates the properties of the alloys to their chemical compositions and microstructures [85, 120, 123]. A required property is obtained by selection of suitable chemical compositions and fabrication processes (e.g., melting, casting, forming and heat treatment) to obtain the desired microstructures. For biomedical materials, such as dental and joint implant materials, corrosion resistance, and mechanical properties (e.g., strength, ductility and work-ability, etc.) are of prime importance for optimum clinical performances [120].

Table 3.13 Mechanical properties of MP35N alloy

Properties		Comments
Hardness, Rockwell B	90	Annealed
Tensile strength, Ultimate	896 MPa	Annealed
Tensile strength, Yield	379 MPa	0.2% YS; Annealed
Elongation at break	65.0 %	Annealed; in 4D
Reduction of area	75.0 %	
Modulus of elasticity	232.8 GPa	Annealed
	234.8 GPa	Cold Worked and Aged
	201.0 GPa Temperature 482°C	Annealed
	201.3 GPa Temperature 482°C	Cold Worked and Aged
Shear modulus	67.8 GPa	Cold Worked and Aged; at 482°C
	70.60 GPa	Annealed; at 482°C
	80.94 GPa	Cold Worked and Aged; at 78°F
	83.36 GPa	Annealed; at 78°F

Table 3.14 Mechanical properties of protasul-21 WF after different processing

Properties	Annealing	Hot plastic forming	Cold plastic forming
$R_{0,2}$ [MPa]	550	700	827
R_m [MPa]	750	1000	1172
A [%]	16	12	12

3.2.3 Titanium and Titanium Alloys

Since 1965, cp titanium has found an increasing application as an implant material in medicine (Table 3.15). Cp titanium has high corrosion resistance and outstanding biocompatibility [1, 2, 46, 47, 71, 110]. One reason for these advantages may be a protective oxide layer, which forms spontaneously on the implant surface as long as oxygen is present. Variable film thickness can be electrochemically created yielding various colors of implants [33].

Biomechanical properties of cp titanium depend mostly on the amount of trace elements present. Four grades of cp titanium are distinguished for medical applications (ISO 5832-2). The grading depends on the amounts of nitrogen, carbon, hydrogen, iron, and oxygen (Tables 3. 16 and 3.17). Increasing amounts of trace elements elevate tensile strength but reduce ductility of cp titanium. Titanium plates, used for osteosynthesis, offer less stress-shielding to bone tissue, because of their low Young's modulus of 68 GPa [110]. Therefore, titanium is an ideal biomaterial for plates and screws used for fracture treatment.

Titanium exists in two allotropic forms. At low temperatures, it has a closed packed hexagonal crystal structure (cph), which is commonly known as α, whereas above 883°C, it has a body centered cubic structure (bcc) termed β. The α to β transformation temperature of pure titanium either increases or decreases based on the nature of the alloying elements. The alloying elements (such as Al, O, N, etc.) that tend to stabilize the α phase are called alpha stabilizers and the addition of these elements increase the beta transus temperature, while elements that stabilize β phase are known as beta stabilizers (V, Mo, Nb, Fe, Cr, etc.) and addition of these elements depress the β transus temperature. Some of the elements that do not have marked effect on the stability of either of the phase but form solid solutions with titanium are termed as neutral elements (Zr and

Sn). It has been shown that the addition of Zr stabilizes the β phase in Ti–Zr–Nb system.

Table 3.15 Physical properties of titanium

Properties			Value
Atomic number			22
Atomic mass			47.90
Crystallographic structure			
	α (hexagonal)	*a* [Å]	4.6832±0.0004
		c [Å]	2.9504±0.0004
	β (cubic)	*a* [Å]	3.28±0.003
Density [g/cm^3]			4.54
Temperature coefficient of heat expansion α at 20°C [K^{-1}]			$8.4 \cdot 10^{-6}$
Thermal conductivity [W/(m·K)]			19.2
Melting temperature [°C]			1668
Evaporation temperature [°C]			3260
Temperature of allotropic transformation [°C]			882.5
Electric resistivity			
High-purity titanium [μΩ·cm]			42
Commercial-purity titanium [μΩ·cm]			55
Young's modulus [GPa]			105
Yield strength [MPa]			692

Table 3.16 Titanium and its alloy grades

Grades	ISO
Cp-Titanium Grade 1	ISO 5832-2
Cp-Titanium Grade 2	ISO 5832-2
Cp-Titanium Grade 4	ISO 5832-2
Titanium alloy Ti Al6V4 (ELI)	ISO 5832-3
Titanium alloy Ti Al6Nb7	ISO 5832-11

Table 3.17 Chemical compositions of different titanium grades

Grade		C	Fe	O	H	N	Al	V	Ti
1	max.	0.08	0.20	0.18	0.01	0.03	—	—	bal.
2	max.	0.08	0.30	0.25	0.01	0.03	—	—	bal.
4	max.	0.08	0.50	0.40	0.01	0.05	—	—	bal.
		0.08	0.25	0.13	0.01	0.05	5.50	3.50	
ELI	max.						6.50	4.50	bal.

ELI – extra low interstitials

The α and β phases also form the basis for normally accepted classification of titanium alloys. Alloys having only α stabilizer and consisting entirely of a phase are known as α alloys. Alloys containing 1–2% of α stabilizers and about 5–10% of β phase are termed as near α alloys. Alloys containing higher amounts of α stabilizers, which results in 10–30% of β phase in the microstructure, are known as $\alpha + \beta$ alloys. Alloys with higher β stabilizers where β phase can be retained by fast cooling are known as metastable β alloys. These alloys decompose to $\alpha + \beta$ on aging. Most of the biomedical titanium alloys belong to $\alpha + \beta$ or metastable β class.

An alternative and potentially more attractive method for enhancing the mechanical performance of commercial-purity titanium has recently been reported by Valiev and coworkers [131]. These investigators have investigated the strengthening of grade 2 commercial-purity titanium utilizing equal channel angular pressing (ECAP) in combination with other deformation processes. Procedures examined include ECAP (8 passes) at 400°C (#1), ECAP + 65% cold rolling (#2), and ECAP + rolling followed by annealing at 300°C, for 1 h (#3). Ultrafine grained (UFG) structures can range from an equiaxed cellular microstructure to a sub-grain structure with a defined boundary structure. In all cases, the micro hardness of severely deformed commercial-purity titanium was superior to that of the original coarse-grained commercial-purity titanium (Table 3.18). The yield and ultimate tensile strengths also exhibit this enhancement, a 140% increase in ultimate tensile strength vis-à-vis coarse-grained commercial-purity titanium being observed. Notably this increase was achieved while maintaining an elongation to failure of 9%.

Table 3.18 Microhardness, tensile mechanical properties and fatigue limit of grade 2 Cp Ti in different states in comparison to Ti–6Al–4V ELI [131]

Structure type	HV (MPa)	UTS (MPa)	YS (MPa)	EL. (%)	RA (%)	Fatigue limit (MPa)
Coarse-grained	1800	460	380	26	60	238 ± 10
UFG #1 (Equiaxed, submicron-grained)	2700	710	625	14	60	403 ± 8
UFG #2 (Fibrous, with high dislocation density)	2821	960	725	10	45	434 ± 5
UFG #3 (subgrained with internal cells)	2850	1100	915	9	40	500 ± 8

(Contd)

Table 3.18 (*Contd*)

Structure type	HV (MPa)	UTS (MPa)	YS (MPa)	EL. (%)	RA (%)	Fatigue limit (MPa)
Ti-6Al-4V ELI (annealed)	—	965	875	10-15	25-47	515

To improve mechanical strength of titanium implants, α + β titanium alloys have been developed (Tables 3.19–3.21, Fig. 3.4). These titanium alloys have the greatest commercial potential and are widely used as load-bearing orthopedic implants due to their relatively good fatigue resistance and biological passivity [36].

Table 3.19 Ti-based biomaterials

No.	Chemical composition	Crystallographic structure
1	Cp Ti — grades 1, 2, 3, 4	α
2	Ti-6Al-4V ELI	α + β
3	Ti-6Al-4V	α + β
4	Ti-6Al-7Nb	α + β
5	Ti-5Al-2,5Fe	β-rich, α + β
6	Ti-5Al-3Mo-4Zr	α + β
7	Ti-15Sn-4Nb-2Ta-0.2Pd	α + β
8	Ti-15Zr-4Nb-2Ta-0.2Pd	α + β
9	Ti-13Nb-13Zr	mostly β
10	Ti-12Mo-6Zr-2Fe	β
11	Ti-15Mo	β
12	Ti-16Nb-10Hf	β
13	Ti-15Mo-5Zr-3Al	β
14	Ti-15Mo-3Nb	β
15	Ti-35.3Nb-5.1Ta-7.1Zr	β
16	Ti-29Nb-13Ta-4.6Zr	β

Table 3.20 Mechanical properties of biomedical Ti-based alloys

Composition	R_m [MPa]	R_e [MPa]	*A* [%]	Z [%]	*E* [GPa]	Structure type
Ti, grade 1	240	170	24	30	102.7	α
Ti, grade 2	345	275	20	30	102.7	α
Ti, grade 3	450	380	18	30	103.4	α
Ti, grade 4	550	485	15	25	104.1	α
Ti-6Al-4V ELI	860–965	795–875	10–15	25–47	101–110	α + β
Ti-6Al-4V (a)	895–930	825–869	6–10	20–25	110–114	α + β
Ti-6Al-7Nb	900–1050	880–950	8–15	25–45	114	α + β
Ti-5Al-2.5Fe	1020	895	15	35	112	α + β

(*Contd*)

Table 3.19 (*Contd*)

Composition	R_m [MPa]	R_e [MPa]	*A* [%]	Z [%]	*E* [GPa]	Structure type
Ti-5Al-1.5B	925–1080	820–930	15–17	36–45	110	
Ti-15Sn-4Nb-2Ta-0.2Pd						
(a)	860	790	21	64	89	
(b)	1109	1020	10	39	103	
Ti-15Zr-4Nb-4Ta-0.2Pd						α + β
(a)	715	693	28	67	94	
(b)	919	806	18	72	99	
Ti-13Nb-13Zr (b)	973–1037	836–908	10–16	27–53	79–84	β
Ti-12Mo-6Zr-2Fe (a)	1060–1100	900–1060	18–22	64–73	74–85	β
Ti-15Mo (a)	874	544	21	82	78	β
Ti-15Mo-5Zr-3Al (b)	1060–1100	1000–1060	18–22	64–73	—	
Ti-15Mo-2,8Nb-0.2Si (a)	979–999	945–987	16–18	60	83	β
Ti-35,3Nb-5,1Ta-7.1Zr	597	547	19	68	55	β
Ti-29Nb-13Ta-4.6Zr (b)	911	864	13	—	80	β

(a) Heat treatment.
(b) Aging treatment.

Table 3.21 Mechanical properties of some Ti-based dental alloys

Composition	Synthesis method	R_m [MPa]	R_e [MPa]	*A* [%]	HV
Ti-20Cr-0.2Si	Casting	874	669	6	318
Ti-25Pd-5Cr	Casting	880	659	5	261
Ti-13Cu-4.5Ni	Casting	703	—	2,1	—
Ti-6Al-4V	Casting	976	847	5,1	—
Ti-6Al-4V	Forming	954	729	10	346
Ti-6Al-7Nb	Casting	933	817	7,1	—
Ti-Ni	Casting	470	—	8	190

In all conventional titanium alloys, α to β transus temperature (known as β transus) plays a central role in evolution of microstructure and is of great technological importance in determining heat treatment and processing schedule. The various microstructures that are developed under different thermomechanical processing conditions are shown in Fig. 3.4 [46]. The alloys processed/heat treated above the β transus temperature result in acicular or lamellar structure and are typically known as β treated structure. When these alloys are mechanically processed below the β transus (α + β phase field) and heat treated in α + β phase region, the microstructure consists of a mixture of equiaxed α and β phases. Depending upon the alloy chemistry, heat treatment temperatures, and cooling rate, volume fraction of equiaxed α and nature of β phases may change. In faster cooled structure, transformed β phase may constitute martensite or α laths along with the retained β, while on slow cooling the transformed β phase may entirely be retained β. In metastable β alloys, the β phase is usually retained on quenching from the β phase field and very fines a precipitates on aging at lower temperatures, which leads to extremely high strength in these alloys [46].

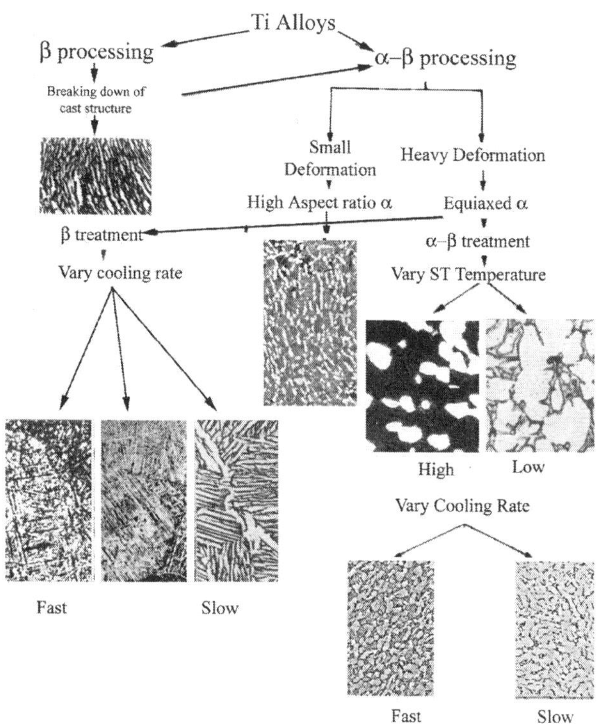

Figure 3.4 Influence of thermomechanical processing on development of various microstructure in alpha beta titanium alloys [46].

Current goals in the development of new Ti-based biomaterials are: (i) to avoid potentially toxic elements, such as vanadium, to further improve biocompatibility; (ii) to produce titanium alloys with a high fatigue strength but a low Young's modulus compared to cortical bone (E = 10–25 GPa), to minimize stress shielding and improve fracture healing. β-titanium alloys partially fulfill these requirements. They were developed already in the 1980s and 1990s, but it usually takes years before new implant materials find their way into clinical application. One of these improved β titanium alloys is $Ti_{12}Mo_6Zr_2Fe$, which is currently used for implant materials. Whether these new implant materials will prove to have superior clinical properties compared to cp titanium and α + β titanium alloys, has to be proven in clinical trials.

Implant sensitivity to titanium alloys is very seldom, despite components such as vanadium which are described to be cytotoxic [145]. *In vivo* studies have shown that Ti–6Al–4V has similar biological compatibility to vanadium-free titanium alloys [40].

Titanium and its alloy are also getting much attention in dental applications. Currently, Ti-based alloy is expected as an alternative candidate to Ag–Pd–Au–Cu alloy. The representative dental titanium alloys reported and their mechanical properties are shown in Table 3.21 [95, 97, 102].

Some alloys have very different chemical compositions from those for implants. Pure titanium and Ti–6Al–4V are the main materials in the dental field as well as in the surgical field. Ti–6Al–7Nb, which has been developed for surgical implants, is also attractive for dental applications [68]. The casting process is dominant in dental applications and especially the elongation which is very low although the strength is kept to be high. Therefore, the development in elongation without reducing strength is investigated in cast titanium alloys.

Recently, Ti–40Zr, Ti–5Al–13Ta, and Ti–43.1 at.% Zr–10.2 at.% Al–3.6 at.% V [37, 67] have been proposed. Titanium alloys are very reactive and have relatively higher melting point comparing with other dental alloys like Au based alloys and Ag based alloys. Therefore, the low melting point titanium alloys and low reactive mold materials are desired for dental precision castings. In dental precision casting, alumina based, and magnesia based mold materials are mainly used [146]. Magnesia based mold material is more suitable for dental precision casting of titanium alloys comparing with alumina based mold materials [68]. Calcia based mold material is also reported to be suitable for dental precision casting [88, 89].

It is important to note that between Ti-based alloys, only TiNi has been put into wide practical use as super elastic and shape memory alloy. Formerly, TiNi shape memory alloy was tried to apply to implants [15, 16, 45]. However, since TiNi contains a large amount of Ni, which causes allergy at high rate, the usage of TiNi shape memory alloys is restricted. Therefore, the research and development of Ni-free super elastic and shape memory titanium alloys composed of non-toxic elements for biomedical applications are increasing.

Recently, the super multifunctional titanium alloy "GUM METAL" has been developed. Gum metal is a multi-functional β-type titanium alloy that has a body centered cubic structure with an ultra-low elastic modulus (40 GPa), high strength (more than 1.100 MPa), high elastic deformability (2.5%), high plastic deformability at room temperature without work hardening, non-linear elasticity without any hysteresis (the difference in stress-strain curve in uploading and down-loading passes causing energy loss in elastic deformation) as well as Invar and Elinvar properties. Gum metal has been put into practical use for glass frames. The chemical composition of gum metal is very similar to that of Ti–Nb–Ta–Zr system alloy for biomedical applications [45, 104].

The super elastic behavior has been observed in Ti–29Nb–13Ta–4.6Zr for biomedical applications [45, 104]. Ti–29Nb–13Ta–4.6Zr has been also put into practical use for glass frames as a brand name of bio-titan. Ti–Nb–Sn system alloy is developing as Ni-free shape memory titanium alloy for biomedical applications. Its martensite transformation temperature (M_s point) decreases with an increasingly amount of Nb or Sn, and the shape memory effect is recognized when the alloy is deformed below austenite transformation temperature (A_f point) which is similar to the case of Ti–Ni shape memory alloy. It has been reported that an elastic strain of 3.5% is obtained at the composition of Ti–18 at.% Nb–4 at.% Sn [100]. The research and development of Ti–Mo–Ga system alloy, Ti–Mo–Ge system alloy, or Ti–Mo–Al system alloy, Ti–Ta system alloy, Ti–Ta–Zr system alloy, Ti–Sc–Mo system alloy as shape memory titanium alloys for biomedical applications are also noticeable [95, 97]. They are all β type titanium alloys.

3.2.4 Noble Metal Alloys

Noble metal alloys are widely employed in dentistry [55, 81, 93, 101, 112]. In the production of dental restorations, gold, and platinum alloys have been preferred because they exhibit desirable properties during casting and hardening, and various porcelains have been developed for use therewith to provide the desired wear properties and aesthetic appearance.

Dental alloys based on Ag and Pd as the main alloying components has been in use since the 1930s [27]. Typical dental alloys within this system contain from 50–57% Ag, 25–33% Pd, 10–15% Cu, and 5–20% Au (in weight %). The main reason for their application is that they cost considerably less than do gold alloys, and, at the same time, they are claimed to have satisfactory service properties as dental restoratives.

The structure of a commercial dental Ag–Pd–Cu–Au casting alloy has been studied by Her *et al.* after various heat treatments [55]. After being annealed at 400°C, 500°C, and 600°C for seven weeks, the alloy consisted of three phases: a Cu- and Pd-rich fcc phase (α_1) with $a = 0.372$ nm, a Ag-rich matrix (α_2) with $a = 0.399$ nm, and an ordered CsCl-type bcc PdCu phase with $a = 0.296$ nm. The PdCu phase was not observed above 600°C, and the proportion of the α_1 phase decreased sharply above 700°C. After being annealed at 900°C, the alloy matrix was partly decomposed at the Cu-enriched grain boundaries. The decomposed areas grew into the grain interior during subsequent precipitation hardening. No segregation of Au was detected after casting, and the element was evenly distributed throughout the alloy structure after all heat treatments.

Ag–Pd–Cu-based casting alloys are used in dentistry as substitutes for the costly high-gold alloys too [61]. Increasing the Pd content of a single-phase PdCu alloy has generally been shown to increase its corrosion resistance [129]. Furthermore, it is common practice to add about 1 wt% Zn to noble and precious dental alloys in order to avoid the oxidation of Cu during melting.

3.3 Ceramic Biomaterials

Ceramic biomaterials became an accepted group of materials for medical applications. The selection and application of synthetic materials for surgical implants has been directly dependent upon the biocompatibility profiles of specific prosthetic devices [20, 32, 69,

75, 127]. Bioceramics have become a diverse class of biomaterials presently including three basic types: bioinert high strength ceramics, bioactive ceramics which form direct chemical bonds with bone or even with soft tissue of a living organism; various bioresorbable ceramics that actively participate in the metabolic processes of an organism with the predictable results (Table 3.22) [127]. Alumina, zirconia, and carbon are termed bioinert. Bioglass and glass ceramics are bioactive. Calcium phosphate ceramics are categorized as bioresorbable. In the search to improve the biocompatibility and mechanical strength of implant materials, attention has been directed toward the potential use of ceramic/ceramic composites, too.

Table 3.22 Ceramics used in biomedical applications

Ceramic	Chemical formula	Comments
Alumina	Al_2O_3	bioinert
Zirconia	ZrO_2	
Carbon	C	
Bioglass	Na_2O CaO P_2O_5-SiO_2	bioactive
Tricalcium phosphate	$Ca_3(PO_4)_2$	bioresorbable
Hydroxyapatite	$Ca_{10}(PO_4)_2(OH)_2$	

Applications of ceramic biomaterials range from bulk ceramic structures as joint and bone replacements to fully or partially biodegradable substrates for the controlled delivery of pharmaceutical drugs, growth factors, and morphogenetically inductive substances [127]. Mechanical properties of some ceramic biomaterials in comparison to natural bone are presented in Table 3.23.

Table 3.23 Mechanical properties of ceramic biomaterials in comparison to natural bone

	Young's modulus *E* [GPa]	Compressive strength σUCS [MPa]	Hardness HV	Fracture toughness [Mpa $m^{1/2}$]	Density [g/cm^3]
Alumina	380	4500	2000–3000	5.0–6.0	>3.9
Zirconia (PS)	150–200	2000	1000–3000	4.0–12.0	6.0
Graphite	20–25	138	NA	NA	1.5–1.9
Pyrolitic carbon	17–28	900	NA	NA	1.7–2.2
Bioglass	75	1000	NA	0.7	2.5
Hydroxyapatite	73–117	600	NA	<1	3.1
Natural bone	3–20	130–180	NA	3–6	NA

PS, partially stabilized; NA, not available.

3.3.1 Bioinert Ceramics

The term bioinert refer to any material that has minimal interaction with its surrounding tissues once placed within human body. This type of bioceramic shows minimal interfacial bonds with the living tissues [53]. Single oxide ceramic, alumina (Al_2O_3), zirconia (ZrO_2), and carbon are typical examples of bioinert ceramic. In a human body, they are expected to be nontoxic, non-allergenic, and non-carcinogenic for a lifetime, which leads to a corresponding range of engineering design philosophies for medical application [82].

3.3.1.1 Alumina

Aluminum oxide, also known as alumina, is a well-proven, biocompatible ceramic that has been used as a dental restorative material for many years. An alumina ceramic has characteristics of high hardness and high abrasion resistance. The reasons for the excellent wear and friction behavior of Al_2O_3 are associated with the surface energy and surface smoothness of this ceramic. There is only one thermodynamically stable phase, i.e., Al_2O_3 that has a hexagonal structure with aluminum ions at the octahedral interstitial sites [50, 62]. Abrasion resistance, strength, and chemical inertness of alumina have made it to be recognized as a ceramic for dental and bone implants. Aluminum oxide has been used as a dental porcelain pigment for some 60 years and as a ceramic restoration substructure for 25 years. Dental patients should rest assured that dental ceramics composed of aluminum oxide are not a source of soluble metallic aluminum in the body.

The new generation alumina ceramics are manufactured by hot isostatic pressing. They have high density (3.98 g/cm^3), high purity, and fine grains (1.8 μm). Fine grain structure of alumina allows achieving very low surface roughness after polishing (less than 0.05 μm R_a), which determines extremely low wear rate and low coefficient of friction.

The benefits of alumina-on-alumina hip prostheses are: low wear rate, high hardness (HV = 2200), high scratch resistance (low three body abrasive wear), low coefficient of friction, excellent surface finish, no ion dissolution, high strength, high fatigue strength, good wetting by the synovial fluid, and biocompatibility.

The main disadvantage of alumina as a material for hip prostheses is its low fracture toughness and strength. The fracture toughness and four-point flexural strength of alumina were considerably improved due to high density and fine grain structure (four-point bending flexural strength of the new generation alumina has reached 630 MPa). Further improvement of the fracture toughness and flexure strength was made in alumina with additions of 17–20 vol.% of zirconia (ZrO_2) and some other oxides (e.g., chromium oxide). Zirconia toughened alumina (ZTA) has flexural strength of 1100 MPa.

The biocompatibility of alumina ceramic has been tested. Takami *et al.* investigated Al_2O_3 ceramic as a blood-contacting material using a standard assessment *in vitro* and *in vivo* analysis [125]. The examined items were systemic toxicity, sensitization (guinea pig maximization test), cytotoxicity (elution test), mutagenicity (Ames test), direct contact hemolysis, and thrombogenicity. These findings indicate that Al_2O_3 is biocompatible material for double-pivot bearings in the centrifugal blood pump.

3.3.1.2 Zirconia

Zirconia ceramics has been used in medical applications since 1985 [35, 36, 107, 108, 134]. Polycrystalline tetragonal zirconia possesses high fracture toughness and flexural strength due to its dense and fine grain structure (Table 3.24). Pure tetragonal zirconia is metastable. Under an influence of temperature and mechanical stress, tetragonal zirconia may transform to the monoclinic form. The phase transformation is associated with volume expansion of approximately 3–4%. This means that components made of pure zirconium oxide would burst due to volume increase of grains and tension. In order to depress possible allotropic transformation of tetragonal zirconia, it is doped with yttria (about 5.15%). Yttrium tetragonal zirconia polycrystals (Y-TZP) based systems are the more recent addition in to the high-strength all-ceramic systems that are used for crowns and fixed partial dentures [108].

Table 3.24 Mechanical properties of TZP [108]

Property	TZP material
Color	White
Chemical compositions	Zirconium oxide and Yttrium oxide 3 mol% Hafnium oxide < 2% Aluminum oxide + Silicone oxide <1% Total 100%
Density g cm^{-3}	>6
Porosity %	<0.1

(*Contd*)

Table 3.24 *(Contd)*

Property	TZP material
Bending strength MPa	900–1200
Compression strength MPa	2000
Fracture toughness K_{IC}	7–10
Coefficient of thermal expansion K^{-1}	11×10^{-6}
Thermal conductivity WmK^{-1}	2
Hardness $HV_{0.1}$	1200

In vitro evaluation of the mutagenic and carcinogenic, capacity of the high-purity zirconia ceramic confirmed that it did not elicit such effects on the cells [25]. In 1990s, zirconium material was used as endodontic posts [70] and as implant abutments [17]. This heralded the use of zirconium in to dentistry. Due to its excellent physical properties, white color, and superior biocompatibility, it is being evaluated as an alternative framework for full coverage all-ceramic crowns and fixed partial dentures (FPD). Dental ceramics can be classified according to their crystalline phase and fabrication technique (Table 3.25). From feldspathic porcelains to zirconia-based all-ceramics, tremendous progress has been made in terms of mechanical performance, with a 10-fold increase in flexural strength and fracture toughness.

Table 3.25 Classification of dental ceramics [32]

	Fabrication technique	Crystalline phase
Metal-ceramics	Sintering	Leucite
	Heat-pressing on metal	Leucite, leucite & fluorapatite
All-ceramics	Sintering	Leucite
	Heat-pressing	Leucite, lithium disilicate
	Dry pressing and sintering	Alumina
	Slip-casting & glass infiltration	Alumina, spinel, alumina–zirconia (12Ce-TZP)
	Soft machining & glass-infiltration	Alumina, alumina–zirconia (12Ce-TZP)
	Soft machining & sintering	Alumina, zirconia (3Y-TZP)
	Soft machining, sintering & heat-pressing	Zirconia/fluorapatite–leucite glass–ceramic
	Hard machining	Sanidine, leucite
	Hard machining & heat treatment	Lithium disilicate

Metal–ceramic systems for dental restorations have been available since the 1960s. They rely on the application and firing of a veneering ceramic onto a metal substructure to produce an esthetically acceptable restoration. Veneering ceramics for metal–ceramic restorations, commonly named feldspathic porcelains, are usually leucite ($KAlSi_2O_6$)-based [138]. Feldspar-derived glass alone exhibits a low coefficient of thermal expansion, around 8.6×10^{-6}/°K [133]. The addition of leucite to feldspar glass led to the production of veneering ceramics with a coefficient of thermal expansion compatible with that of the metal substructure.

With the success of zirconia as a dental restorative material, manufactured 3Y-TZP abutments and implants have recently been introduced on the dental market. Low-temperature degradation of 3Y-TZP involves microstructural changes such as grain pull-out, microcracking, and surface roughening [31, 34]. It has also been demonstrated that surface finish and residual surface stresses strongly influence the response of 3Y-TZP to low-temperature degradation [35]. Ceria-stabilized zirconia/alumina nanocomposites for dental applications have been shown to exhibit high flexural strength (1422 ± 60 MPa), high reliability, and an excellent resistance to low-temperature degradation [91].

3.3.1.3 Carbon

Over the past years, various carbon materials have been investigated in many areas of medicine [6, 7, 130]. Carbon occurs in different lattice forms, from amorphous through intermediate to crystalline, e.g., hexagonal or regular lattice diamond. The broad range of applications of carbon materials in medicine is also the outcome of many preparation methods. Carbon implant materials can be produced by pyrolysis of solid, liquid, and gaseous carbon precursors. The pyrolysis method applied to carbon compounds allows the production of carbon of different structure and microstructure in the form of thin films or thick layers, solid pyrolytic carbon, powders, and in the form of fibers. The different forms of carbon that have received attention are artificial graphites, glasslike carbon, carbon fibers, pyrolytic carbon, and vapor-grown carbon coatings. Among pure forms of carbon, the pyrolytic carbons and carbon coatings on various substrates have found clinical applications in the cardiovascular, orthopedic, and dental areas as well as in long-term transcutaneous applications [10–12, 18, 77]. Most of the studies on carbon fibrous implants confirm that carbon fibers do not inhibit tissue growth, and thus can act as a scaffold for tissue proliferation [41].

On the other hand, carbon nanotubes (CNTs) are very prevalent in today's world of medical research and are being highly researched in the fields of efficient drug delivery and biosensing methods for disease treatment and health monitoring [56, 79]. CNTs possess exceptional mechanical, thermal, and electrical properties, facilitating their use as reinforcements or additives in various materials to improve the properties of the materials. In the medical field, biomaterials are expected to be developed using CNTs for clinical use. Biomaterials often are placed adjacent to bone. The use of CNTs is anticipated in these biomaterials applied to bone mainly to improve their overall mechanical properties, for applications such as high-strength arthroplasty prostheses or fixation plates and screws that will not fail [116]. The use of CNTs in drug delivery and biosensing technology has the potential to revolutionize medicine. Functionalization of single-wall nanotubes (SWNTs) has proven to enhance solubility and allow for efficient tumor targeting/drug delivery. Research shows that functionalized carbon nanotubes are non-cytotoxic and preserve the functionality of primary immune cells [38].

3.3.2 Bioactive Ceramics

A bioactive material is a material that obtains a specific biological response at the interface of the material, which would result in the formation of a bond between the tissues and the material [49, 52, 54, 142]. A bioactive ceramic undergoes chemical reactions in the body, but only at its surface. Upon implantation, surface-reactive ceramics form strong bonds with the closest tissue. The surface reactive implants respond to local pH changes by releasing Ca^{2+}, Na^{+}, and K^{+} ions and lead to bonding of tissues at the interfaces [51]. The ion exchange reaction between the bioactive implant and the surrounding body fluids, in some cases, results in the formation of a biologically active carbonated apatite (CHA) layer on the implant that is a mimic to the mineral phase of bones. Common bioactive ceramics used in orthopedic surgery are bioglass, ceravital, and A-W glass ceramic. However, the mechanical properties of these bioactive ceramic are generally weaker than bioinert ceramics. Only A-W glass ceramic has higher mechanical strength than cortical bone. Bioactive ceramics are most frequently used as bone defect fillers. They are supplied in the forms of block, porous material, and granules [49]. The glass ceramic has good biocompatibility, bioactivity, and no toxicity, making it useful as a biomaterial in artificial bone and dental implants.

The base components in most bioactive glasses and glass ceramic made by traditional high-temperature melting, casting and sintering are SiO_2, Na_2O, CaO, and P_2O_5. The first and well studied composition is 45S5 Bioglass, which contains 45% SiO_2, 24.5% Na_2O, 24.5% CaO, and 6% P_2O_5 (in weight percent). The compositions and some mechanical properties of several typical bioactive glasses are given in Table 3.26.

Table 3.26 Composition (wt %) and mechanical properties of bioactive glasses [20]

Component	45S5 Bioglass	45S5, 4F Bioglass	45B15S5 Bioglass	52S4.6 Bioglass	55S4.3 Bioglass
SiO_2	45	45	30	52	55
P_2O_5	8	6	6	6	6
CaO	24.5	14.7	24.5	21	19.5
Na_2O	24.5	24.5	24.5	21	19.5
CaF_2	—	9.8	—	—	—
B_2O_3	—	—	15	—	—
Class	A	A	A	A	B
BS [MPa]	40–60	—	—	—	—
E [GPa]	30–50	—	—	—	—

BS, Bending strength; *E*, Young's modulus.

A series of glasses in four component system with a constant 6 wt% P_2O_5 were studied [49]. The compositional dependence of the bioactive bonding-boundary for this system is shown in Fig. 3.5. In the region A, the glasses are bioactive and bond to bone. Glasses in region B behave as nearly inert materials and result in a fibrous capsule at the implant-tissue interface. Glasses in the region C are resorbed within 10–30 days in tissue. Compositions in region D are not technically practical and have not been implanted.

The low strength of a monophase bioactive glass restricts its clinical applications to non-load bearing situation. A way to attempt to solve this problem is to prepare a glass-ceramic by a process of crystallization of glass [42, 49, 51]. Several kinds of bioactive glass-ceramics have been developed, including bioactive A/W glass ceramic. The compositions and mechanical properties of these glass ceramics are given in Table 3.27.

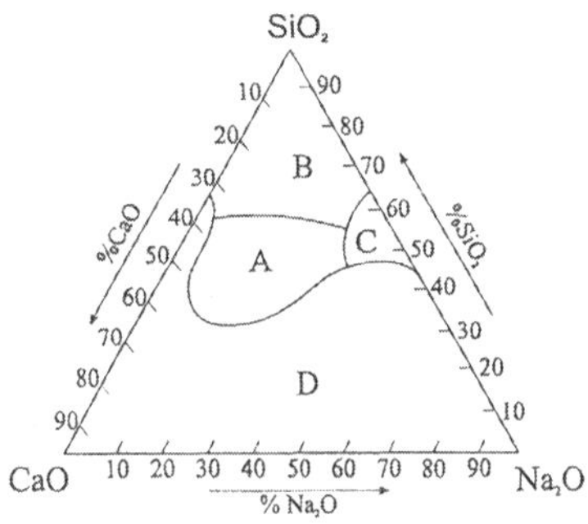

Figure 3.5 Compositional dependence (in weight percent) of bone bonding and soft-tissue bonding of bioactive glasses and glass-ceramics [20].

Table 3.27 Composition (wt%) and mechanical properties of A/W and Ceravital glass ceramics [20]

Component	A/W glass ceramic	KG Cera Ceravital	Mina 13 Ceravital	KGy213 Ceravital	M8/1 Ceravital
SiO_2	34.2	46.2	46	38	50
$Ca(PO_3)_2$	—	25.5	16	13.5	7.1
CaO	44.9	20.0	33	31	—
P_2O_5	16.3	—	—	—	—
Na_2O	—	4.8	—	4	5
MgO	4.6	2.9	5	—	—
CaF_2	0.5	—	—	—	—
K_2O	—	0.4	—	—	—
Al_2O_3	—	—	—	7	1.5
Ta_2O_5	—	—	—	5.5	—
TiO_2	—	—	—	1	—
B_2O_3	—	—	—	—	4
$Al(PO_3)_3$	—	—	—	—	2.4
SrO	—	—	—	—	20
La_2O_3	—	—	—	—	6
Gd_2O_3	—	—	—	—	4
BS [MPa]	215	—	—	70–88	—
E [GPa]	35	—	—	—	—

BS, bending strength; *E*, Young's modulus.

3.3.3 Bioresorbable Ceramics

Bioresorbable implants are designed to degrade gradually with time and be replaced with natural tissues [60]. It leads to tissue regeneration, instead of their replacement. The rate of degradation varies from one material to another. Calcium phosphate ceramics (CPC) are bioresorbable ceramics (Table 3.28).

Table 3.28 Calcium phosphate ceramics (CPC)

Acronym	Name	Formula	Ca/P
VCPA	Monocalcium phosphate anhydrate	$Ca(H_2PO_4)_2$	0.5
DCPA	Dicalcium phosphate anhydrate	$CaHPO_4$	1.0
OCP	Octacalcium phosphate	$Ca_8(HPO_4)_2(PO_4)_45H_2O$	1.33
TCP	Alpha tricalcium phosphate	$Ca_3(PO_4)_2$	1.5
TCP	Beta tricalcium phosphate	$Ca_3(PO_4)_2$	1.5
ACP	Amorphous calcium phosphate	$Ca_x(PO_4)_y nH_2O$	1.1–1.5
HA	Hydroxyapatite	$Ca_{10}(PO_4)_6(OH)_2$	1.67
cd-HA	Calcium deficient hydroxyapatite	$Ca_9(HPO_4)(PO_4)_2OH$	1.5
TetCP	Tetra calcium phosphate	$Ca_4(PO_4)_2O$	2.0

The most widely used calcium phosphate based bioceramics are hydroxyapatite (HAP) and β-tricalcium phosphate (β-TCP). Hydroxyapatite has the chemical formula $Ca_{10}(PO_4)_6(OH)_2$, the Ca/P ratio being 1.67 and possesses a hexagonal structure. It is the most stable phase of various calcium phosphates. It is stable in body fluid and in dry or moist air up to 1200°C and does not decompose and has shown to be bioactive due to its resorbable behavior.

Hydroxyapatite is the natural mineral phase of bone and teeth and, hence, is compatible with the body environment as well as to match closely some of the physical properties of bone. For biomedical purposes, the carbonated apatite and fluorapatite are the materials of interest because of assumed similarity to bony apatite and decreased solubility in aqueous solutions respectively [90]. It is important to note, that carbonated fluorapatite–gelatin nanocomposites were grown by the double-diffusion technique within a gelatin gel [114]. Recent studies have shown that irradiation of dental enamel by carbon dioxide (CO_2) laser irradiation, with and without fluoride, can inhibit the acid-induced surface dissolution of the carbonated apatite (CAP) mineral of the tooth [105].

β-tricalcium phosphate (β-TCP) is represented by the chemical formula $Ca_3(PO_4)_2$, the Ca/P ratio being 1.5. β-TCP shows an X-ray pattern consistent with a pure hexagonal crystal structure, although the related α-TCP is monoclinic. β-TCP turns into α-TCP around 1200°C; the latter phase is considered to be stable in the range 700 to 1200°C. β-TCP is highly soluble in body fluid.

To achieve an optimum resorbability of the material, studies have mainly focused on the biphasic calcium phosphate ceramics composed of HAP and TCP. Several results suggest that the resorbability of biphasic ceramics is largely determined by the HAP/TCP ratio.

Takatoshoi *et al.* examined the biocompatibility of calcium phosphate bone cement for clinical use by an *in vitro* assay system [126]. The experimental material was the bone replaceability calcium phosphate cement (BRCPC, Norian PDC, Shofu). The obtained results indicate that BRCPC has bioactivity such as osteogenesis. On the other hand, BRCPC does not have cytotoxicity. Recently, Bogdanski *et al.* investigated nickel–titanium shape memory alloys (NiTi-SMA) [9]. These alloys have mechanical properties (superelasticity, shape memory effect) that make them very interesting for clinical applications. Due to the high nickel content of the material, there are concerns about possible immunological reactions of the surrounding tissue. The proliferation of osteoblast-like cells (MG-63) and the cytokine release (interleukin (IL)-6, IL-8, tumor necrosis factor (TNF)-a) from polymorphonuclear neutrophil (PMN) leukocytes as well as peripheral blood mononuclear cells (PBMC) in response to different NiTi specimen was studied: (a) NiTi coated with calcium phosphate (octacalcium phosphate + hydroxyapatite; OCP/HAP) from an supersaturated aqueous solution, consisting of a dense layer of sharp-edged platelets; (b) NiTi coated with calcium phosphate (hydroxyapatite) by high-temperature plasma-spraying, consisting of a rough dense layer of globular particles; (c) NiTi after geometrical structuring (microdrilling). In the first two cases, the aim was to improve the biocompatibility while retaining the mechanical properties. In the third case, the effect of mechanical treatment (damage of the passivating TiO_2 surface layer) was investigated. In comparison to non-coated samples, the OCP/HAP-coated SMA led to an increase in the release of cytokines, in contrast, the HAP-coated samples led to a decrease in cytokine release from PBMC. The attachment, viability, and the proliferation of MG-63-cells in the near vicinity of microdrilled structures were not different from non-strained surface areas. These results show that cell-biological *in vitro* techniques are suitable for the prediction of tissue reactions toward implant coatings or mechanically altered NiTi-surfaces. The surface morphology (platelets versus globular particles) appears to play a significant role that supersedes that of the chemical composition of the surface [9].

3.4 Polymers

Polymers used in today's dentistry, commonly known as "dental resins", are used in all specialties of the profession, whether restorative, prosthodontics, or surgical (Table 3.29) [4, 84, 86, 140].

Table 3.29 Requirements for a denture polymers [4, 80, 92, 106, 140]

Properties	Characterization
Biocompatibilty	Biocompatibility is the ability of a polymer material or a device to remain biologically inert during its functional period. Possible effects of biomaterials on the living environment due to a lack of blood compatibility are thrombogenicity, and the induction of hemolysis. In addition, the biomaterial must not be carcinogenic, immunogenic, antileukotactic, or mutagenic. In turn, the environment should not cause degradation or corrosion of the biomaterial that would result in loss of physical and mechanical properties.
Physical properties	A denture polymer should possess adequate resilience and strength to biting, chewing, impact forces, and excessive wear under mastication. It should be stable under all conditions of service, including thermal and loading shocks. Mechanical properties increase considerably with an increase in number average molecular weight.
Aesthetic properties	The resin should exhibit sufficient translucency and transparency (hue, chroma and, value) to match the adjacent structures and tissues. It should be capable of being pigmented or tinted to camouflage the surroundings. Once fabricated, it should maintain the appearance and color and not change subsequently.
Chemical stability	The biomaterial should be chemically stable and not deteriorate inside the oral cavity by inducing some chemical reaction or an adverse event. It should preferably polymerize to completion, without leaching any residual monomers.
Rheometric properties	The flow behavior in polymers involves elastic and plastic deformation (viscous flow) and elastic recovery when stresses are released. Molecular weight, chain length, number of cross linkages, temperature, and applied force greatly determine the typical behavior. Plastic flow is irreversible and causes permanent polymer deformation, compared to elastic recovery in certain polymers, when applied stress is removed. Biopolymers exhibit complex combined elasticplastic deformation called as visco-elastic recovery.

(*Contd*)

Table 3.29 *(Contd)*

Properties	Characterization
Thermal properties	Polymers normally show a large variation in their properties with temperature. At sufficiently low temperatures, amorphous polymers are hard and glass-like, compared to softer and more flexible, when a critical temperature is reached usually the glass transition temperature (T_g). The T_g of a plastic is one of the very important set points in determining whether the polymer is thermosetting or thermoplastic and hence our desired clinical properties are affected.

Because of its esthetic properties, resin can be used for the reproduction of lost tooth structure. There are two types of direct filling resins. The older type is an unfilled polymethyl methacrylate (PMMA). Such a resin is often referred to as an acrylic resin. The other type is the composite resin. Requirements for a denture polymers are included in Table 3.29 [4, 80, 92, 106, 140].

3.4.1 Acrylic Resins

Acrylic resin is a general term for any one of the plastics (resin) generated through chemical reaction by applying polymerization initiator and heat to a monomer. Polymethyl methacrylate, one of the more rigid variations of acrylic resin, was first marketed in Germany in 1927. A similar form of the resin known as Plexiglas® began to be manufactured in the United States in 1936. One of the main characteristic features of PMMA is its high transparency. With its high weather resistance, it has been known to last over 30 years; it does not easily turn yellow or crumble by sunlight. Acrylic (unfilled) resins are used as temporary crown material. Temporary crowns are placed to protect the crown preparation and provide patient comfort during the time the permanent crown is being constructed. Acrylic resins are composed from powder and liquid. The powder is composed of a polymethyl methacrylate, a polymer, and contains some inert coloring pigments. The liquid is a monomethyl methacrylate, a monomer.

Acrylic resins are used in individual impression trays and orthodontic devices, in addition to dentures and artificial crowns. PMMA is well known for its property of being a bone cement (for fixing hip implants) and of its use in making acrylic glass (as a base for artificial fingernails and varnish).

In respect to the diverse applications of polymers in denture science, higher molecular weight methacrylate polymers such as polyethylmethacrylate (PEMA) and polybutylmethacrylate (PBMA) have found uses as soft denture liners [141].

Acrylic resins are available in several shades to match tooth shading (esthetics). They have a low thermal conductivity. These resins are not easily washed out by the acids of the oral cavity (low solubility). Acrylic resins are also resilient, which allows them to be used in stress-bearing areas. Acrylic resins exhibit a moderate shrinkage from 3 to 8 percent. Acrylic resins have a low resistance to wear.

3.4.2 Composite Resins

Composite resins are the most commonly used material for all permanent anterior restorations [14, 44, 80]. These resins make excellent restorative materials because of their good resistance to wear and their excellent esthetics. Different fillers (glass, silica, borosilicates, and fused quartz), binders and processing techniques (rods, fibers, and matte) are being experimented to develop the ideal denture polymer, suiting clinical requirements [3]. Several technological solutions have been proposed and are under testing. Polyethylene woven fibers, braided fibers and unidirectional fibers are under experimentation [28, 39, 122]. Fiberflex based on Kevlar, developed by Dupont, is under consideration as a unidirectional fiber reinforced denture material.

Composite resins have excellent esthetic properties and good resistance to wear because of the filler. They also have an acceptable compressive strength. Thermal expansion is at a minimum — the manufacturer claims that the thermal expansion is close to the normal expansion of tooth structure. Solubility and shrinkage are low.

A new fluoride-containing, methacrylate-based denture lining material has been introduced [48]. Because of the fluoride content, this material is expected to demonstrate low water sorption and solubility and high stain resistance, but these attributes remain untested.

Composite resins are widely used today instead of amalgams for restorative dental practice. The mechanical properties of the cured composite resins are shown to be dependent on the curing process [78]. It is thus highly desirable to monitor the curing status of the resin in real time as to achieve the optimum performance.

Synthetic polymer poly-L-lactide (PLLA) is often used as polymeric material and has some features, which make it useful for combining with HAp [58]. Fibers of PLLA may strengthen HAp and its good bioresorbility provides space for tissue extension. Composite biomaterial HAp/PLLA embodies good features of each of these biomaterials. Differences in porosity, microstructure, compressive consistency as well as bioresorbility of HAp/PLLA may be achieved by using PLLA with different mole masses [59]. The most appropriate way of recognizing the usability of biocomposite is its implantation into the organism.

Recently, much effort has been directed at the fabrication of carbon nanotubes (CNTs)/polymer composites, and the characterization of their physical properties. Among them, composites comprising CNTs and the biocompatible polymers are of special interest due to their potential for specific biomedical applications [147]. Zhang *et al.* report the preparation of the MWCNT/poly(L-lactide) composite and the corresponding spectroscopic (Raman) and the microscopic (SEM, TEM) characterization [147]. The electronic transport, thermal properties, and biocompatibility of this composite have also been investigated. The Raman spectroscopic analysis suggests the interaction between PLLA and MWCNT occurs mainly through the hydrophobic $C{-}CH_3$ functional groups. The DC conductivity of the composite increases as the MWCNT loading is increased. Such behavior can be described by a percolation mechanism in which a percolation threshold at about 14 wt% MWCNT loading is observed with the maximum end conductivity of $0.1\ S\cdot cm^{-1}$. The DSC study of the PLLA/MWCNT composite reveals that the MWCNTs in the composite have the effect of inducing crystallization and plasticizing the polymer matrix. The results from the cell culture test suggest that the presence of MWCNT in the composite inhibits the growth of the fibroblast cells.

Chen *et al.* have fabricated the biocompatible poly(lactic acid) (PLA) nanofibers by using electrospinning and then adopted to blend with other nanomaterials such as nano-TiO_2, which have been widely applied in biomedical and bioengineering fields, as a potential drug carrier or protective container for biologically active drug molecules [21]. The observations indicate that by combining this nanocomposite with the drug accumulation of anticancer drug daunorubicin, a broad-spectrum anticancer agent, the blends of nano-TiO_2 and PLA nanofibers could afford promising scaffolding for noncovalent attachment of multiple drugs, which could efficiently enhance the drug uptake in target leukemia cells. These new nanocomposites have good biocompatibility, ease of surface chemistry modification, and very high surface area, which may afford the possibility for their promising application in pharmacology and biomedical engineering areas.

3.5 Composite Biomaterials

Composite materials may be defined as those materials that consist of two or more fundamentally different components that are able to act synergistically to give properties superior to those provided by either component alone [72, 94, 128]. Natural composites include bone, wood, dentin, cartilage, and skin. In biomaterials, it is important that each constituent of the composite be biocompatible. Some applications of composites in biomaterial applications are: (i) dental filling composites, (ii) reinforced methyl methacrylate bone cement and ultrahigh molecular weight polyethylene, and (iii) dental/orthopedic implants with porous surfaces.

Composites made of bioinert and bioactive ceramics are produced to achieve two important features, bioactivity and mechanical strength [43, 113]. Alumina ceramic can form composites with hydroxyapatite that are bioactive. The properties of composites depend very much upon structure, shape, volume fraction, and the interface among the constituents, as well as upon the chemical composition and their final microstructure.

For, example, the dental composite resins consist of a polymer matrix and stiff inorganic inclusions [26]. The inorganic inclusions confer a relatively high stiffness and high wear resistance on the material. Available dental composite resins use quartz, barium glass, and colloidal silica as fillers. Fillers have particle size from 0.04 μm to 13 μm, and concentrations from 33 to 78% by weight. In view of the greater density of the inorganic filler phase, a 77% weight percent of filler corresponds to a volume percent of about 55%. The matrix consists of a polymer, typically BIS-GMA. Polymerization can be initiated by a thermochemical initiator such as benzoyl peroxide, or by a photochemical initiator (benzoin alkyl ether) which generates free radicals when subjected to ultraviolet light. Dental composites have a Young's modulus in the range of 10–16 GPa, and the compressive strength from 170 to 260 MPa (Table 3.30).

Table 3.30 Properties of bone, teeth and some biomaterials [72]

Material	Young's modulus *E* [GPa]	Density ρ (g/m^3)	Strength (MPa)
Hard Tissue			
Tooth, bone, human compact bone, longitudinal direction	17	0.8	130
Tooth dentin	18	2.1	138
Tooth enamel	50	2.9	(compression)
Polymers			
Polyethylene (UHMW) 10.94 30 (tension)	1	0.94	30 (tension)
Polymethyl methacrylate, PMMA 3 1.1 65 (tension)	3	1.1	65 (tension)
PMMA bone cement 21.18 30 (tension)	2	1.18	30 (tension)
Metals			
316L Stainless steel (wrought)	200	7.9	1000 (tension)
Co–Cr–Mo (cast)	230	8.3	660 (tension)
Co–Ni–Cr–Mo (wrought)	230	9.2	1800 (tension)
Ti–6Al–4V	110	4.5	900 (tension)
Composites			
Graphite-epoxy (unidirectional fibrous, high modulus)	215	1.63	1240 (tension)
Graphite-epoxy (quasi-isotropic fibrous)	46	1.55	579 (tension)
Dental composite resins (particulate)	10–16	—	170–260 (compression)
Foams			
Polymer foams	10^{-4}–1	0.002–0.8	0.01–1 (tension)

For the successful achievement of three-dimensional scaffolds, several characterization criteria are required. They can be divided into four categories: (i) morphology (e.g., porosity, pore size, surface area); (ii) mechanical properties (e.g., compressive and tensile strength); (iii) bulk properties (e.g., degradation and its relevant mechanical properties); and (iv) surface properties (e.g., surface energy, chemistry, charge) [63]. Recently, a strong and bioactive ceramic scaffold consisting of a porous zirconia body coated with apatite double layers (fluorapatite (FA) as an inner layer and hydroxyapatite (HA) as an outer layer) was successfully fabricated [66]. The authors investigate the *in vivo* performance of the engineered bioceramic scaffolds using a rabbit calvarial defect model. In particular, the porosity and pore size of the scaffolds are varied in order to observe the geometrical effects of the scaffolds on their bone formation behaviors. The scaffolds supported on a zirconia framework can be produced with an extremely high porosity (~84–87%), while retaining excellent compressive strength (~7–8 MPa), which has been unachievable in the case of pure apatite scaffolds (~74% porosity with ~2 MPa strength). All of the specimens show a good healing response without adverse tissue reactions. Good healing is shown at 4 weeks post-surgery with the ingrowth of new bone into the macropore-channels of the scaffolds. The apatite-coated zirconia scaffolds show good bone forming ability and are considered to be a promising scaffolding material for bone regeneration since they possess a high level of both mechanical and biological properties [73].

Dental implants for load bearing applications require the use of materials that are both bioactive and have significant mechanical strength [13]. There are no existing materials that readily fit these two criteria. While titanium and titanium-based alloys have excellent mechanical properties and are generally well tolerated in a physiological environment, they have negligible capacity for osteointegration. Two approaches have been taken to improve the osteointegration: a surface modification approach and a composite approach. In the surface modification approach, the surface of the primary material is modified by chemical or electrochemical methods to make the surface bioactive, or a bioactive coating is applied using, for example, plasma deposition techniques. In the composite approach, the primary material is combined with bioactive materials, such as bioglass or bioceramics. Conventional powder metallurgy methods [23, 98], or plasma-assisted processes [30, 87] are commonly used for fabricating the composites.

The biocomposites prepared from powder mixtures of titanium (α-Ti), hydroxyapatite (HA), and bioactive glass (BG) (SiO_2–CaO–P_2O_5–B_2O_3–MgO–TiO_2–CaF_2) were investigated, too [99]. The results showed that complex reactions among the starting materials mainly depended on the initial Ti/HA ratios as well as the sintering temperatures.

Although titanium (Ti) and its alloys are the preferred metal materials for orthopedics and dentistry because of their high strength and good biocompatibility [132, 136], they are bioinert biomaterials and cannot directly bond to the bone. Moreover, HA coatings to improve the surface bioactivity of titanium and its alloys often flake off as a result of poor ceramic/metal interface bonding, which may make the surgery fail [19, 144]. The problems mentioned above can be solved by the fabrication of HA–Ti composites. Recently the studies on the fabrication and characterization of HA-based composites reinforced with Ti particles have been reported [22, 23, 98]. The significant toughening effect by energy-absorbing mechanism due to the plastic deformation of ductile Ti particles at the tips of cracks was found in both HA-20 vol%Ti composite and HA-40 vol%Ti composite [22]. Ti-20 vol% HA composite with a relative density of 97.86% was fabricated by a hot pressing technique. The phase constitution of Ti-20 vol% HA composite is similar to that of HA-based composite with Ti and HA as the predominant phases. TEM observation shows the bonding state of Ti/HA interface is good, however, there exists an interfacial transition zone between Ti and HA. Elastic modulus and Vicker's hardness of Ti-20 vol% HA composite are 102.6 and 3.41 GPa respectively. In comparison with pure Ti metal fabricated under the same conditions, the composite has a much lower bending strength (170.1 MPa) and fracture toughness ($3.57\ \mathrm{MPa \cdot m^{1/2}}$), which are only about 17.5 and 12% of those of pure Ti metal, nevertheless it can meet the basic strength and toughness demands of replacing hard tissue in heavy load-bearing applications. Both Ti matrix in the composite and pure Ti metal fabricated by hot-pressing process present quasi-cleavage fracture. The *in vivo* studies indicate that Ti-20 vol% HA composite has good biocompatibility and can integrate with bone. The osteointegration ability of the composite is better than that of pure titanium, especially in the early stage after the implantation, which may be due to the presence of HA ceramic in the Ti-matrix composite.

Bioactivity of the Ti–HA composite in a simulated body fluid (SBF) was investigated [23, 98]. Main crystal phases of the as-fabricated composite are found to be Ti_2O, $CaTiO_3$, CaO, α-Ti, and a TiP-like phase. When the composite is immersed in the simulated body fluid for a certain time, a poor-crystallized, calcium-deficient, carbonate-containing apatite film will form on the surface of the composite. It can be concluded that the biocomposite, fabricated from HA and Ti powders by powder metallurgy technique, has the ability to induce apatite nucleation and growth on its surface from the SBF. Among all the crystal phases in the as-fabricated composite, Ti_2O has the ability to induce apatite formation. Furthermore, the dissolve of CaO phase also provides favorable conditions for the apatite nucleation and growth.

References

1. Aghelli, H., Malmström, J., Hanarp, P., and Sutherland, D.S. (2006). Nanostructured biointerfaces, *Mat. Sc. Eng. C*, **26**, pp. 911–917.
2. Ahmed, T., Lomg, M., Silvestri, J., Ruiz, C., and Rack, H.J. (1996). A new low modulus, biocompatible titanium alloy, in *Titanium'95* (eds. Blenkinsop, P.A., Evans, W.J., and Flower, H.M.), The Institute of Materials, pp. 1760.
3. Bae, J.M., Kim, K.N., Hattori, M., Hasegawa, K., Yoshinari, M., Kawada, E., and Oda, Y. (2001). The flexural properties of fiber-reinforced composites with light-polymerized polymer matrix, *Int. J. Prosthodont.*, **14**, pp. 33–39.
4. Bhola, R., Bhola, S.M., Liang, H., and Mishra, B. (2010). Biocompatible denture polymers – a review, *Trends Biomater. Artif. Organs*, Vol **23**, pp.129–136.
5. Biehl, V., and Breme, J. (2001). Metallic Biomaterials, *Mat.-wiss. u. Werkstofftech.* **32**, pp. 137–141.
6. Blazewicz, M., Blazewicz, S., and Wajler, C. (1994). Mechanical and implant behaviour of chemically modified carbon braids, *Ceramics Intern.*, **20**, pp. 99–103.
7. Blazewicz, M. (2001). Carbon materials in the treatment of soft and hard tissue injuries, *Eur. Cells Mater.*, **2**, pp. 21–29.
8. Boccaccini, A.R., and Blaker, J.J. (2005). Bioactive composite materials for tissue engineering scaffolds, *Expert Rev. Med. Dev.*, **2**, pp. 303–317.
9. Bogdanski, D., Epple, M., Esenwein, S.-A., Muhr, G., Petzoldt, V., Prymak, O., Weinert, K., and Koller, M. (2004). Biocompatibility of calcium phosphate-coated and of geometrically structured nickel–titanium (NiTi) by *in vitro* testing methods, *ESOMAT: European Symposium on Martensitic Transformation*, ROYAUME-UNI, **378**, pp. 527–531.
10. Bokros, J.C. (1977). Carbon biomedical devices, *Carbon*, **15**, pp. 355–371.
11. Bokros, J.C., Lagrange, L.D., and Schoen, F. (1972). Control of structure carbon for use in bioengineering, in *Chemistry and Physics of Carbon* (ed. Walker, P.L.), Marcel Dekker, New York, **9**, pp. 03–171.
12. Bokros, J.C. (1983). Carbon in medical devices, *Ceramics Int.*, **9**, pp. 3–7.
13. Bonfield, W. (1988). Hydroxyapatite-reinforced polyethylene as an analogous material for bone replacement, in *Bioceramics: Materials Characteristics vs in vivo Behavior* (eds Ducheyne, P., and Lemons, J.E.), Annals of New York Academy of Science, New York, **523**, pp. 173–177.
14. Bowen, R.L. (1963). Properties of silica reinforced polymer for dental restorations, *J. Ame. Dent. Ass.*, **66**, pp. 57–64.

15. Breme, J., and Helsen, J. (1998). Selection of materials, in *Metals as Biomaterials* (eds. Helsen, J.A., and Breme, J.), John Wiley & Sons Ltd., Chichester, pp. 1– 37.

16. Breme, J., and Helsen, J. (1998). Selection of materials, in *Metals as Biomaterials* (eds. Helsen, J.A., and Breme, J.), John Wiley & Sons Ltd., Chichester, pp. 37–71.

17. Brodbeck, U. (2003). The ZiReal post: A new ceramic implant abutment, *J. Esthet. Restor. Dent.*, **15**, pp. 10–23.

18. Brittberg, M., Faxen, E., and Peterson, L. (1994). Carbon fiber scaffolds in the treatment of early knee osteoarthritis. A prospective 4-year follow up of 37 patients, *Clin. Orthop.*, **307**, pp. 155–164.

19. Brossa, F., Cigada, A., Chiesa, R., Paracchini, L., and Consonni, C. (1994). Post-deposition treatment effects on hydroxyapatite vacuum plasma spray coatings, *J. Mater. Sci. Mater. Med.*, **5**, pp. 855–857.

20. Cao W.P., and Hench, L.L. (1996). Bioactive materials, *Ceramics Intern.*, **22**, pp. 493–507.

21. Chen, C., Lv, G., Pan, C., Song, M., Wu, C., Guo, D., Wang, X., Chen, B., and Gu, Z. (2007). Poly(lactic acid) (PLA) based nanocomposites – a novel way of drug-releasing, *Biomed. Mater.*, **2**, pp. L1–L4.

22. Chu, C., Lin, P., Dong, Y., Xue, X., Zhu, J., and Yin, Z. (2002). Fabrication and characterization of hydroxyapatite reinforced with 20 vol % Ti particles for use as hard tissue replacement, *J. Mater. Sci. Mater. Med.*, **13**, pp. 985–992.

23. Chu, C., Xue, X., Zhu, J., and Yin, Z. (2006). Fabrication and characterization of titanium-matrix composite with 20 vol% hydroxyapatite for use as heavy load-bearing hard tissue replacement, *J. Mat. Sc. Mat. Med.*, **17**, pp. 245–251.

24. Cogliano, V.J. (1999). *IARC Monographs on the Evaluation of Carcinogenic Risks to Humans*, IARC, Lyon; Surgical Implants and other Foreign Bodies, vol. 74.

25. Covacci, V., Bruzzese, N., Maccauro, G., Andreassi, C., Ricci., G.A., Piconi, C., Marmo, E., Marmo, E., Burger, W., and Cittadini, A. (1999). *In vitro* evaluation of mutogenic and carcinogenic power of high purity zirconia ceramic, *Biomaterials*, **20**, pp. 371–376.

26. Craig, R. (1981). Chemistry, composition, and properties of composite resins, in *Dental Clinics of North America* (ed. Horn, H.), Saunders, Philadelphia, PA.

27. Craing, R.G., and Peyton, F.A. eds. (1975). *Restorative Dental Materials, 5th ed.*, St. Louis, MO: The C.V. Mosby Co., p. 318.

28. Culy, G., and Tyas, M. (1998). Direct resin bonded fiber reinforced anterior bridges, *Aust. Dent. J.*, **43**, pp. 1–4.

29. Davis, H.E., and Leach, J.K. (2008). Hybrid and Composite Biomaterials for Tissue Engineering, *in Topics in Multifunctional Biomaterials and Devices* (ed. Ashammakhi, N.), University of California, chapter 10.

30. De Groot, K., Geesink, R., Klein, C.P.A.T., and Serekian, P. (1987). Plasma sprayed coatings of hydroxyapatite, *J. Biomater. Res.*, **21**, pp. 1375–1381.

31. Denry, I.L., and Holloway, J.A. (2006). Microstructural and crystallographic surface changes after grinding zirconia-based dental ceramics, *J. Biomed. Mater. Res., Appl. Biomater., B*, **76**, pp. 440–448.

32. Denry, I., and Holloway, J.A. (2010). Ceramics for Dental Applications: A Review, *Materials*, **3**, pp. 351–368.

33. Deplancke, J.L., Degrez, M., Fontana, A., and Winand, R. (1982). Self-colour anodizing of titanium, *Suf. Technol.*, **16**, pp. 153–162.

34. Deville, S., Grémillard, L., Chevalier, J., and Fantozzi, G. (2005). A critical comparison of methods for the determination of the aging sensitivity in biomedical grade yttria-stabilized zirconia, *J. Biomed. Mater. Res., Appl. Biomater., B*, **72**, pp. 239–245.

35. Deville, S., Chevalier, J., and Grémillard, L. (2006). Influence of surface finish and residual stresses on the ageing sensitivity of biomedical grade zirconia, *Biomaterials*, **27**, pp. 2186–2192.

36. Dhir, G., Berzins, D.W., Dhuru, V.B., Periathamby, A.R., and Dentino, A. (2007). Physical properties of denture base resins potentially resistant to candida adhesion, *J. Prosthodont.*, **16**, pp. 465–472.

37. Doi, H., Yoneyama, T., Kobayashi, E., and Hamanaka, H. (1998). Mechanical properties and corrosion resistance of Ti–5Al–13Ta alloy castings, *J. Jpn. Soc. Dent. Mater. Dev.*, **17**, pp. 247–252.

38. Dumortier, H., Lacotte, S., Pastorin, G., Marega, R., Wu, W., Bonifazi, D., Briand, J., Prato, M., Muller, S., and Bianco, A. (2006). Functionalized carbon nanotubes are non-cytotoxic and preserve the functionality of primary immune cells, *Nano Letters*, **6**, pp. 1522– 1528.

39. Eskitascioglu, C., Eskitascioglu, A., and Belli, S. (2004). Use of polyethylene ribbon to create a provisional fixed partial denture after immediate implant placement, *J. Prosthet. Dent.*, **91**, pp. 11–14.

40. Fini, M., Giavaresi, G., Torricelli, P., Krajewski, A., Ravaglioli, A., Belmonte, M.M., Biagini, G., and Giardino, R. (2001). Biocompatibility and osseointegration in osteoporotic bone, a preliminary *in vitro* and *in vivo* study, *J. Bone Joint Surgery, B*, **83**, pp. 139–143.

41. Forster, I.W., Ralis, Z.A., McKibbin, B., and Jenkins, D.H.R. (1978). Biological reaction to carbon fiber implants, *Clin. Orthop. Rel. Res.*, **131**, pp. 299–307.

42. Fox, T.G., and Flory, P.J. (1950). The glass transition, *J. Appl. Phys.*, **21**, pp. 581–591.

43. Freed, L., Vunjak-Novakovic, G., Biron, R., Eagles, D., Lesnoy, D., Barlow, S., and Langer, R. (1994). Biodegradable polymer scaffolds for tissue engineering. *BioTechnology*, **12**, pp. 689–693.

44. Freilich, M., Meiers, J., Duncan, J., Eckrote, K., and Goldberg, A. (2002). Clinical evaluation of fiber reinforced fixed bridges, *J. Ame. Dent. Assoc.*, **133**, pp. 1524–1534.

45. Fukunaga, K., and Niinomi, M. (2001). Microstructural observation of deformed structure of new β-type titanium alloy, Ti–29Nb–13Ta–4.6Zr, for biomedical applications, *Proc. Fall Meet JIM*, p. 308.

46. Geetha, M., Singh, A.K., Asokamani, R., and Gogia, A.K. (2009). Ti based biomaterials, the ultimate choice for orthopedic implants – a review, *Progr. Mat. Sc.*, **54**, pp. 397–425.

47. Guehennec, L., Souedan, A., Layrolle, P., and Amouriq, Y. (2007). Surface treatments of titanium dental implants for rapid osseointegration, *Dental Materials*, **23**, pp. 844–854.

48. Hayakawa, I., Akiba, N., Keh, E.S., and Kasuga, Y. (2006). Physical properties of a new denture lining material containing a fluoroalkyl methacrylate polymer, *J. Prosthetic Dentistry*, **96**, pp. 53–58.

49. Hench, L.L. (1988). Bioactoive ceramics, *Ann. New York Ac. Sc.*, **523**, pp. 54–71.

50. Hench, L.L. (1991). Bioceramics: From Concept to Clinic, *J. Am. Ceram. Soc.*, **74**, pp. 1487–1510.

51. Hench, L.L., and Wilson, J. eds (1993). An introduction to bioceramics, in *Bioceramics*, Word Scientific Publishing Co. Pte, Ltd, Singapore, pp. 1–24.

52. Hench, L.L. (1998). Bioactive materials: the potential for tissue regeneration, *J. Biomed. Mater. Res.*, **41**, pp. 511–518.

53. Hench L.L. (1998) Bioceramics. *J. Am. Ceram. Soc.*, **81**, *pp.* 1705–1728.

54. Hench, L.L. (2006). The story of bioglass, *J. Mater. Sc., Mat. Med.*, **17**, pp. 967–978.

55. Her, H., Jrgensen, R., Srbrden, E., and Suoninen, E. (1982). Precipitations in a Dental Ag–Pd–Cu–Au Alloy, *J. Dent. Res.*, **61**, pp. 673–677.

56. Hilder, T.A., and Hill, J.M. (2008). Carbon nanotubes as drug delivery nanocapsules, *Curr. Appl. Phys.*, **8**, pp. 258–261.

57. Hodgson, A.W.E, Kurz, S., Virtanen, S., Fervel, V., Olsson, C.O.A., and Mischler, S. (2004). Passive and transpassive behaviour of CoCrMo in simulated biological solutions, *Electrochimica Acta*, **49**, pp. 2167–2178.

58. Ignjatovic, N., Tomic, S., Dakic, M., Miljkovic, M., Plavšic, M., and Uskokovic, D. (1999). Synthesis and properties of hydroxyapatite/poly-Llactide composite biomaterials, *Biomaterials*, **20**, pp. 809–816.

59. Ignjatovic, N., Savic, V., Najman, S., Plavšic, M., and Uskokovic, D. (2000). A study of HAp/PLLA composite as a substitute for bone powder, using FT-IR spectroscopy, *Biomaterials*, **22**, pp. 571–575.

60. Jarcho, M. (1981). Calcium phosphate ceramics as hard tissue prosthetics, *Clin. Orthop. Relat. Res.*, **157**, pp. 259–278.

61. Karnowsky, M.M. (1978). Ageing of a Pd-35 wt% Ag-25 wt% Cu solid solution alloy, *J. Met. Sci.*, **13**, pp. 2339–2346.

62. Kelly J.R. (1997). Ceramics in restorative and prosthetic dentistry. *Ann. Rev. Mater. Sci.*, **27**, pp. 443–68.

63. Khang, G., Lee, S.J., Kim, M.S., and Lee, H.B. (2006). Biomaterials: tissue engineering and scaffold, in *Encyclopedia of Medical Devices and Instrumentation, 2nd ed.* (ed. Webster, J.), Wiley Press, New York.

64. Khatak, H.S., and Raj, N. (2010). *Corrosion of austenitic stainless steel: Mechanism, mitigation and monitoring*, Woodhead Publishing Limited, Cambridge, UK.

65. Kim, S.S., Ahn, K.M., Park, M.S., Lee, J.H., Choi, C.Y., and Kim, B.S. (2007). A poly(lactide-coglycolide)/hydroxyapatite composite scaffold with enhanced osteoconductivity, *J. Biomed. Mater. Res. A*, **80**, pp. 206–215.

66. Kim, H.W. (2008). Bone formation on the apatite-coated zirconia porous scaffolds within a rabbit calvarial defect, *J. Biomater. Appl.*, **22**, pp. 485–504.

67. Kobayashi, E., Doi, H., Yoneyama, T., Hamanaka, H., Matsumoto, S., and Kudaka, K. (1995). Evaluation of mechanical properties of dental casting Ti–Zr based alloys, *J. Jpn. Soc. Dent. Mater. Dev.*, **14**, pp. 321–328.

68. Kobayashi, E., Doi, H., Takahashi, M., Nakano, T., Yoneyama, T., and Hamanaka, H. (1995). Castability and mechanical properties of Ti–6Al–7Nb alloy dental-cast, *J. Jpn. Soc. Dent. Mater. Dev.*, **14**, pp. 406–413.

69. Kokubo, T. (ed) (2005). *Bioceramics and their clinical applications*, Woodhead Publishing Limited, Cambridge.

70. Koutayas, O. S., and Kern, M (1999). All-ceramic posts and cores: The state of the art, *Quintessence Int.*, **30**, pp. 383–392.

71. Kramer, K.H. (1999). Implants for sugery – a survey on metallic materials, in *Materials for Medical Engineering* (Stallforth, H., and Revell, T.), Euromat, Wiley-VCH, Germany, **2**, p. 5.

72. Lakes, R. (2000). Composite Biomaterials, in *The Biomedical Engineering Handbook*, 2nd ed. (ed. Bronzino, J.D., and Raton, B.), CRC Press LLC., Boca Raton.

73. Laurencin, C., Khan, Y., El-Amin S.F. (2006). Bone graft substitutes *Expert Rev. Med. Dev.*, **3**, pp. 49–57.

74. Tateishi, T. (ed) (2007). Microstructures and mechanical properties of biomedical Co–Cr–Mo alloys with combination of N addition and Cr-enrichment, in *Biomaterials in Asia: 1st Asian Biomaterials Congress* (Lee, S.H., Nomura, N., and Chiba, A.), Tsukuba, Japan.

75. Lemons, J.E. (1996). Ceramics: past, present, and future, *Bone*, **19**, pp. S121–S128.

76. Lim, I.A.L. (2004). Biocompatibilty of stent materials. MUR Jornal, **11**, p. 34.

77. Louis, J.P., and Dabadie, M. (1996). Fibrous carbon implants for the maintenance of bone volume after tooth avulsion: first clinical results, *Biomaterials*, **11**, pp. 525–528.

78. Lovell, L.G., Lua, H., Elliotta, J.E., Stansburyb, J.W., and Bowman, C.N. (2001). The effect of cure rate on the mechanical properties of dental resins, *Dent. Mater.*, **17**, pp. 504–511.

79. Lu, F., Gu, L., Meziani, M.J., Wang, X., Luo, P.G., Veca, L.M., Cao, L., and Sun, Y. (2009). Advances in bioapplications of carbon nanotubes, *Adv. Mater.*, **21**, pp. 139–152.

80. Maachi, R.L., and Craig, R.G. (1969). Physical and mechanical properties of composite restorative materials, *J. Ame. Dent. Ass.*, **78**, pp. 328–334.

81. Matthys, C.M., and Sommerling-Van Peteghem, R.S. (1987). The use of non-precious metal alloys for dental crowns and bridges, *Rev. Belge Med. Dent.*, **42**, pp. 114–119.

82. Marti A. (2000). Inert bioceramics (Al_2O_3, ZrO_2) for medical application, *Injury*, **31**, pp. 33–36.

83. Marti, A. (2000). Cobalt-base alloys used in bone surgery, *Injury Int. J. Care Injured*, **31**, pp. 18–21.

84. Marxkors, R., and Meiners, H. (1993). *Taschenbuch Der Zahnarztlichen Werkstoffkunde, 4th ed.*, Hanser, Munchen.

85. Massalski, T.B. (1965). Alloying behavior and effects in concentrated solid solutions, Gordon Breach Science, New York.

86. May, K.B., Razzoog, M.E., Koran, A., and Robinson, E. (1992). Denture base resins: comparison study of color stability, *J. Prosthet. Dent.*, **68**, pp. 78–82.

87. Mcpherson, R., Gane, N., and Bastow, T. (1995). Structure characterization of plasma-sprayed hydroxyapatite coating, *J. Mater. Sci.: Mater. Med.*, **6**, pp. 327–334.

88. Miyazaki, T., Tani, Y., Tamaki, Y., Suzuki, E., and Miyaji, T. (1987). Application of CaO to dental investments, *J. Jpn. Soc. Dent. Mater. Dev.*, **6**, pp. 437–440.

89. Miyazaki, T., Tani, Y., Tamaki, Y., Suzuki, E., and Miyaji, T. (1987). Casting of titanium with calcia investment (Part 2). Improvement of mold treatment and finess on castings, *J. Jpn. Soc. Dent. Mater. Dev.*, **6**, pp. 633–638.

90. Monroe, E.A., Votava, W., Bass, D.B., and Mc Mullen, J. (1971). New calcium phosphate ceramic material for bone and tooth implants, *J. Dental Res.*, **50**, pp. 860 –861.

91. Nawa, M., Nakamoto, S., Sekino, T., and Niihara, K. (1998). Tough and strong Ce-TZP/Alumina nanocomposites doped with titania, *Ceram. Int.*, **24**, pp. 497–506.

92. Nielsen, L.E. (1962). *Mechanical properties of polymers*, Reinhold, NewYork.

93. Niemi, L., and Her, H. (1984). The Structure of a Commercial Dental Ag–Pd–Cu–Au Casting Alloy, *J. Dental Res.*, **63**, pp. 149–154.

94. Niespodziana, K., Jurczyk, K., Jakubowicz, J., and Jurczyk, M. (2010). Fabrication and properties of titanium - hydroxyapatite nanocomposites, *Mater. Chem. Phys.*, **123**, pp. 160–165.

95. Niinomi, M. (1998). Mechanical properties of biomedical titanium alloys, *Mater. Sci. Eng. A*, **243**, pp. 231–236.

96. Niinomi, M., Fukui, H., Hattori, T., Kyo, K., and Suzuki, A. (2002). Development of high biocompatible Ti alloy, Ti–29Nb–13Ta–4.6Zr, *Mater. Jpn.*, **41**, pp. 221–223.

97. Niinomi, M. (2003). Recent research and development in titanium alloys for biomedical applications and healthcare foods, *Science and Technology of Adv. Mater.*, **4**, pp. 445–454.

98. Ning, C.Q., and Zhou, Y (2002). *In vivo* bioactivity of a biocomposite fabricated from HA and Ti powders by powder metallurgy method, *Biomaterials*, **23**, pp. 2909–2916.

99. Ning, C.Q., and Zhou, Y. (2004). On the microstructure of biocomposites sintered from Ti, HA and bioactive glass, *Biomaterials*, **25**, pp. 3379–3387.
100. Nitta, K., Watanabe, S., Masahashi, M., Hosoda, H., and Hanawa, S. (2001). Ni-free Ti–Nb–Sn shape memory alloys, in *Structural Biomaterials for the 21st Century*, (ed. Niinomi, M., Okabe, T., Taleff, E.M., Lesuer, D.R., and Lippard, H.F.), TMS, Warrendale, PA, pp. 25–34.
101. Ohta, M., Hisatsune, K., and Yamane, M. (1979). Age Hardening of Ag-Pd-Cu Dental Alloy, *J. Less Common Met.*, **65**, pp. 11–21.
102. Okuno, O. (1996). Titanium alloys for dental applications, *J. Jpn. Soc. Biomater.*, **14**, pp. 267–273.
103. Ornhagen, C., Nilsson, J.O., and Vannevik, H. (1996). Characterization of a nitrogen-rich austenitic stainless steel used for osteosynthesis devices, *J. Biomed. Mater. Res.*, **31**, pp. 97–103.
104. Otomo, H., Niinomi, M., Akahori, T., Fukui, H., and Suzuki, A. (2003). Relationship between microstructure and cold working ratio of a new β-type titanium alloy, *Proc. Fall Meet. Jpn Inst. Met.*, pp. 324.
105. Phan, N.D., Fried, D., Featherstone, J.D.B. (1999). Laser-induced transformation of carbonated apatite to fluorapatite on bovine enamel, in *SPIE proceedings series - Lasers in dentistry. Conference N°5, San Jose CA, vol. 3593*, SPIE, Bellingham, WA, pp. 233–240.
106. Phillips, R.M. (1973). *Skinners science of dental materials* 7th ed., W.B. Saunders Company, Philadelphia.
107. Piconi, C., and Maccauro, G. (1999). Zirconia as ceramic biomaterial, *Biomaterials*, **20**, pp. 1–25.
108. Pilathadka, S., Vahalová, D., and Vosáhlo, T. (2007). The zirconia: a new dental ceramic material. An overview, *Prague Medical Report*, **108**, pp. 5–12.
109. Pohler, E,M. (2000). Unalloyed titanium for implants in bone surgery, *Injury*, **31**, pp. D7–D13.
110. Rack, H.J., and Qazi, J.J. (2006). Titanium alloys for biomedical applications, *Mater. Sci. Eng. C*, **26**, pp. 1269–1277.
111. Ratner, B.D., Hoffman, A.S., Schoen, F.J., and Lemons, J.E. (2004). *Biomaterials science: an introduction to materials in medicine*, Academic Press, Orlando.
112. Raub, E., and Worwag, G. (1955). Die Silber-Palladium-Kupfer-Legierungen, *Z Metallkunde*, **46**, pp. 52–57.
113. Rodriguez-Lorenzo, L.M., Salinas, A.J., Vallet-Regi, M and Raman, J.S. (1996). Composite biomaterials based on ceramic polymers. I.

Reinforced systems based on Al_2O_3/PMMA/PLLA, *J. Biomed Mater. Res.*, **30**, pp. 515–522.

114. Rosseeva, E.V., Buder, J., Simon, P., Schwarz, U., Frank-Kamenetskaya, O.V and Kniep, R. (2008). Synthesis, characterization, and morphogenesis of carbonated fluorapatite–gelatine nanocomposites: A complex biomimetic approach toward the mineralization of hard tissues, *Chem. Mater.*, **20**, pp. 6003–6013.

115. Sachlos, E., Gotora, D., and Czernuszka, J.T. (2006). Collagen scaffolds reinforced with biomimetic composite nano-sized carbonate-substituted hydroxyapatite crystals and shaped by rapid prototyping to contain internal microchannels, *Tissue Eng.*, **12**, pp. 2479–2487.

116. Saito, N., Usui, Y., Aoki, K., Narita, N., Shimizu, M., Ogiwara, N., Nakamura, K., Ishigaki, N., Kato, H., Taruta, S., and Endo, M. (2008). Carbon nanotubes for biomaterials in contact with bone, *Curr. Med. Chem.*, **15**, pp. 523–527.

117. Sakaguch, N., Niinomi, M., and Akahori, T. (2004). Tensile Deformation Behavior of Ti–Nb–Ta–Zr Biomedical Alloys, *Mater. Trans.*, **45**, pp. 1113–1119.

118. Sakaguchi, N., Niinomi, M., Akahori, T., Takeda, J., and Toda, H. (2005). Relationships between tensile deformation behavior and microstructure in Ti–Nb–Ta–Zr system alloys, *Mat. Sc. Eng. C*, **25**, pp. 363–369.

119. Sander, C.G., Leeuwenburgh, J.A., Jansen, J.M., Wouter, A.D., Rouwkema, J., van Blitterswijk, C.A., Kirkpatrick, C.J., and Williams D.F. (2008). Trends in biomaterials research: An analysis of the scientific programme of the World Biomaterials Congress 2008, *Biomaterials*, **29**, pp. 3047–3052.

120. Shi, L.Q., Nortwood, D.O., and Cao, Z.W. (1993). Alloy design and microstructure of a biomedical Co–Cr alloy, *J. Materials Science*, **28**, pp. 1312–1316.

121. Shulman, L.B., and Driskell, T,D. (1997). Dental implants: a historical perspective., in *Implants in Denistry* (ed. Block, M.S., and Kent, J.N.), Saunders, Philadelphia, pp. 2–9.

122. Shuman, I. (2000). Replacement of tooth with a direct fiber reinforced direct bonded restoration, *Gen. Dent.*, **48**, pp. 314–318.

123. Smith, W.F. (1981). Structure and properties of engineering alloys, McGraw-Hill, New York.

124. Sumita, M., Hanawa, T., and Teoh, S.H. (2004). Development of nitrogen-containing nickel-free austenitic stainless steels for metallic biomaterials — review, *Mat. Sc. Eng. C*, **24**, pp. 753–760.

125. Takami, Y., Nakazawa, T., Makinouchi, K., Glueck, J., and Nosé, Y. (1997). Biocompatibility of alumina ceramic and polyethylene as materials for pivot bearings of a centrifugal blood pump, *J. Biomed. Mat. Res.*, **36**, pp. 381–386.

126. Takatoshi, N., Noriyasu, H., Hiroya, K., Naoya, S., Fumiaki, I., Kazuhiro, G., and Takashi, A. (2004). Biocompatibility of bone replaceability calcium phosphate cement (BRCPC) with human pulp cells *in vitro*, *Jap. J. Conserv. Dentistry*, **47**, pp. 476–482.

127. Thamaraiselvi, T.V., and Rajeswari, S. (2004). Biological evaluation of bioceramic materials – a review, *Trends Biomater. Artif. Organs*, **18**, pp. 9–17.

128. Tulinski, M., and Jurczyk, M. (2009). Mechanical ans corrosion properties of Ni-free austenitic stainless steel/hydroxyapatite nanocomposities, *Solid State Phenomena*, **151**, pp. 213–216.

129. Vaidyanathan, T.K., and Prasad, A. (1981). *In vitro* corrosion and tarnish analysis of the Ag-Pd binary system, *J. Dent. Res.*, **60**, pp. 707–715.

130. Vallana, F., Pasquino, E., Rinaldi, S., Galloni, M., Gatti, A.M., Modica, F, and Benech, A. (1993). Carbofilm: Present and future applications in biological devices, *Ceramics Intern.*, **19**, pp. 169–174.

131. Valiev, R.Z., Stolyarov, V.V., Rack, H.J., Lowe, T.C., and Shrivastava, S. (2004). *Medical Device Materials*, ASM, Cleveland, OH, pp. 362.

132. Van Noort, R. (1987). Titanium: The implant material today, *J Mater. Sci.*, **22**, pp. 3801–3811.

133. Vergano, P.J., Hill, D.C., and Uhlmann, D.R. (1967). Thermal expansion of feldspar glasses, *J. Am. Ceram. Soc.*, **50**, pp. 59–60.

134. Vult von Steyern, P. (2005). All-ceramic fixed partial dentures. Studies on aluminum oxide and zirconium dioxide based ceramic systems, *Swed. Dent. J.*, **173**, pp. 1–69.

135. Wahl, D.A., Sachlos, E., Liu, C.Z., and Czernuszka, J.T. (2007). Controlling the processing of collagenhydroxyapatite scaffolds for bone tissue engineering, *J. Mater. Sci. Mater. Med.*, **18**, pp. 201–209.

136. Wang, K. (1996). The use and properties of titanium and titanium alloys for medical applications in the USA, *Mater. Sci. Eng. A*, **213**, pp. 134–137.

137. Weinand, C., Gupta, R., Huang, A.Y., Weinberg, E., Madisch, I., Qudsi, R.A., Neville, C.M., Pomerantseva, I., and Vacanti, J.P. (2007). Comparison of hydrogels in the *in vivo* formation of tissue-engineered bone using mesenchymal stem cells and beta-tricalcium phosphate, *Tissue Eng.*, **13**, pp. 757–765.

138. Weinstein, M., Weinstein, L.K., Katz, S., and Weinstein, A.B. (1962). Fused porcelain-to-metal teeth, US Patent 3,052,982, USA.

139. Williams, D.F., Black, J., and Doherty, P.J. (1992). Second consensus conference on definitions in biomaterials, in *Biomaterial–tissue interfaces, Advances in biomaterials*, vol. 10 (Doherty, P.J., Williams, R.F., and Lee, A.J.C.), Elsevier, Amsterdam.

140. Williams, D.J. (1971). *Polymer science and engineering*, Prentice Hall, New Jersey.

141. Wilson, H.J., and Tomlin, H.R. (1969). Soft lining materials: some relevant properties and their determination, *J. Prosth. Dent.*, **21**, pp. 244–250.

142. Wilson, J., Stanley, H.R., Clark, A.E., Hench, L.L., and Gross, U. (1988). Development of bioglass for dental implantation, *Phillip J. Restaurative Zahnmedizin*, **5**, pp. 242–246.

143. Woodard, J.R., Hilldore, A.J., Lan, S.K., Park, C.J., Morgan, A.W., Eurell, J.A.C., Clark, S.G., Wheeler. M.B., Jamison, R.D., and Johnson, A.J.W. (2007). The mechanical properties and osteoconductivity of hydroxyapatite bone scaffolds with multi-scale porosity, *Biomaterials*, **28**, pp. 45–54.

144. Yang, C.Y., Wang, B.C., Chang, E., and Wu, B.C. (1995). Bond degradation at the plasma-sprayed HA coating/Ti–6Al–4V alloy interface: an *in vitro* study, *J. Mater. Sci. Mater. Med.*, **6**, pp. 258–265.

145. Yang, I.H., Kim, S.Y., Rubash, H.E., and Shanbhag, A.S. (1999). Fabrication of submicron titanium-alloy particles for biological response studies, *J. Biomed. Mat. Res., Part B: Appl. Biomat.*, **48**, pp. 220–223.

146. Yoshitani, S., Niinomi, M., Fukunaga, K., Fukui, H., Takeuchi, T., and Katsura, S. (2000). Dental precision castability of biocompatible β-type titanium alloy, *Cur. Adv. Mater. Proc.*, **13**, pp. 1345.

147. Zhang, D., Madhuvanthi, A., Cech, J., Siegmar, R., and Curran, S.A. (2006). Poly(L-lactide) (PLLA)/multiwalled carbon nanotube (MWCNT) composite: Characterization and biocompatibility evaluation, *The J. Phys. Chem., B*, **110**, pp. 12910–12915.

Chapter 4

Nanotechnology

Mieczyslaw Jurczyk[a,*] and Mieczyslawa U. Jurczyk[b]

[a]*Poznan University of Technology, Institute of Materials Science and Engineering, M. Sklodowska-Curie Sq. 5, 60-695 Poznan, Poland*

[b]*Division Mother's and Child's Health, University of Medical Sciences, Polna 33 Street, 60-535 Poznan, Poland*

[*]mieczyslaw.jurczyk@put.poznan.pl

Nanotechnology is seen as one of the most important fields of innovation and technology today [6, 8, 59]. The size is in the range of 1–100 nm the theories of classical and quantum mechanics are no long valid and a rich variety of unexpected properties are possible. New applications for materials can be created with novel or significantly enhanced properties. Products produced from these materials exhibit unique properties [6]. Nanoparticles are produced in two main ways (Fig. 4.1). The traditional way of producing fine particles has been "top-down" referring to the reduction through attrition and various methods of comminution in the traditional sense. During the past few years and increasingly, methods of production using "bottom-up" techniques are being increasingly utilized [39]. Bottom up approach refers to the buildup of a material from the bottom: atom by atom, molecule by molecule, or cluster by cluster.

Bionanomaterials for Dental Applications
Edited by Mieczyslaw Jurczyk

ISBN 978-981-4303-83-5 (Hardcover), 978-981-4303-84-2 (eBook)
www.panstanford.com

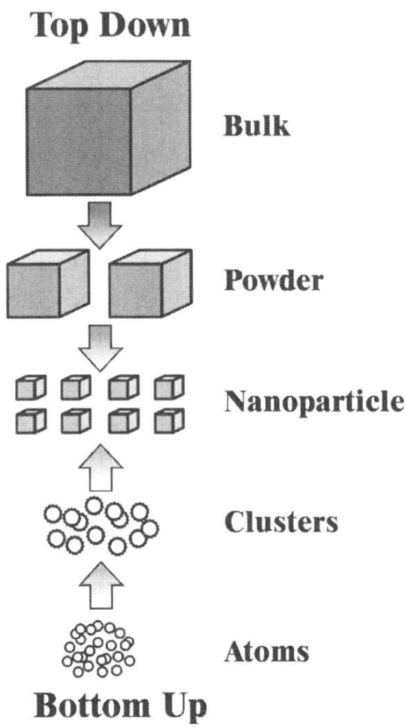

Figure 4.1 Schematic representation of the building up of nanostructures.

Both approaches play a very important role in modern industry and most likely in nanotechnology as well [59]. There are advantages and disadvantages in both approaches. High energy ball milling is a typical top-down method in making nanoparticles. The biggest problem with top-down approach is the imperfection of surface structure and significant crystallographic damage to the processed patterns. Regardless of the defects produced by top-down approach, they will continue to play an important role in the synthesis of nanostructures.

Although the bottom-up approach is frequently referred in nanotechnology, it is not a new concept. Currently, it plays an important role in the fabrication and processing of nanostructures. It is important to note, that the produced materials can have significantly different properties, depending on the route chosen to fabricate them. Apart from direct atom manipulation, there are various widely known methods for producing nanomaterials: mechanical, wet chemical, form-in-place, and gas-phase synthesis [55, 56].

4.1 Synthesis of Nanopowders

4.1.1 Top-Down Approach

Many different methods are available for producing small grained nanostructures. Mainly they are based on the production of fine grained powders (down the nanometer scale) and a subsequent Hot Isostatic Pressing (HIP) for consolidation. Mechanical processes include mechanical grinding, high-energy ball milling, mechanical alloying, and reactive milling [2, 45]. The advantages of these techniques are that they are simple, require low-cost equipment, and provided that a coarse feedstock powder can be made, the powder can be processed. However, there can be difficulties such as agglomeration of the powders and contamination from the process equipment. Mechanical processes are commonly used only for in organics and metals/alloys.

In high intensity ball milling, high impact collisions are used to reduce microcrystalline materials down to nanocrystalline structures without chemical change. A relatively new technique termed Mechanochemical Processing (MCP) technology is a novel solid-state process for the manufacture of a wide range of nanopowders [31]. Dry milling is used to induce chemical reactions through ball/powder collisions that result in nanoparticles formed within a salt matrix. Particle size is defined by the chemistry of the reactant mix, milling, and heat treatment conditions. Particle agglomeration is minimized by the salt matrix, which is then removed by a simple washing procedure.

4.1.1.1 Milling processes

In this book, we concentrate generally on nanoprocessing conducted entirely in a solid state. This is a class of methods based on processes conducted in high-energy ball mills. Such methods are based on high-energy grinding and milling of materials. They provide top-down approach toward manufacturing nanomaterials, bionanomaterials, and/or bionanocomposites [22, 25, 34, 35, 49].

Use of mechanical milling for synthesis of new alloys, compounds, and nanomaterials was not envisioned until recent decades. The primary objectives of milling have always been mixing or blending, change in particle shape, and size reduction. The milling in ball mills combines both crushing/shearing and impact forces combined in various proportions, depending on the equipment used. A variety of ball mills has been developed for different purposes including tumbler mills, attrition mills, shaker mills, vibratory mills, planetary mills, etc. For example, high-energy milling forces can be obtained using high frequencies and small amplitudes of vibration. Shaker mills (e.g., SPEX model 8000) which are preferable for small batches of powder (approximately 5–7 grams are sufficient for research purposes) are highly energetic and reactions can take place one order of magnitude faster than with other types of mill. Since the kinetic energy of the balls is a function of their mass and velocity, dense materials (steel or tungsten carbide) are preferable to ceramic balls.

Figure 4.2 shows the SPEX Model 8,000 high-energy mill manufactured by the CertiPrep Company (Metuchen, NJ, USA). The vials (Fig. 4.3) are made of hardened ferritic steel, but vials made of zirconia, alumina, agate, and hard-metal tungsten carbide are also used to greatly limit contamination by grinding media, mainly iron [44]. The milling that yields nanometric structure takes place about 10 times faster in a SPEX ball mill than in a planetary ball mill. The milling time needed to produce fine powders and nanostructure depends obviously on the nature of material, but also on other parameters, such as the weight of the milling balls and the ratio of the weight of balls to the weight of powder (BPR). The ball to powder weight ratio can vary from 1 to 100. The time needed for milling to the same fines of a powder usually decreases with the increase in the BPR. Agglomeration and cold welding of metal particulates to the reactor wall and the milling balls or to form balls made of the milled powder itself are common phenomena that occur during milling of pure ductile metals in vials filled with inert gas, such as argon.

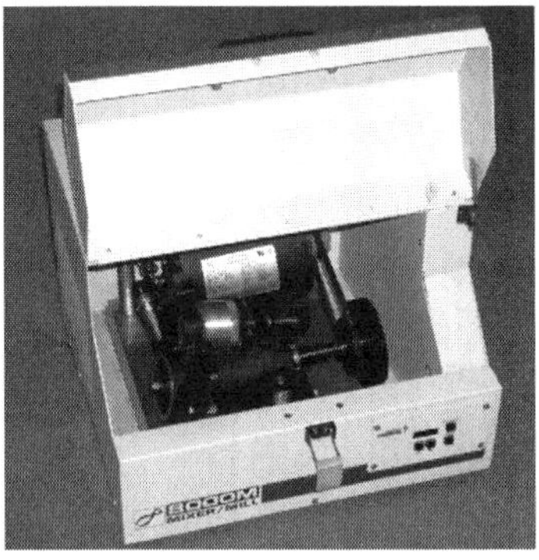

Figure 4.2 SPEX ball mill.

Figure 4.3 Steal vial for SPEX mill.

4.1.1.2 Nanoprocessing

Five major processing methods can be employed during milling of materials in high-energy ball mills: (i) high energy ball milling (HEBM), (ii) mechanical alloying (MA), (iii) mechanochemical activation synthesis (MCAS), (iv) mechanochemical synthesis (MCS), and (v) mechanical amorphization (MAM) [54].

High-energy ball milling is nanotechnology top down approach for the synthesis of nanoparticles. In the HEBM process, brute force is applied to material, whether it is a metal, a pre-alloyed intermetallic, or a solid chemical (stoichiometric) compound. Short milling times can break thin chemically passive surface coatings (e.g., surface oxides) and expose fresh, clean chemically active metallic surface. Such milling can also introduce defects into solid compounds. It results in an increased chemical activity of milled media toward both gasses and chemical reactions in solutions and electrolytes. This process is often termed mechanochemical activation synthesis.

A process wherein mixture of elemental metal powders, or powders of metal and nonmetal are milled long enough to trigger alloying of elemental powders in a solid-state process is termed mechanical alloying. Since substantial rearrangement of chemical species must take place during alloying, the latter requires long times of milling, usually more than twice the time needed for preparation of nanopowders by mechanical milling. Even longer milling results in the term, mechanical amorphization of a crystalline solid.

Reactive mechanical milling can be realized by reacting two chemical compounds A and B in a solid-state synthesis that yields a distinct chemical compound C. This is termed mechanical synthesis or mechanosynthesis. Reactive milling can also be conducted by milling metal powders in ball mills filled with a reactive gas.

4.1.1.3 Mechanical alloying

Mechanical alloying, the main mechanical process, was developed in the 1970s at the International Nickel Co as a technique for dispersing nanosize inclusions into nickel-based alloys (Fig. 4.4) [3]. During the last years, the MA process has been successfully used to prepare a variety of alloy powders including powders exhibiting supersaturated solid solutions, quasicrystals, amorphous phases, and nano-intermetallic compounds [25, 45]. MA technique has been proved a novel and promising method for alloy formation [26, 36, 42, 44].

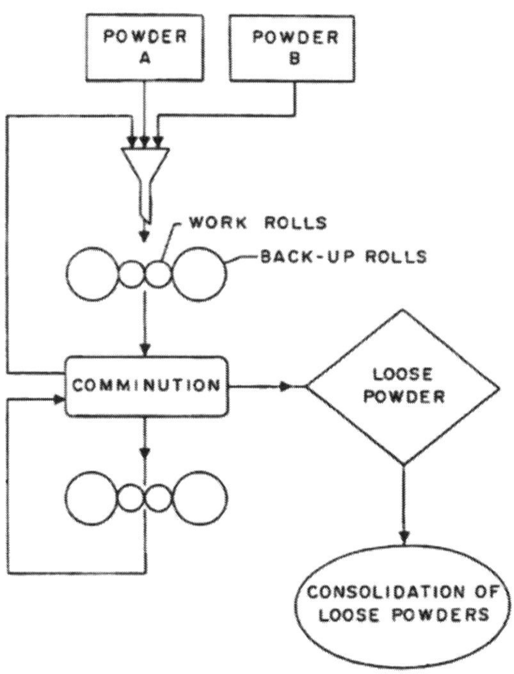

Figure 4.4 Mechanical alloying process [3].

Mechanical alloying is conducted through milling of two elemental metal powders. Longer milling, order of hours, leads to atomic-level mixing of the metals and produces an alloy consisting of these metals. Until the advent of this new, non-equilibrium, and low-temperature solid-state processing method, metal alloys could only be manufactured by melt casting and metal foundry practices.

The raw materials used for MA, are available commercially high purity powders that have sizes in the range of 1–100 μm. During the mechanical alloying process, the powder particles are periodically trapped between colliding balls and are plastically deformed. Such a feature occurs by the generation of a wide number of dislocations as well as other lattice defects. Furthermore, the ball collisions cause fracturing and cold welding of the elementary particles, forming clean interfaces at the atomic scale. Further milling lead to an increase of the interface number and the sizes of the elementary component area decrease from millimeter to submicrometer lengths. Concurrently to this decrease of the elementary distribution, some nanocrystalline intermediate phases are produced inside the particles or at its surfaces. As the milling duration develops, the content fraction of such intermediate compounds increases leading to a final product whose properties are the function of the milling conditions.

The sequence of concurrent mechanical and chemical events can be written as follows [54]: mechanical stressing → severe plastic deformation → formation of a submicron lamellar microstructure → cold interdiffusion of metal atoms between lamellae or nanograins (cold welding) → fracture → formation of nanostructure → extended solid solubility → mechanical alloying with formation of thermodynamically stable and /or metastable phases → amorphization.

It was shown, that mechanical alloying has produced amorphous phases in metals. But differentiation between a "truly" amorphous, extremely fine grained, or a material in which very small crystals are embedded in an amorphous matrix in so produced materials has not been easy on the basis of diffraction basis [45]. Only the supplementary investigations by neutron diffraction can unambiguously confirmed that the phases produced by MA are truly amorphous. The milled powder is finally heat treated to obtain the desired microstructure and properties. Annealing leads to grain growth and release of microstrain.

Nanocrystalline materials exhibit quite different properties from both crystalline and amorphous materials, due to structure, in which extremely fine grains are separated by what some investigators have characterized as "glass-like" disordered grain boundaries. The mechanism of amorphous phase formation by MA is due to a chemical solid state reaction, which is believed to be caused by the formation of a multilayer structure during milling [45]. When a mixture of elemental powders is milled, the formation of the amorphous phase is due mainly to an ultimate interdiffusion of atoms that occurs at fresh surfaces and interfaces created by mechanical milling. This interdiffusion is promoted by defects and chemical disorder in the crystalline structure.

The basic process of mechanical alloying is illustrated in Fig. 4.5. Microcrystalline powders are placed together with a number of hardened steel or tungsten carbide (WC) coated balls in a sealed container which is shaken. The most effective ratio for the ball to powder masses is ten.

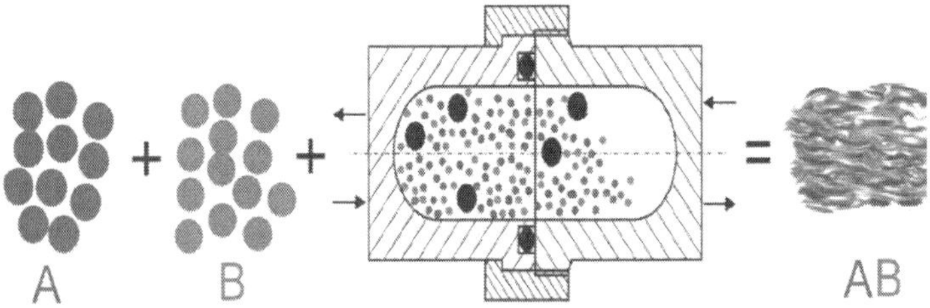

Figure 4.5 Schematic cross-sectional representation of MA process for synthesizing nanometer-sized powders (SPEX 8000 mixer mill).

Furthermore, for all nanocrystalline materials prepared by a variety of different synthesis routes, surface, and interface, contamination is a major concern. In particular, during mechanical attrition, contamination by the milling tools (iron) and atmosphere (trace elements of O_2 and other rare gases) can be a problem (see Fig. 4.4) [24, 43]. By minimizing the milling time and using the purest, most ductile metal powders available, a thin coating of the milling tools by the respective powder material can be obtained which reduces Fe contamination tremendously. Atmospheric contamination can be minimized or eliminated by sealing the vial with a flexible "O" ring after the powder has been loaded in an inert gas glove box.

The cleanliness of the surface of nanocrystalline alloys prepared by mechanical alloying method was studied by X-ray photoelectron spectroscopy (XPS) and Auger electron spectroscopy (AES) [24]. In Fig. 4.6, we show the element specific Auger intensities of the nanocrystalline $LaNi_{4.2}Al_{0.8}$ sample as a function of the sputtering time, converted to depth. Data for microcrystalline $LaNi_{4.2}Al_{0.8}$ sample are included, too.

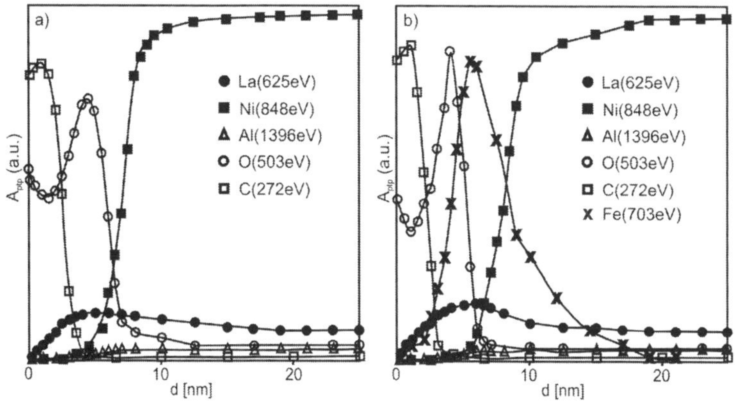

Figure 4.6 AES spectrum of microcrystalline (a) and nanocrystalline (b) $LaNi_{4.2}Al_{0.8}$ alloy vs. sputtering time, as converted to depth. The sample surface is located on the left-hand side.

As it can be seen, there is relatively high concentration of carbon and oxygen immediately on the surface, which could be due to carbonates or adsorbed atmospheric CO_2. The carbon concentration strongly decreases towards the interior of the sample. At the metal interface itself, only oxygen is present, making it very likely that only an oxide layer is formed, and no other compounds, the latter growing apparently with a smaller probability. We have found that at the oxide–metal interface, mainly lanthanum and nickel atoms are present. Taking into account that the escape-depth of the Auger electron from nickel and aluminum atoms is about 2 nm, the concentration of these elements on the metallic surface is significantly lower compared to the average bulk composition. Therefore, the lanthanum atoms, which segregate to the surface form a La based oxide layer under atmospheric conditions. The oxidation process is depth limited such that an oxide-covering layer with a well-defined thickness is formed by which the lower lying metal is prevented from further oxidation. In this way, one can obtain a self-stabilized oxide–metal structure. Very similar behavior was observed for the microcrystalline Co thin films oxidized under atmospheric conditions. From the peak-to-peak amplitude, we have calculated a maximum atomic concentration of oxygen inside the sample as 2 at.%. The concentration of carbon impurities inside the sample was below 0.5 at.%.

Figure 4.6b shows the element specific Auger intensities of the nanocrystalline $LaNi_{4.2}Al_{0.8}$ sample as a function of the sputtering time, converted to depth. Similarly to the results obtained for the microcrystalline sample, there is relatively high concentration of carbon and oxygen immediately on the surface. The carbon concentration strongly decreases towards the interior of the sample. We have found that at the oxide–metal interface, only iron impurities and lanthanum atoms are present. As the escape-depth of the Auger electrons from nickel and aluminum atoms is about 2 nm, we conclude that these elements are practically not present on the metallic surface. In other words, lanthanum atoms and iron impurities strongly segregate to the surface and form an oxide layer under atmospheric conditions. The lower lying Ni atoms form a metallic subsurface layer and are responsible for the observed high hydrogenation rate in accordance with earlier findings. The above segregation process is stronger compared to that observed for the microcrystalline sample. The presence of the significant amount of iron atoms in the surface layer of the nanocrystalline $LaNi_{4.2}Al_{0.8}$ alloy could be explained by Fe impurities trapped in the MA powders from erosion of the milling media. The amount of the Fe impurities considerably decreases in the subsurface layer of the sample. From the peak-to-peak amplitude we have estimated a maximum atomic concentration of oxygen as ~2 at.% in the interior of the sample. Similarly to the results reported for the polycrystalline $LaNi_{4.2}Al_{0.8}$ alloy, the concentration of carbon impurities inside the sample was below 0.5 at.%.

Concluding, the contamination with Fe based wear debris in mechanically synthesized nanomaterials can generally be reduced to less than 1–2 % and oxygen contamination to less than 300 ppm. But, contamination through the milling atmosphere can have a positive impact on the milling conditions if one wants to prepare metal or ceramic nanocomposites with one of the metallic elements being chemically highly reactive with the gas (or fluid) environment. On the other side, main advantage of top-down approach can be high production rates of nano powders.

4.1.2 Bottom-Up Approach

4.1.2.1 Wet chemical processes

These include colloidal chemistry, hydrothermal methods, sol-gels, and other precipitation processes. Generally, solutions of different ions are mixed in well-defined quantities and under controlled conditions of heat, temperature, and pressure to promote the formation of insoluble compounds, which precipitate out of solution. Among the advantages of the established wet chemistry methods, are good control over microstructure and particle morphology.

The sol-gel process involves the evolution of inorganic networks through the formation of a colloidal suspension (sol) and gelation of the sol to form a network in a continuous liquid phase (gel) [28]. The precursors for synthesizing these colloids consist usually of a metal or metalloid element surrounded by various reactive ligands. The starting material is processed to form a dispersible oxide and forms a sol in contact with water or dilute acid. Removal of the liquid from the sol yields the gel, and the sol/gel transition controls the particle size and shape. Calcination of the gel produces the oxide.

Sol-gel processing refers to the hydrolysis and condensation of alkoxide-based precursors such as $Si(OEt)_4$(tetraethyl orthosilicate, or TEOS). The reactions involved in the sol-gel chemistry based on the hydrolysis and condensation of metal alkoxides $M(OR)_z$ can be described as follows:

$$MOR + H_2O \rightarrow MOH + ROH \text{ (hydrolysis)} \tag{4.1}$$

$$MOH + ROM \rightarrow M\text{-}O\text{-}M + ROH \text{ (condensation)} \tag{4.2}$$

The production of glasses by the sol-gel method permits preparation of glasses at far lower temperatures than is possible by using conventional melting. It also makes possible synthesis of compositions that are difficult to obtain by conventional means because of problems associated with volatilization, high melting temperatures, or crystallization. In addition, the sol-gel approach is a high-purity process that leads to excellent homogeneity. The most successful applications utilize the composition control, microstructure control, purity, and uniformity of the method combined with the ability to form various shapes at low temperatures. Finally, the sol-gel approach is adaptable to producing films and fibers as well as bulk pieces.

Films and coatings were the first commercial applications of the sol-gel process. The development of sol-gel-based optical materials has also been quite successful, and applications include monoliths (lenses, prisms, and lasers), fibers (waveguides), and a wide variety of optical films. Other important applications of sol-gel technology utilize controlled porosity and high surface area for catalyst supports, porous membranes, and thermal insulation.

4.1.2.2 Form in place processes

These include lithography, vacuum deposition processes such as physical vapor deposition (PVD) and chemical vapor deposition (CVD), and spray coatings. These processes are more geared to the production of nanostructured layers and coatings, but can be used to fabricate nanoparticles by scraping the deposits from the collector [55].

Physical vapor deposition (PVD) is fundamentally a vaporization coating technique, involving transfer of material on an atomic level. It is an alternative process to electroplating. The process is similar to chemical vapor deposition (CVD) except that the raw materials/precursors, i.e. the material that is going to be deposited starts out in solid form, whereas in CVD, the precursors are introduced to the reaction chamber in the gaseous state. It incorporates processes such as sputter coating and pulsed laser deposition (PLD).

PVD processes are carried out under vacuum conditions. The process involved four steps: evaporation, transportation, reaction, and deposition. During the first stage, a target consisting of the material to be deposited is bombarded by a high energy source such as a beam of electrons or ions. This dislodges atoms from the surface of the target, "vaporizing" them. The next process simply consists of the movement of "vaporized" atoms from the target to the substrate to be coated and will generally be a straight line affair.

In some cases, coatings will consist of metal oxides, nitrides, carbides, and other such materials. In these cases, the target will consist of the metal. The atoms of metal will then react with the appropriate gas during the transport stage. For the above examples, the reactive gases may be oxygen, nitrogen and methane. PVD coatings are deposited for numerous reasons. Some of the main ones are: improved hardness and wear resistance, reduced friction, and improved oxidation resistance.

In the case of chemical vapor deposition, feed gases are reacted in a chamber and the resulting species are attracted to a substrate. Once again the reaction products can be controlled and not only in terms of composition but also in terms of how they are deposited. The substrate effectively provides a template from where the deposited coating can grow in a very well controlled manner. Electro-deposition involves a similar process; however, the controlled coating is deposited from solution by the application of an electric field.

4.1.2.3 Gas phase synthesis

These include gas condensation processing (GPC), flame pyrolysis, electro-explosion, laser ablation, high-temperature evaporation, and plasma synthesis techniques. Gas condensation was the first technique used to synthesize nanocrystalline metals and alloys [9–11]. In this technique a metallic or inorganic material is vaporized using thermal evaporation sources such as Joule heated refractory crucibles, electron beam evaporation devices or sputtering sources in an atmosphere of 1–50 mbar He (or another inert gas like Ar, Ne, Kr). Cluster form in the vicinity of the source by homogenous nucleation in the gas phase and grow by coalescence and incorporation of atoms from the gas phase. The cluster or particle size depends critically on the residence time of the particles in the growth regime and can be influenced by the gas pressure, the kind of inert gas, i.e. He, Ar or Kr, and on the evaporation rate/vapor pressure of the evaporating material.

The average particle size of the nanoparticles increases with an increasing gas pressure, vapor pressure, and mass of the inert gas. A rotating cylindrical device cooled with liquid nitrogen was employed for the particle collection. The nanoparticles in the size range from 2–50 nm are extracted from the gas flow by thermophoretic forces and deposited loosely on the surface of the collection device as a powder of low density and no agglomeration. Subsequently, the nanoparticles are removed from the surface of the cylinder by means of a scraper in the form of a metallic plate. In addition to this cold finger device, several techniques known from aerosol science have now been implemented for the use in gas condensation systems such as corona discharge, etc. These methods allow for the continuous operation of the collection device and are better suited for larger scale synthesis of nanopowders. However, these methods can only be used in a system designed for gas flow, i.e. a dynamic vacuum is generated by means of both continuous pumping and gas inlet via mass flow controller. A major advantage over convectional gas flow is the improved control of the particle sizes. Depending on the flow rate of the He-gas, particle sizes are reduced by 80% and standard deviations by 18%.

Evaporation can be done from refractory metal crucibles (W, Ta, or Mo). If metals with high melting points or metals which react with the crucibles, are to be prepared, sputtering, i.e. for W and Zr, or laser or electron beam evaporation has to be used. Synthesis of alloys or intermetallic compounds by thermal evaporation can only be done in the exceptional cases that the vapor pressures of the constituents' elements are similar. As an alternative, sputtering from an alloy or mixed target can be employed.

However, control of the composition of the elements has been difficult and reproducibility is poor. Nanocrystalline oxide powders are formed by controlled postoxidation of primary nanoparticles of a pure metal (e.g., Ti to TiO_2) or a suboxide (e.g., ZrO to ZrO_2). Although the gas condensation method including the variations has been widely employed to prepare a variety of metallic and ceramic materials, quantities have so far been limited to a laboratory scale. Recently, it should be mentioned that the scale-up of the gas condensation method for industrial production of nanocrystalline oxides by a company called nanophase technologies has been successful.

Laser ablation has been extensively used for the preparation of nanoparticles and particulate films. In this process, a laser beam is used as the primary excitation source of ablation for generating clusters directly from a solid sample in a wide variety of applications. Laser vaporization cluster beams were introduced by Smalley and coworkers to overcome the limitations of oven sources [15]. In this method, a high energy pulsed laser with an intensity flux exceeding 10^7 W/cm^3 is focused on a target containing the material to be made into clusters. The resulting plasma causes highly efficient vaporization since with current, pulsed lasers one can easily generate temperatures at the target material greater than 10^4 K. This high temperature vaporizes all known substances so quickly that the rest of the source can operate at room temperature. Typical yields are 10^{14}–10^{15} atoms from a surface area of 0.01 cm^2 in a 10^{-8} s pulse. The local atomic vapor density can exceed 10^{18} atom/cm^3 (equivalent to 100 Torr pressure) in the microseconds following the laser pulse. The hot metal vapor is entrained in a pulsed flow of carrier gas (typically He) and expanded through a nozzle into a vacuum. The cool, high-density helium flowing over the target serves as a buffer gas in which clusters of the target material form, thermalize to near room temperature and then cool to a few K in the subsequent supersonic expansion.

In a recent investigation utilizing a novel atomization system, (LINA-SPARK™), LSA, based on laser spark atomization of solids has been developed that seems to be very versatile for different materials [19, 20]. Briefly, the LSA is capable of evaporating material at a rate of about 20 μg/s from a solid target under argon atmosphere. The small dimensions of the particles and the possibility to form thick films make the LSA quite an efficient tool for the production of ceramic particles and coatings, and also an ablation source for analytical applications such as the coupling to induced coupled plasma emission spectrometry. ICP, the formation of the nanoparticles, has been explained following a liquefaction process which generates an aerosol, followed by the cooling/solidification of the droplets which results in the formation of fog.

Both RF and DC plasmas are being used successfully to make a wide range of materials [11]. In plasma arcing, the very high temperatures associated with the formation of an arc or plasma is used to effectively separate the atomic species of the feedstock, which quickly recombine outside the plasma to form nanosize particles, which may have novel compositions. The heat source is very clean and controllable and the temperatures in the plasmas can reach in excess of 9000°C, which means that even highly refractory materials can be processed.

Short-range ordering in metal alloys may be formed during deposition of atoms or atom clusters from gas phases. The short-range order of atoms may even be preserved in amorphous metallic alloys. The approach to build up nanocrystals through such self-organization of atoms is well recognized in the preparation of nanomaterials and termed the bottom-up nanotechnology approach.

4.2 Nanopowder Consolidation

Nanosintering is the sintering of nanometer size powders. For certain nanomaterial applications, sintering of powders while retaining grain sizes in the nanometer range becomes a critical processing step. Fully dense specimens with nanosize features are most important for structural, magnetic, electric, electronic, and medical applications. The focus in the consolidation of out-of-equilibrium powders has been the retention of the metastable condition of the initial structures.

The sintering process of powder materials with particle in the nanometer range was studied in many laboratories [13, 14, 33]. For example, high-energy ball milled and heat treated $Nd_2(Fe, Co, Zr)_{14}B/\alpha\text{-}Fe$ and $Nd(Fe, Mo)_{12}N_x/\alpha\text{-}Fe$ magnetic nanopowders have been compacted by hot pressing to form bulk magnets [23]. Studies of nanopowder densification have lead to a better understanding of numerous sintering issues such as powder agglomeration, surface condition or contamination, pore role in sintering, and grain growth. Thermodynamic and kinetic aspects of metastable powder densification, sintering mechanisms, and scaling laws applicable to nanopowder sintering were examined, as well [13, 58].

4.2.1 Consolidation Methods

When powder particle size decreases below the micron range, the consolidation efforts are faced with additional problems related to powder agglomeration, interparticle friction, and contamination. In spite of these new challenges, all known consolidation methods have been applied for full densification of both ceramic and metal nanopowders, including conventional sintering. Non-conventional sintering methods include: microwave sintering, shock or dynamic consolidation, and field-assisted sintering. The main purpose of using these methods is to enhance densification and prevent grain growth. The very short high temperature exposure during consolidation provides the best means to retain fine grain size or out of equilibrium conditions such as amorphous structures or supersaturated solid solutions [46]. Examples of dense materials that have retained nanosize grains <100 nm by conventional sintering are provided in Table 4.1 [33, 41].

A large variety of pressure assisted methods has been applied for consolidation of nanopowders ranging from low (<100 MPa) to high pressure methods (>0.5 GPa) such as piston cylinder, diamond anvil, and torsion-pressure method (sometimes referred to as severe plastic deformation consolidation — SPDC). To illustrate the pressure effects, Hahn fully sintered TiO_2 at 998–1098°C (~0.35 T_m) by applying 1 GPa while retaining the morphologically metastable structure (i.e., with no grain growth) [17].

Table 4.1 Densities and grain sizes in selected nanopowders densified by conventional sintering

		Sintering parameters:			Final properties:	
Material:	Init. Gr. size(nm)	Temp. (K)	Time (hrs)	Atm.	Density (%)	Gr. size (nm)
ZrO_2	6 to 9	1400	1.3	air	full	80
ZrO_2-3 mol.% Y_2O_3	<10	1373	1	NR	99.9	80
TiO_2	~6	873	NR	NR	99	<60

4.2.2 Nanosintering

Thermodynamically, nanopowders are unstable due to large surface area. Nanoparticles adopt different surface energies than regular ones, for instance, by a different local atomic arrangement at the surface. TEM studies showed that nanoparticles have a faceted appearance with anisotropic surface energies (e.g., in γ-Al_2O_3 [7]). Kinetically, sintering of nanopowders is significantly enhanced. Sintering of nanoparticles indicated depressed sintering onset temperatures (0.2–0.3 T_m) as compared to conventional powders (0.5–0.8 T_m). Molecular dynamics (MD) simulations indicated extremely fast sintering of nanoparticles [58]. Surface diffusion cannot explain this behavior. Therefore, some other sintering mechanisms have been suggested: dislocation motion, grain rotation, viscous flow, and grain boundary slip [14, 33]. The rapid shrinkage was attributed to fast dislocation activity driven by the contact Hertzian stresses that exceed the ideal shear strength [58]. After the neck forms, the adjacent particles rotate to achieve minimum grain boundary energy. Such rotation process was confirmed by TEM studies of nanoparticles [37, 56].

Full densification of nanopowders is completed at temperatures lower than that for conventional powders. To rationalize this decrease in the sintering temperature, different scaling laws have been applied [1, 18]. Considering Herring's law, the sintering temperature dependence on the particle size may be calculated.

Reasonable agreement of experimental and calculated temperatures was found in TiO_2 and Al_2O_3 assuming certain diffusion mechanisms [32].

4.2.3 Pore Effects and Grain Growth

The influence of pores on grain coarsening has been well documented theoretically and experimentally, mostly in ceramics. Open pores in nanopowders inhibit grain growth in a similar way that pores prevent grain coarsening in ordinary grain sized ceramics. The pinning action of the pores is, however, difficult to predict. Gupta established an empirical direct relationship between density and grain size in the intermediate stage sintering [16]. In a pore controlled grain growth model, Liu and Patterson found a linear relationship between the inverse of grain size and the pore surface area per unit volume [29].

A modified sintering law that directly accounts for pore size effects on densification rate was proposed by Mayo [33]:

$$\frac{1}{\rho(1-\rho)}\frac{d\rho}{dt} \propto \frac{1}{d^n}\frac{1}{r}\exp\left(\frac{-Q}{RT}\right) \quad (4.3)$$

where p is density, d is the particle size, n is a constant dependent on the sintering mechanism, r is the pore radius, Q is the activation energy, R is the gas constant, and T is the absolute sintering temperature. This equation predicts that the highest densification rate occurs for the finest pore size. A small pore size is also critical in controlling the final grain size based on the pore pinning effect. For these purposes, a small and uniform pore population is desired in the green compact.

Based on detailed microstructural studies, the average grain size can be measured for the sintered samples as a function of temperature [53]:

$$d^n - d_0^n = kt \quad (4.4)$$

where d_0 is grain size before heat treatment, t is time and $k = k_0 \exp(-E/(kT))$. For ZrO_2 nanopowders experimental coefficient $n = 3$ and activation energy $E = 4.2$ eV.

4.3 Severe Plastic Deformation

Top-down approaches involve the refinement of coarse grains to ultra-fine grains by severe plastic deformation (SPD) techniques that subject the work-piece to high-accumulated strains [30]. With increasing straining of the material, the size of both microstructural levels decreases in size. At the same time, the disorientation — difference in crystallographic orientation — is increasing. In order to obtain the smallest microstructural sizes, plastic strains more than 600–800% are necessary. A number of severe plastic deformation (SPD) methods for producing bulk ultra fine grain metals/alloys have been developed [30, 50]. The significant feature of these processes is that the external dimensions of the work-piece do not change significantly during the processing.

Examples of these methods include equal channel angular extrusion or pressing (ECAE or ECAP) [40, 50–52, 57], cyclic extrusion compression method (CEC) [38], high pressure torsion (HPT) [27], twist extrusion (TE) [5], friction stir processing (FSP) [47], and multi directional forging (MDF), also known as multiaxial compression/forging (MAC/F) [12]. In addition, there are several methods of producing ultra fine grain sheet metals, such as accumulative roll bonding (ARB) [48], and repeated corrugation and straightening (RCS) [21]. From different variants of SPD techniques, only a few have industrial potential.

4.3.1 Equal Channel Angular Pressing

Equal Channel Angular Pressing (ECAP) is one of the discontinuous processes of severe plastic deformation. In principal, the tool consists of two intersecting channels of the same cross section that meet at an angle 2α (Fig. 4.7).

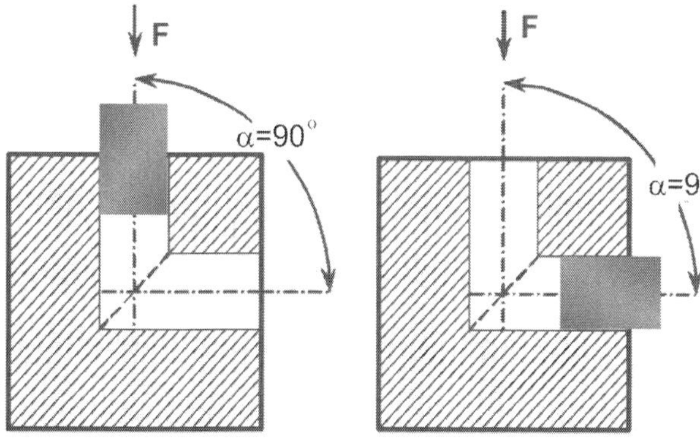

Figure 4.7 Schematic ECAE/P process.

The geometry of this tool provides that the material is deformed by simple shear at ideal, frictionless, and conditions. The cross section of the specimen remains about equal before and after a processing step, thus it is possible to subject one specimen several times to ECAP in order to reach highest degrees of plastic deformation. A circular or squared cross section of the channel provides the possibility of a materials processing at different routes that are distinguished by their different combinations of sample rotation around the channel axes between consecutive processing steps.

4.3.2 Cyclic Extrusion Compression Method

Richert J. and Richert M. came up with the idea of cyclic extrusion compression method (CEC) [38]. CEC involves the cyclic flow of metal between the alternating extrusion and compression chambers (Fig. 4.8). The deformation effect could obviously be achieved with the frame/die fixed and the movable punches or vice versa.

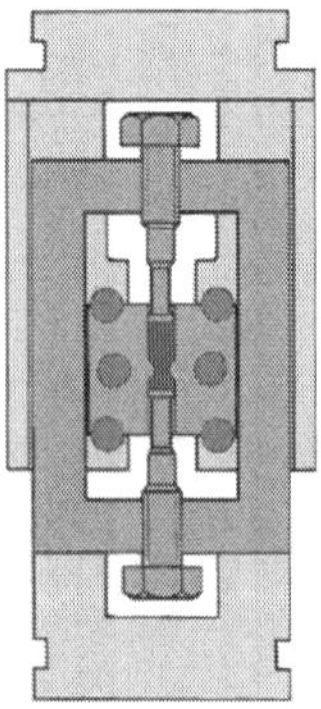

Figure 4.8 Reciprocating extrusion compression method.

4.3.3 High Pressure Torsion

High Pressure Torsion (HPT) is one technique of severe plastic deformation and provides a continuous shear deformation of the metallic material. In Fig. 4.9, a sketch of a HPT tool is shown. In principal, a disc shaped sample is deformed by pure shear between two anvils that are rotating with respect to each other. The necessary torsion-momentum is provided by friction forces at contact areas anvil-material.

Figure 4.9 High-pressure torsion technique.

4.3.4 Other Severe Plastic Deformation Methods

Twist Extrusion (TE) process was developed in 1999 [4]. This process can change the structure of materials, significantly improving some of their physical and mechanical properties. TE works by extruding a prism specimen through a matrix whose profile consists of two prism-like regions separated by a twist passage. The extruded material undergoes an intense shift, with the property that the final cross-section of the specimen is identical to the initial cross-section. This property allows for a repeated extrusion, that accumulates the value of deformation.

There are currently three main application areas of TE: obtaining ultra-fine grained crystalline and nano-crystalline structures in bulk specimens; increasing the plasticity of secondary non-ferrous metals and alloys, which allows one to significantly broaden the range of production; and obtaining bulk specimens by consolidating porous materials which allows one to create substantially different, new compositions with unique characteristics.

Friction Stir Processing (FSP) provides the ability to thermo-mechanically process selective locations on the structure's surface and to some considerable depth (>25 mm) to enhance specific properties.

Multi-Axial Compressions/Forgings involves the deformation of a rectangular cross section samples through a series of compressions so that the initial dimensions of the billet are retained. The loading direction is changed through 90° between successive compressions. MAG/F's are effective in producing fine grain structure, but are deficient due to the non-uniform strain distribution along the billet cross-section. However, this non-uniformity can be eliminated by very good lubrication of the billet and through a large number of compression/forgings steps.

References

1. Alymov, M.I., Maltina, E.I., and Stepanov, Y.N. (1992). Model of initial stage of ultrafine metal powder sintering, *Nanostruct. Mat.*, **4**, pp. 737–742.
2. Benjamin, J.S. (1976). Mechanical alloying, *Sci. Am.*, **234**, pp. 40–57.
3. Benjamin, J. S. (1997). *Mechanical alloying process*, United States Patent 5688303, USA.
4. Beygelzimer, Y., Varukhin, V., Synkov, S., Sapronov, A., and Synkov, V. (1999). New techniques for accumulating large plastic deformations using hydroextrusion, *Fizika i Tekhnika Vusokih Davlenii* (*High Pressure Physics and Technology*, in Russian), **6**, pp. 499–508.
5. Beygelzimer, Y., Varyukhin, V., Orlov, D., Efros, B., Stolyarov, V., and Salimgareyev, H. (2002). Ultrafine grained materials II, in *Proceedings of a Symposium held during the 2002 TMS Annual Meeting I, Seattle, Washington, February 17-21, 2002* (ed. Zhu, Y.T., Langdon, T.G., Mishra, R.S., Semiatin, S.L., Saran, M.J., and Lowe, T.C.), TMS, Warrendale PA, pp. 297–304.
6. Bleeker, R.A., Troilo, L.M., and Ciminello, D.P. (2004). Patenting nanotechnology, *Materials Today*, **2**, pp. 44–48.
7. Bonevics, J.E., and Marks, L.D. (1992). The sintering behavior of ultrafine alumina particles, *J. Mater Res.*, **7**, pp. 1489–1500.
8. Dowling, A.P. (2004). Development of nanotechnologies, *Materials Today*, **7**, pp. 30–35.
9. Fernández, A., Reddy, E.P., Rojas, T.C., and Sánchez-López, J.C. (1999). Application of the gas phase condensation to the preparation of nanoparticles, *Vacuum*, **52**, pp. 83–88.
10. Festag, G., Steinbrueck, A., Csaki, A., Moeller, R., and Fritzsche, W. (2007). Single particle studies of the autocatalytic metal deposition onto surface bound gold nanoparticles reveal a linear growth, *Nanotechnology*, **18(1)**.
11. Francois, G. (2001). *Advances in Plasma Physics Research*, Volume I, Nova Science Publishers, New York.

12. Ghosh, A.K. (1988). *Method for producing a fine grain aluminum alloy using three axes deformation*, U.S. Patent 4,721,537, USA.

13. Groza, J.R. (2002). Nanocrystalline powder consolidation methods, in *Nanostructured Materials: Processing, Properties, and Potential Applications*, (ed. Koch, C.C.), Noyes Publ. Westwood, New York, pp. 115–178.

14. Groza, J.R. (1999). Nanosintering, *Nanostruct. Mater.*, **12**, pp. 987–992.

15. Guo, T., Nikolaev, P., Thess, A., Colbert, D.T., and Smalley, R.E. (1995). Catalytic growth of single-walled nanotubes by laser vaporization, *Chem. Phys. Let.*, **243**, pp. 49–54.

16. Gupta, K. (1972). Possible correlation between density and grain size during sintering, *J. Am. Ceram. Soc.*, **55**, pp. 276–277.

17. Hahn, H. (1993). Microstructure and properties of nanostructured oxides, *Nanostruct. Mater.*, **2**, pp. 251–265.

18. Herring, C. (1950). Effect of change of scale on sintering phenomena, *J. Appl. Phys.*, **21**, pp. 301–303.

19. Houriet, R., Vacassy, R., Hofmann, H., and Vogel, W. (1998). Thin film growth using ablation of ceramics with a Lina-spark atomizer, *Mater. Res. Soc. Symp. Proc.*, **526**, pp. 117–122.

20. Houriet, R., Vacassy, R., and Hofmann, H. (1999). Synthesis of powders and films using a new laser ablation technique, *Nanostruct. Mater.*, **11**, pp. 1155–1163.

21. Huang, J.Y., Zhu, Y.T., Jiang, H., and Lowe, T.C. (2001). Microstructure and dislocation configuration in nanostructured Cu processed by repetitive corrugation and straightening, *Acta Mater.*, **49**, pp. 1497–1505.

22. Jurczyk, K., Niespodziana, K., Stopa, J., and Jurczyk, M. (2009). Titanium-ceramic dental bionanomaterials, *J. Clin. Periodontol.*, **36**, Suppl. 9, pp. 73–74.

23. Jurczyk, M., and Gwan, P.G. (1995). Magnets produced by hot pressing $Nd_2(Fe, Co, Zr)_{14}B/\alpha$-Fe and $Nd(Fe, Mo)_{12}Nx/\alpha$-Fe powders, *J. Alloys Compd.*, **230**, pp. L1–L3.

24. Jurczyk, M., Smardz, K., Rajewski, W., and Smardz, L. (2001). Nanocrystalline $LaNi_{4.2}Al_{0.8}$ prepared by mechanical alloying and annealing and its hydride formation, *Mater. Sci. Eng. A*, **303**, pp. 70–76.

25. Jurczyk, M., and Nowak, M. (2008). Nanomaterials for hydrogen storage synthesized by mechanical alloying, in *Nanostructured Materials in Electrochemistry* (ed. Eftekhari Ali), Wiley, pp. 349–386.

26. Jurczyk, M., Smardz, L., Okonska, I., Jankowska, E., Nowak, M., and Smardz, K. (2008). Nanoscale Mg-based materials for hydrogen storage, *Int. J. Hydrogen Energy*, **33**, pp. 374–380.
27. Kim, H.S. (2001). Finite element analysis of high pressure torsion processing, *J. Mat. Proc. Technol.*, **113**, pp. 617–621.
28. Klein, L. (1994). *Sol-Gel Optics: Processing and Applications*, Springer Verlag.
29. Liu, Y., and Patterson, B.R. (1993). Grain growth inhibition by porosity, *Acta Met. Mater.*, **41**, pp. 2651–2656.
30. Love, T.C., and Valiev, R.Z. (eds) (2000). *Investigations and Applications of Severe Plastic Deformation*, Kluwer Academia Pub., Dordrecht.
31. Lü, L., and Lai, M.O. (1998). *Mechanical Alloying*, Kluwer Academic, USA.
32. Messing, G.L., and Kumagai, M. (1994). Low temperature sintering of α alumina seeded boehmite gels, *Am. Ceram. Bull.*, **73**, pp. 88–91.
33. Mayo, M. (1996). Processing of nanocrystalline ceramics from ultrafine particles, *Int. Mater. Rev.*, **41**, pp. 85–115.
34. Niespodziana, K., Jurczyk, K., Jakubowicz, J., and Jurczyk, M. (2010). Fabrication and properties of titanium–hydroxyapatite nanocomposites, *Mater. Chem. Phys.*, **123**, pp. 160–165.
35. Niespodziana, K., Jurczyk, K., Miklaszewski, A., and Jurczyk, M. (2010). Hybrid Ti-ceramic bionanomaterials for medical engineering, *Phys. Status Solidi C*, **7**, pp. 1363–1366.
36. Okonska, I., Iwasieczko, W., Jarzebski, M., Nowak, M., and Jurczyk, M. (2007). Hydrogenation properties of amorphous 2Mg + Fe/x wt% Ni materials prepared by mechanical alloying (x = 0, 100, 200), *Int. J. Hydrogen Energy*, **32**, pp. 4166–4190.
37. Rankin, J., and Sheldon, B.W. (1995). In situ TEM sintering of nanosized ZrO_2 particles, *Mater. Sci. Eng. A*, **204**, pp. 48–53.
38. Richert, J., and Richert. M. (1986). A new method for unlimited deformation of metals and alloys, *Aluminium*, **62**, pp. 604–607.
39. Scheu, M., Veefkind, V., Verbandt, Y., Galan, M.G., Absalom, R., and Förster, W. (2006). Mapping nanotechnology patents: The EPO approach, *World Patent Inf.*, **28**, pp 204–211.
40. Segal, V.M. (1977). *The method of material preparation for subsequent working*, USSR Patent No. 575892, USA.
41. Skandan, G. (1995). Processing of nanostructured zirconia ceramics, *Nanostruct. Mater.*, **5**, pp. 111–126.

42. Smardz, K., Smardz, L., Okonska, I., Nowak, M., and Jurczyk, M. (2008). XPS valence band and segregation effect in nanocrystalline Mg_2Ni-type materials, *Int. J. Hydrogen Energy*, **33**, pp. 387–392.

43. Smardz., L., Smardz, K., Nowak, M., and Jurczyk, M. (2001). Structure and electronic properties of $La(Ni, Al)_5$ alloys, *Cryst. Res. Technol.*, **36**, pp. 1385–1392.

44. Smardz, L., Jurczyk, M., Smardz, K., Nowak, M., Makowiecka, M., and Okonska, I. (2008). Electronic structure of nanocrystalline and polycrystalline hydrogen storage materials, *Renewable Energy*, **33**, pp. 201–210.

45. Suryanarayna, C. (2001). Mechanical alloying, *Progr. Mater. Sci.*, **46**, pp. 1–184.

46. Suryanarayana, C., and Koch, C.C. (1999). Nanostructured materials, in *Non-Equilibrium Processing of Materials* (ed. Suryanarayana, C.), Elsevier Science Pub., Oxford, pp. 313–346.

47. Thomas, W.M., Edward, N.D., Needham, J.C., Murch, M.G., Temple-Smith, P., Dawes, A., and Christopher, J. (1991). *Friction stir welding*, GB Patent Application 9125978.8. (1995), US Patent 5460317, USA.

48. Tsuji, N., Saito, Y., Utsunomiya, H., and Tanigawa, S. (1999). Ultra-fine grained bulk steel produced by accumulative roll-bonding (ARB) process, *Scripta Mater.*, **40**, pp. 795–800.

49. Tulinski, M., and Jurczyk, M. (2009). Mechanical and corrosion properties of Ni-free austenitic stainless steel/hydroxyapatite nanocomposities, *Solid State Phenomena*, **151**, pp. 213–216.

50. Valiev, R.Z., Islamgaliev, R.K., and Alexandrov, I.V. (2000). Bulk nanostructured materials from severe plastic deformation, *Prog. Mater. Sci.*, **45**, pp. 103–189.

51. Valiev, R.Z., Zehetbauer, M.J., Estrin, Y., Höppel, H.W., Ivanisenko, Y., Hahn, H., Wilde, G., Roven, H.J., Sauvage, X., and Langdon, T.G. (2007). The innovation potential of bulk nanostructured materials, *Adv. Eng. Mater.*, **9**, pp. 527–533.

52. Valiev, R.Z., Semenova, I.P., Latysh, V.V., Rack, H., Lowe, T.C., Petruzelka, J., Dluhos, L., Hrusak, D., and Sochova, J. (2009). Nanostructured titanium for biomedical applications, *Adv. Eng. Mater.*, **10**, B15–B17.

53. Vaben R., Stöver, D. (1999). Processing and properties of nanophase ceramics, *J. Mat. Proc. Tech.*, **92–93**, pp. 77–84.

54. Varin, R.A., Czujko, T., and Wronski, Z.S. (2009). *Nanomaterials for Solid State Hydrogen Storage*, Springer, New York.

55. Wang, Z.L., Liu, Y., and Zhang, Z. (eds) (2003). *Handbook of Nanophase and Nanostructured Materials*, Kluwer Academic/Plenum Publishers, New York, pp. 1–25

56. Yeadon, M., Yang, J.C, Ghaly, M., Olynick, D. L., Averbach, R. S., and Gibson, J. M. (1997). Sintering of sputtered copper nanoparticles on (001) copper substrates, in *Nanophase and Nanocomposite Materials 11* (ed. Komarneni, S., Parker, J.C., and Wollenberger, H.J.,), MRS, Pittsburgh, PA, 1997, pp. 179–184.

57. Zehetbauer, M.J., and Valiev, R.Z. (eds) (2004). *Nanomaterials by Severe Plastic Deformation*, Wiley-VCH, Weinheim.

58. Zhu, H., and Averbach, R.S. (1995). Sintering process of two nanoparticles: a study by molecular dynamics simulations, *Phil. Mag. Lett.*, **73**, pp. 27–33.

59. Zweck, A., Bachmann, G., Luther, W., and Ploetz, C. (2008). Nanotechnology in Germany: from forecasting to technological assessment to sustainability studies, *J. Cleaner Prod.*, **16**, pp. 977–987.

Chapter 5

Corrosion of the Metallic Biomaterials and Implants

Jaroslaw Jakubowicz
Institute of Materials Science and Engineering,
Poznan University of Technology, Sklodowska-Curie 5 Sq., 60-965 Poznan, Poland
jaroslaw.jakubowicz@put.poznan.pl

5.1 Corrosion of the Ti-Based Alloys

Corrosion is an electrochemical process that results in the deterioration of the metal properties, due to its contact with aggressive environments. Metal destruction by corrosion occurs either through the transfer of the metal ions directly into corrosive solution or by continuous dissolution of a surface metal oxide. Uniform corrosive attack is less destructive than localized, causing catastrophic and suddenly damage of the corroding material, like in pitting corrosion. Rapid dissolution of a small localized area may occur through breakdown of passivity, initiation, propagation, and repassivation. The Ti and its alloys are mostly used in implants, because of their high corrosion resistance, due to the stable Ti-oxide formation caused by more cathodic behavior (Fig. 5.1) [55],

Bionanomaterials for Dental Applications
Edited by Mieczyslaw Jurczyk

ISBN 978-981-4303-83-5 (Hardcover), 978-981-4303-84-2 (eBook)
www.panstanford.com

biocompatibility, and mechanical properties. Other alloys, like stainless steel are still in use.

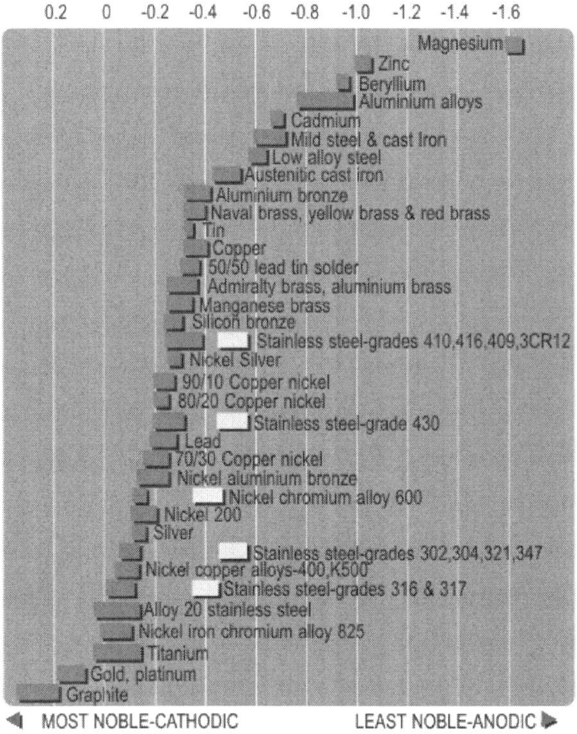

Figure 5.1 Corrosion susceptibility of various metals and alloys [55].

Corrosion resistance of biomaterials plays a key role in theirs successful application in human body environments. All metals and alloys undergo corrosion when they are in contact with body fluids, which are very aggressive due to the presence of chloride ions and proteins. The passive oxide film shields the implant material from the corrosion process and additionally provides protection before releasing the harmful compounds from the implant or its surface into the bone.

The oxide thickness can be increased by anodic oxidation, resulting in different types of oxide (anatase, brukite, and rutile) and theirs thickness. The oxide thickness determines the surface color.

The blood plasma is a highly aggressive environments for metals and alloys, because the presence of a high concentration of chloride ions (Table 5.1) and their ability to induce localized corrosion. Another factor is the body temperature, close to 37°C, which is enough to accelerate electrochemical reactions or even changes in the mechanism of corrosion, which can be different at room temperature [52]. Body fluids also contain different types of biomolecules and cells, which may attach to the biomaterial surface and affect the surface reactions [52].

Table 5.1 Typical chemical composition of human blood plasma [52]

Ion	Concentration ($mmol \cdot L^{-1}$)
Na^+	142.0
K^+	5.0
Mg^{2+}	1.5
Ca^{2+}	2.5
Cl^-	103.0
HCO_3^-	27.0
HPO_4^{2-}	1.0
SO_4^{2-}	0.5

The corrosion can be classified into the following types (Fig. 5.2) [6, 10, 22, 51]:

(i) Uniform corrosion is defined as a chemical or electrochemical reaction that proceeds uniformly over the exposed surface.
(ii) Galvanic corrosion occurs due to the galvanic coupling of two or more different metals in the voltage range (Fig. 5.1). Less corrosion-resistant metal become anode and undergo corrosive attack.
(iii) Crevice corrosion causes the regions of stagnant solutions, which attacks the material through enhanced local ions flux.
(iv) Pitting is a localized corrosion that results in breakdown of the protecting passive layer and holes formation on the surface.
(v) Intergranular corrosion occurs through the grain boundaries, because of theirs higher reactivity (anode area) with respect to the grains (cathode area).
(vi) Selective corrosion (dealloying) results in removal of one element (more anodic with respect to the second one) from a solid alloy by corrosion process.
(vii) Erosion corrosion occurs when a relative movement between a corrosive fluid and the material surface is present, which

enhances the materials removal. This type of corrosion includes both cavitation and fretting corrosion.

(viii) Hydrogen corrosion occurs in the hydrogen atmosphere or in the presence of internal hydrides deteriorating metal properties, enhancing metal brittleness.

(ix) Stress-corrosion occurs together with cracking. This type of corrosion is caused by the simultaneous presence of tensile stress and corrosive medium, resulting in both mechanical properties deterioration and materials removal.

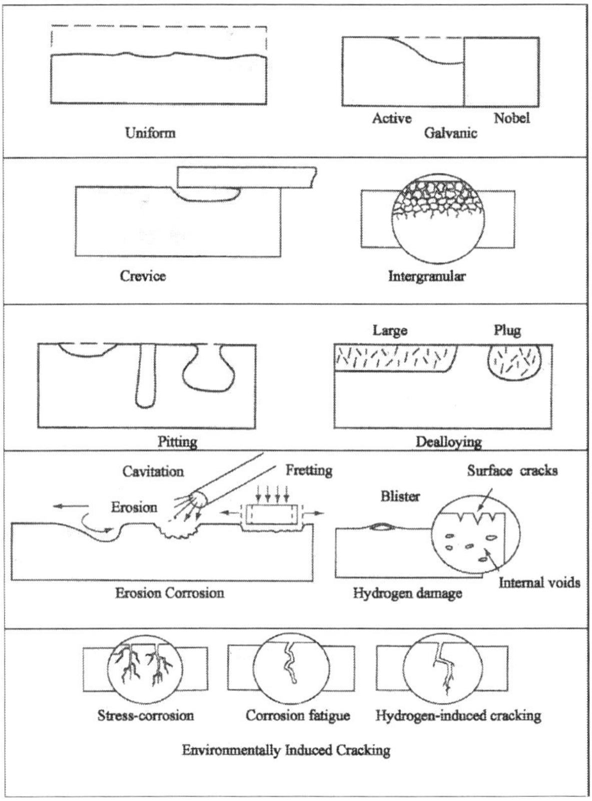

Figure 5.2 Schematic representation of various forms of corrosion [51].

Example images of pitting and crevice corrosion in Ti is shown in Fig. 5.3 and Fig. 5.4 [2]. The corrosion behavior can be determined by time-consuming gravimetric method, when the corrosion rate is estimated after a longtime material dipping in corrosive solution/

atmosphere; or can be estimated in fast electrochemical method, where the corrosion current reflects the corrosion rate of the material. In the electrochemical method, a polarization curve of the material immersed in the corrosive electrolyte is drawn (Fig. 5.5). On the basis of the Tafel tangents, intersection is given a corrosion current and corrosion potential. In the passive range of the corrosion curve, the current density indicates the tendency of passivation and at higher potentials the tendency of pitting corrosion. Generally the lower current densities indicate the higher corrosion resistance.

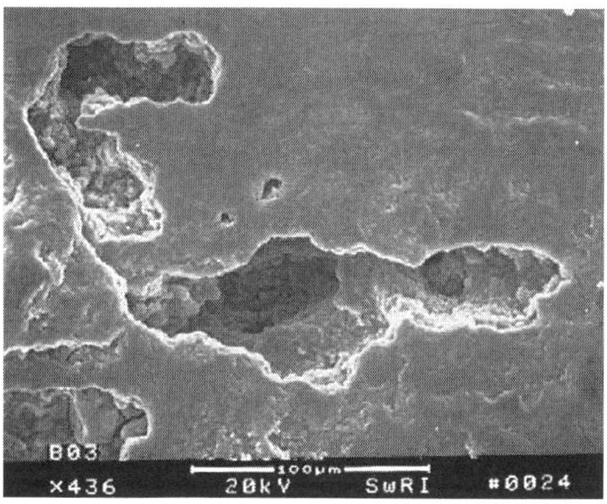

Figure 5.3 SEM image showing morphology of pitting attack after cyclic potentiodynamic polarization testing of Ti Grade 2 in deaerated 1M NaCl at 95°C [2].

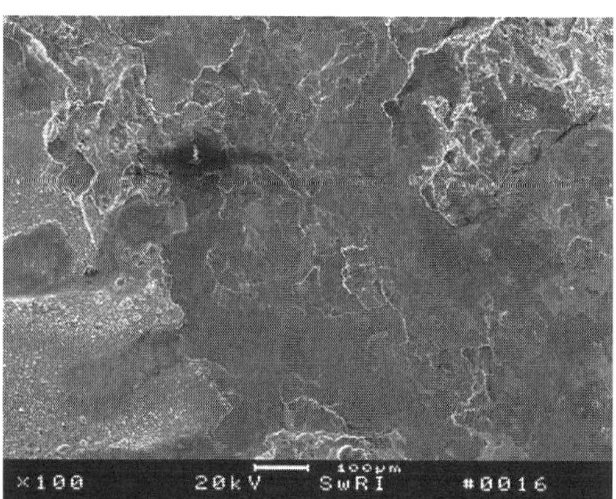

Figure 5.4 SEM image of crevice corrosion on Ti Grade 2 after polarization at 0 V SCE in 1M chloride at 95°C [2].

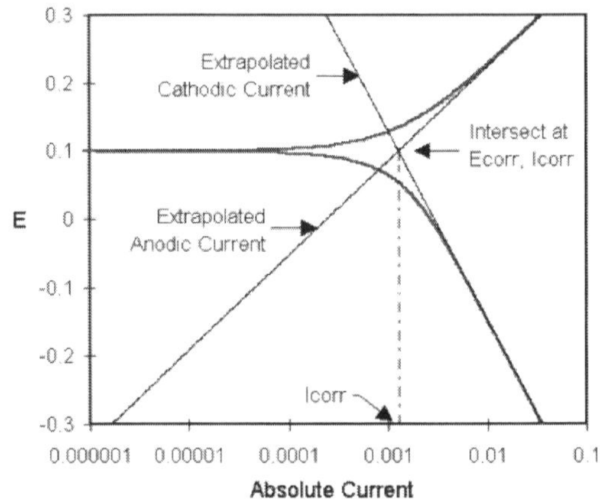

Figure 5.5 Anodic and cathodic part of the polarization curve with Tafel extrapolations, presenting corrosion current and potential [56].

Among the few metals used for implants preparation, Co, Ni, Cu, and V are knows as toxic, while Zr, Ti, Nb, Ta, and Pt exhibited excellent biocompatibility and belong to the vital group in tissue reaction [36]. Al has causal relations with neurotoxicity and senile dementia of the Alzheimer type [23, 27, 36, 54]. The corrosion behavior and biocompatibility of these materials plays a key role for theirs successful implant applications.

Titanium and its alloys are the most corrosion-resistant of the alloys. This is assured by the very high stability of the TiO_2 passive film that forms on the alloy surface in the oxygen-containing atmosphere. Typical thickness of the native oxide is only a few nanometers, but is sufficient to protect against corrosive dissolution. The TiO_2 is thermodynamically stable in the pH range between 2 and 12, and only species, such as HF or H_2O_2, lead to its substantial dissolution [41, 53]. In acidic solutions, the Al-containing alloys such as Ti–6Al–4V, Ti–6Al–7Nb are significantly more susceptible for dissolution than commercially pure Ti (cp-Ti) [39, 53].

The Ti–6Al–4V alloy is one of the most commonly used in the implant applications. The titanium, aluminum, and vanadium are released into the cell tissue by the dissolution of the TiO_2 passive film and wear corrosion process. Hence, the more stable the TiO_2 passive layer on the Ti-based implant alloys is required for the better the corrosion resistance. Implanted alloy corrode in the biological body fluids due to the loosening of implant, infection, wear or fretting corrosion, and metabolism [11]. Unfortunately, TiO_2 film is sensitive for the fluoride ions, always present in oral environments [12, 13, 30], as well as often in tooth gels.

Because the vanadium is a toxic element, vanadium-free Ti alloys such as Ti–6Al–7Nb and Ti–13Nb–13Zr have been developed and replaced vanadium in Ti–6Al–4V [35]. The Ti–6Al–7Nb alloy can be a better alternative to Ti–6Al–4V because of its corrosion resistance and resistance to loss of mechanical properties with changes in pH in simulated body fluid environment [3, 24].

The microstructures of the Ti–6Al–4V consist of two phases α and β (Fig. 5.6). Due to a two-phase microstructure, Ti–6Al–4V is more susceptible to corrosion across the grain boundaries because of galvanic cell formation. Choubey *et al.* [3] investigated corrosion resistance (in simulated body fluids — Hank's solution) of the Ti–6Al–4Nb, Ti–6Al–4Fe and Ti–5Al–2.5Fe alloys and compared their properties with the parent Ti–6Al–4V. All the alloys were α-β-type alloys. Vanadium, niobium, and iron are β-stabilizers, while aluminum is an α-stabilizer; α was the dominant phase in all these alloys. The addition of niobium increases the grain size considerably and results in Widmanstätten-type structure. These new V-free alloys passivates immediately after immersion in the corrosive solution [3]. Addition of alloying elements like niobium or iron does not significantly affect corrosion current density in the passive range. The passive range for Ti–6Al–4V, Ti–6Al–4Fe and Ti–6Al–4Nb is comparable, whereas for Ti–6Al–4Nb is lower. The corrosion rate of Ti–5Al–2.5Fe, Ti–6Al–4V, Ti–6Al–4Fe, and Ti–6Al–4Nb is comparable and is not drastically deteriorated by the Fe substitution [3].

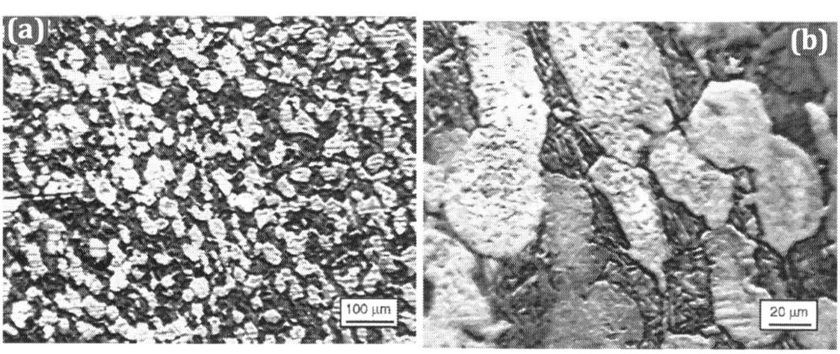

Figure 5.6 Optical micrographs of the Ti–6Al–4V surface [3].

Narayanan and Seshadri [32] investigated corrosion resistance of the Ti–6Al–4V alloy with additionally oxidized surface produced by anodization in 1M H_3PO_4 and 1M H_3PO_4 with dissolved calcium and phosphorus additions. The as prepared porous oxide coating have 90–280 nm thickness and is thicker for 1M H_3PO_4 electrolyte than those produced from electrolyte containing dissolved calcium and phosphorus. The corrosion resistance of as prepared surfaces was investigated in simulated body fluid (8.74 g/L NaCl, 0.35 g/L $NaHCO_3$, 0.06 g/L Na_2HPO_4 0.06 g/L NaH_2PO_4). The coatings produced using 1M H_3PO_4 have lower corrosion rate (because is thicker and less defective) than the coatings produced from electrolyte containing dissolved calcium and phosphorus (Fig. 5.7) [32].

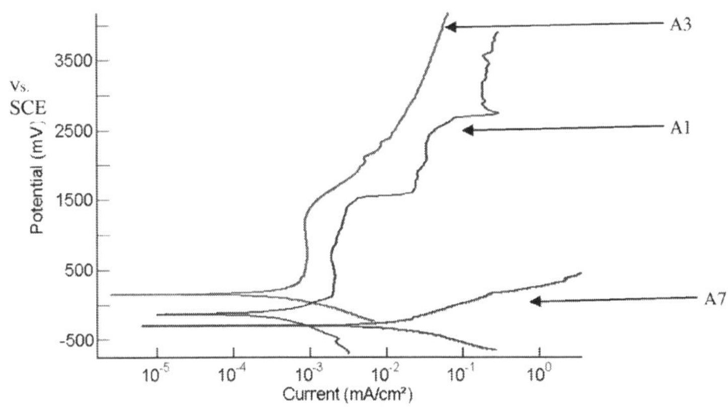

- I_{corr} for samples etched in Ca+P containing solution A1, A3, and A7 is 0.29, 0.42, and 3.68 μA/cm², respectively.
- I_{corr} for samples etched only in H_3PO_4 solution P1, P3, and P7 is 0.014, 0.016, and 0.250 μA/cm², respectively.

Figure 5.7 Potentiodynamic polarization curves of coatings on Ti–6Al–4V produced from electrolyte containing Ca and P; A1 (sample etched through 15 min) and A3 (sample etched through 3 h) show passivation, while A7 (sample etched through 24 h) shows continuous dissolution [32]; (P1, P3, P7 samples etched in H_3PO_4 for 15 min, 3 h, 24 h, not shown here). See also Color Insert.

The V-free alloys were investigated by Okazaki *et al.* [36] too. They prepared a new Ti–15Zr–4Nb–4Ta–0.2Pd–0.2O–0.05N (Ti–Zr-type) and Ti–15Sn–4Nb–2Ta–0.2Pd–0.2O–0.005N (Ti–Sn-type) complex alloys and investigated theirs corrosion resistance, corrosion fatigue, and cytocompatibility. The Ti–15Zr–4Nb–4Ta–0.2Pd–0.2O–0.05N alloy exhibited a higher corrosion resistance in physiological saline solution. The addition of 0.02% O and 0.05% N to Ti–Zr-type alloy improved the mechanical properties at room temperature and corrosion fatigue strength [36]. The comparison of various Ti-alloys on the relative growth ratio of L929 and MC3T3-E1 cells is shown in Fig. 5.8. The relative growth ratio of L929 and MC3T3-E1 cells for the Ti-Zr and Ti-Sn-type alloys are nearly equal to 1 and are slightly higher than those of pure Ti grade 2 and Ti–6Al–4V alloy. The effect of the metallic ions released from these alloys on the relative growth ratios of L929 and MC3T3-E1 cells is shown too (Fig. 5.8b). The relative growth ratios are similar to those of the alloys for both L929 and MC3T3-E1 cells [36].

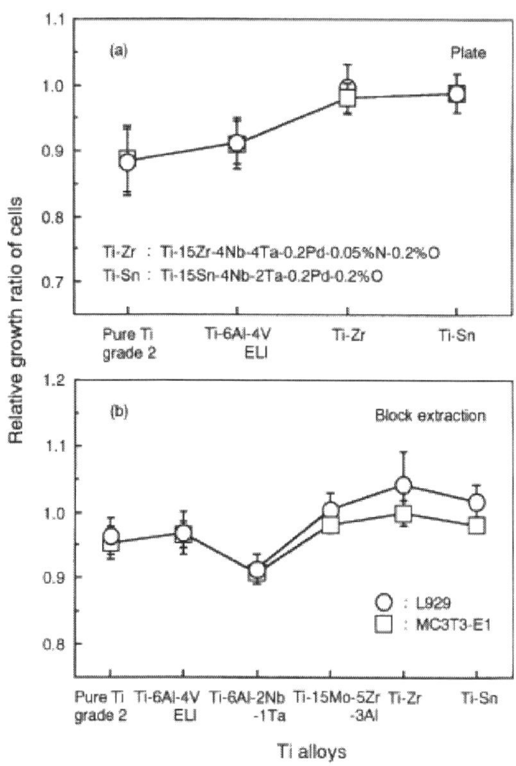

Figure 5.8 Cytocompatibility on the various Ti alloy plates (a) and cytocompatibility of alloy blocks extraction with Eagle's MEM for L929 cells and ultra pure water for MC3T3-E1 cells (b) [36].

Lopez *et al.* [29] investigated corrosion behavior (in Hank's solution consisting 8 g/L NaCl, 1 g/L glucose, 0.4 g/L KCl, 0.35 g/L $NaHCO_3$, 0.14 g/L $CaCl_2$, 0.1 g/L $MgCl_2 \cdot 6H_2O$, 0.06 g/L $Na_2HPO_4 \cdot 2H_2O$, 0.06 g/L KH_2PO_4 and 0.06 g/L $MgSO_4 \cdot 7H_2O$) of three V-free Ti alloys: Ti–7Nb–6Al, Ti–13Nb–13Zr, and Ti–15Zr–4Nb. The presence of Nb in all alloys stabilizes the β phase. The Ti–13Nb–13Zr sample is richer in β phase than Ti–7Nb–6Al and Ti–15Zr–4Nb alloys. The Ti–13Nb–13Zr and Ti–15Zr–4Nb have very low passivation current densities, lower than those of the Ti–7Nb–6Al as well as Ti–6Al–4V, and these materials exhibit an excellent resistance to pitting corrosion [29]. In the Ti alloys, Al and Zr form stable protective passive layers. The higher concentration of Zr in the passive layer of the Zr-containing alloys, measured by Lopez *et al.* [29], as compared with the Al concentration in the passive layer of Ti–7Nb–6Al, explain the lower current densities observed in the first case.

The thermal oxidation of the Ti–6Al–4V alloy was found to produce corrosion and wear resistant surface layers [9]. Oxide thickness and oxygen diffusion zone depth steadily increases with increasing oxidation temperature and time (Fig. 5.9). Oxide layers are composed from anatase and rutile. Additional significant increase in the surface hardness (from 3500 to 9000 HV) was achieved due to formation of a hard oxide layer and an oxygen diffusion zone [26]. Oxidation at 600°C for 60 h produced the most corrosion and corrosion-wear resistant surface (in 5M HCl, 0.9% NaCl, respectively) of the Ti–6Al–4V alloy.

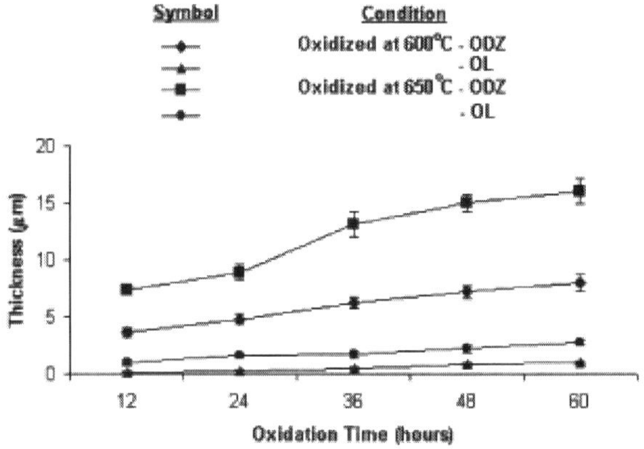

Figure 5.9 The effect of oxidation time and temperature on thickness of OL (oxide layer) and ODZ (oxygen diffusion zone) [26].

Kahan *et al.* [25] investigated the wear of the Ti-alloys in the presence of corrosive environments. The Ti–13Nb–13Zr alloy is more corrosion resistant than Ti–6Al–7Nb and Ti–6Al–4V whereas in the presence of protein, the corrosion resistance of Ti–13Nb–13Zr and Ti–6Al–7Nb is reduced with respect to Ti–6Al–4V [25]. In the presence of the corrosion and wear, the corrosion resistance of Ti–13Nb–13Zr is higher than Ti–6Al–7Nb or Ti–6Al–4V in phosphate-buffered-saline. In the presence of proteins, the corrosion resistance of Ti–13Nb–13Zr and Ti–6Al–7Nb is higher than that of Ti–6Al–4V [25]. The wear of Ti–13Nb–13Zr is lower than that of Ti–6Al–7Nb and Ti–6Al–4V with or without the presence of proteins in a corrosive environment and hence overall degradation when both corrosion and wear processes occurs is lowest for Ti–13Nb–13Zr and highest for Ti–6Al–4V. The presence of proteins reduces the degradation of all three alloys [25]. The increased corrosion resistance of the Ti–13Nb–13Zr is due to the fact that Nb and Zr are less soluble than Al and V, and that the passive oxide layer on the surface of the alloy is more inert, because it consists of surface rutile structure [4, 25, 43].

The toxic effect of Ni on the tissue in the Ti-alloys (with Ni) can be reduced by anodization [47]. Anodization of Ni–Ti alloy in acetic acid improves corrosion resistance in Hanks' solution with respect to untreated NiTi (Fig. 5.10). The Ni/Ti atomic ratio at the surface is very much reduced after anodization, with a value of 0.04 versus that

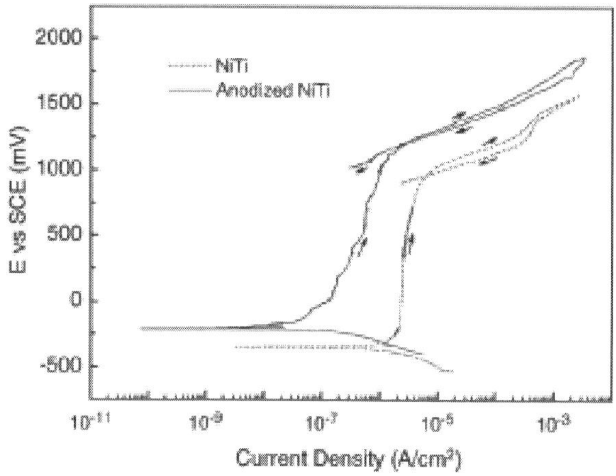

Figure 5.10 Cyclic polarization curves of bare NiTi and anodized NiTi; Hank's solution at pH = 7.4 and 37°C [47].

of 0.30 for bare Ni–Ti [47]. Rondelli and Vicentini [38] investigated the Ni–Ti dental wires in artificial saliva (Fig. 5.11). The Ni–Ti alloy has a wide passive range that extends up to very high potentials, and absence of hysteresis loop, which indicate high resistance to pitting corrosion. For comparison, for the cobalt alloy no pitting occurs at very high potential, whereas the stainless steel exhibited pitting of about 400 mV versus SCE (Fig. 5.11). The Ti–Ni alloys have sufficient resistance against pitting corrosion, comparable to the cobalt-based alloy, and much higher than the stainless steel wire [38].

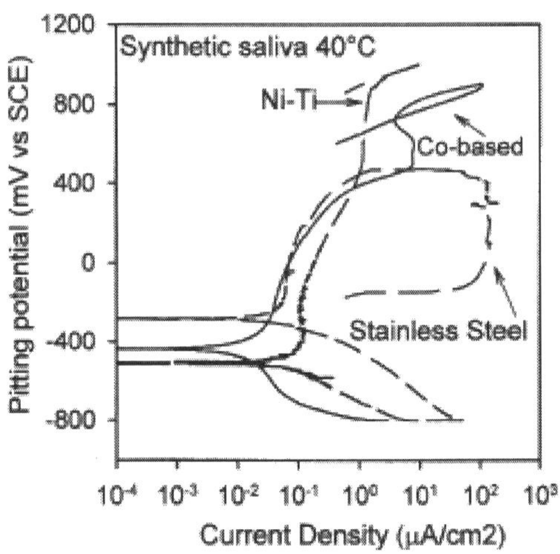

Figure 5.11 Potentiodynamic tests of orthodontic wires in artificial saliva at 40°C [38].

Preparation of Ti-alloys with highly cathodic elements, such as Ag, Pt, and Pd could improve corrosion resistance of the implant alloys. Shim *et al.* [48] and Zhang *et al.* [57] reported that Ti–Ag alloys have better corrosion resistance than that of pure titanium and they predicted that Ti–Ag alloys are less sensitive to fluoride ions, which is extremely important in dental implant applications. Zhang *et al.* [57] investigated the corrosion behavior of the Ti–Ag alloys in artificial saliva with fluoride ions. Figure 5.12 shows the potentiodynamic polarization curves of the commercially pure Ti (CP Ti) and the Ti–Ag alloys in the artificial saliva solutions. In the fluoride-free solution (Fig. 5.12a), both CP Ti and the Ti–Ag alloys show passivation behavior without obvious active–passive transition. The passivation and corrosion current density are lower for Ti–Ag alloy with comparison to the CP Ti [57]. In the solution containing 0.2% NaF (Fig. 5.12b), the corrosion potential of both CP Ti and Ti–Ag decreases and the passivation current and corrosion current densities increases, with respect to the fluoride-free solution. These imply a reduced corrosion resistance. In comparison with CP Ti, the Ti–Ag alloys exhibit better corrosion resistance in fluoride containing solution too [57]. The pure Ti exhibit active–passive transition pass in the fluoride-containing solution. The presence of F^- in the solution retards the formation of passive film on the surface of the Ti. The summarized corrosion parameters of the CP Ti and Ti–Ag alloys, indicating higher corrosion resistance of the Ti–Ag in artificial saliva are presented in Table 5.2 [57]. In the fluoride solution, Na_2TiF_6 or TiF_6^{2-} is formed, both of which are more stable than $Ti0_2$, but unfortunately, they do not exhibit protective, and the corrosion resistance of the Ti alloys decrease in a fluoride-containing solution [2, 57].

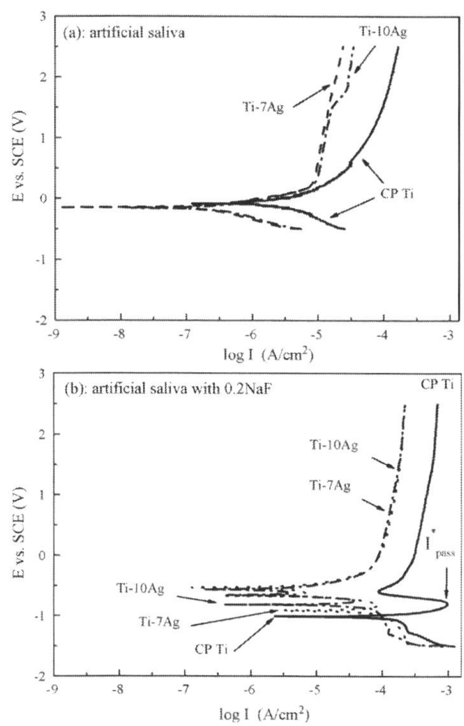

Figure 5.12 Potentiodynamic polarization curves of commercially pure Ti and Ti–Ag alloys in artificial saliva solutions [57].

Table 5.2 Corrosion parameters of pure Ti and Ti–Ag alloys in artificial saliva without/with 0.2% NaF [57]

Artificial saliva	Material	R_p ($\Omega\cdot cm^2$)	E_{corr} (mV)	I_{corr} ($\mu A\cdot cm^{-2}$)
Without NaF	CP Ti	13,903 ± 6102	-68 ± 36	2.11 ± 0.88
	Ti–7Ag	107,210 ± 62370	-87 ± 33	0.33 ± 0.16
	Ti–10Ag	103,639 ± 38661	-143 ± 7	0.29 ± 0.99
With NaF	CP Ti	334 ± 14	-980 ± 24	78 ± 4
	Ti–7Ag	535 ± 79	-870 ± 50	46 ± 22
	Ti–10Ag	540 ± 120	-830 ± 80	73 ± 5

Schiff *et al.* [44] compared the Ni–Ti, Ni–Ti–Co, and Ti–6Al–4V alloys with the titanium, regarding the corrosion resistance in Fusayama-Meyer artificial saliva (0.4 g/L KCl, 0.4 g/L NaCl, 0.906 g/L $CaCl_2\cdot 2H_2O$, 0.690 g/L $NaH_2PO_4\cdot 2H_2O$, 0.005 g/L $Na_2S\cdot 9H_2O$, 1 g/L Urea). For the pure titanium and Ti–6Al–4V, corrosion resistance decreases with medium changes: acidified, fluoridated, or fluoridated–acidified (Table 5.3, Fig. 5.13, Fig. 5.14) [44]. For the Ni–Ti and Ni–Ti–Co alloys, the corrosion current density obtained in Fusayama-Meyer saliva, fluoridated saliva, and acidified saliva media is higher which indicated their lower corrosion resistance [44]. For that reason, patients with dental prosthetic reconstructions or titanium implants should not use fluoridated tooth gels. The corrosion rate in fluoridated environments of the Ti–6Al–4V is similar to the Ti. The corrosion resistance of Ni–Ti and Ni–Ti–Co is lower, especially in neutral and acid saliva environments [44].

The implant surface, after contact with much more aggressive fluoridated–acidified saliva medium is characterized by localized pitting (Fig. 5.15) [44], whereas in normal pH medium pitting does not occur.

Table 5.3 Densities of corrosion currents values (10^{-6} $A\cdot cm^{-2}$) of the samples in different media determined by the Stern-Geary method [44]

I_{corr} (10^{-6} $A\cdot cm^{-2}$)	Ti grade 2	Ti–6Al–4V	Ni–Ti	Ni–Ti–Co
Fusayama Meyer saliva	1 ± 0.2	1.5 ± 0.2	20 ± 2	15 ± 2
Acidified saliva	3 ± 0.3	3.5 ± 0.3	45 ± 4	30 ± 3
Fluoridated saliva	10.5 ± 2	9 ± 0.2	50 ± 5	60 ± 5
Fluoridated-acidified saliva	500 ± 55	480 ± 50	250 ± 30	300 ± 25

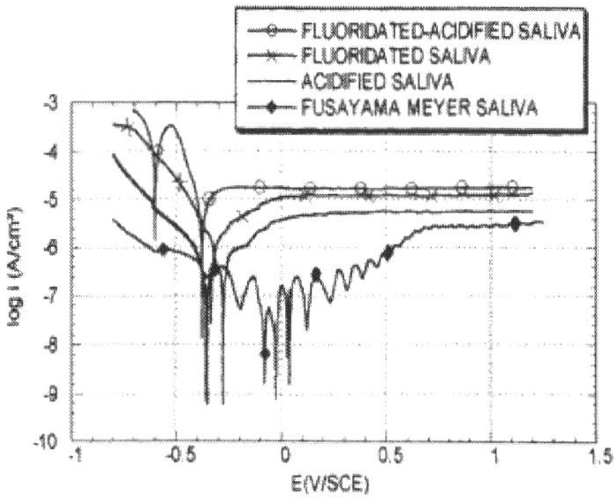

Figure 5.13 Polarization curves for Ti in different media [44].

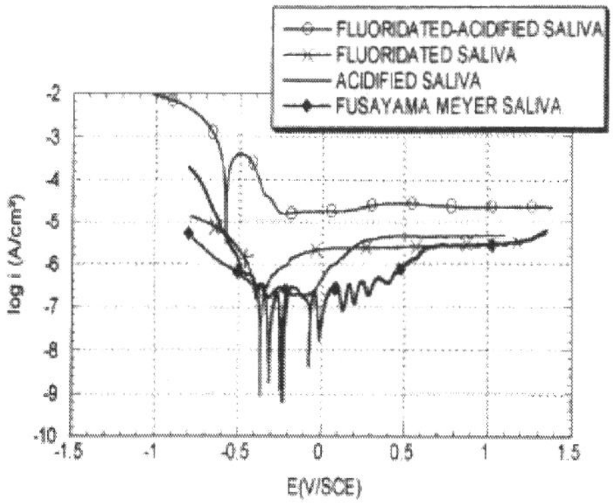

Figure 5.14 Polarization curves for Ti–6Al–4V alloy in different media [44].

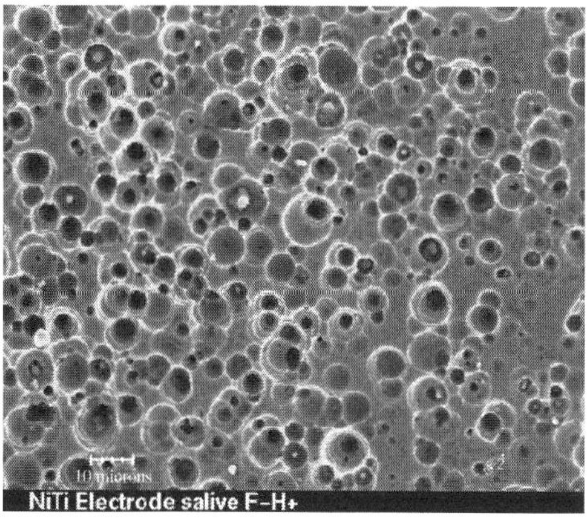

Figure 5.15 SEM image (×1000) of Ni–Ti Sentalloy after corrosion test in fluoridated–acidified saliva [44].

The influence of fluoride and chloride ions on the corrosion behavior of Ni–Ti orthodontic wires was studied by Li *et al.* [28]. The results indicates that Ni–Ti alloy is primarily susceptible to localized corrosion when exposed to a solution containing chloride, while it is susceptible to general corrosion when subjected to a solution containing fluoride.

The effects of chloride and fluoride on NiTi wires are listed in Table 5.4. Ni–Ti orthodontic wires exhibit better behavior in chloride than fluoride solution. Ni–Ti in fluoride solutions does not passivate due to the large magnitude of the anodic current density. Ni–Ti orthodontic wires are susceptible to pitting corrosion in chloride solutions. The combination of fluoride and chloride significantly deteriorate the corrosion resistance of Ni–Ti alloys, with comparison to those in individual fluoride or chloride solutions [28].

Table 5.4 Results of potentiodynamic measurements of NiTi alloy in solutions [28]

Solution type	Concentration (mol L^{-1})	E_{corr} (mV/CE)	I_{corr} (μA cm^{-2})	$I_{passive}$ (μA cm^{-2})
NaF	0.0238	−430 (18)	96.5 (30)	1.865 (0.396) × 10^3
	0.0476	−442 (20)	162 (22)	5.227 (0.116) × 10^3
	0.0714	−428 (36)	160 (18)	8.080 (0.453) × 10^3
	0.1538	−439 (36)	215 (13)	16.27 (0.907) × 10^3
NaCl	0.0238	−171 (37)	0.026 (0.014)	0.623 (0.339)
	0.0476	−106 (27)	0.049 (0.007)	0.940 (0.142)
	0.0714	−105 (925)	0.013 (0.003)	0.520 (0.075)
	0.1538	−161 (3)	0.035 (0.005)	0.353 (0.083)
NaF + NaCl	0.0119 NaF + 0.0119 NaCl	−235 (9)	2.1 (1.4)	185 (978)
	0.0238 NaF + 0.0238 NaCl	−390 (11)	84.1 (24.4)	3.38 (1.252) × 10^3
	0.0714 NaF + 0.0714 NaCl	−448 (11)	140 (33)	14.79 (8.287) × 10^3

Note: Standard deviations are given in parentheses.

The surface morphology after the cyclic polarization test in the mixed solutions containing 0.0714 mol/L NaF and 0.0714 mol/L NaCl indicated, that both pitting and general (uniform) corrosion occurs (Fig. 5.16) [28].

Increase of fluoride concentration drastically deteriorates the corrosion resistance and dissolution rate significantly increases (Fig. 5.17) [2]. Fluoride was found to dramatically increase the corrosion current above a critical concentration of about 0.5 mM.

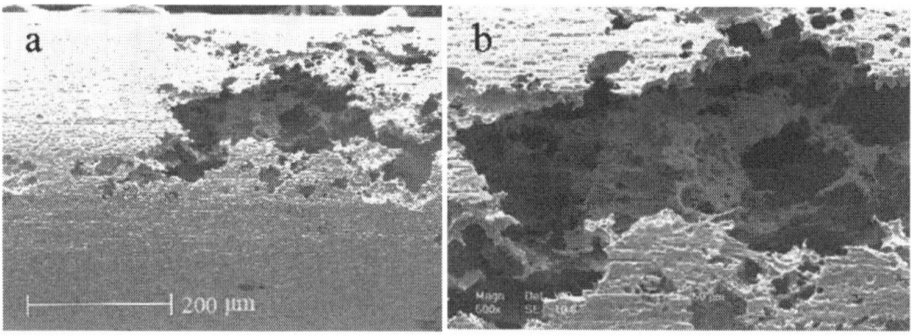

Figure 5.16 SEM images of NiTi wires in mixed solution of chloride and fluoride: (a) low magnification and (b) high magnification [28].

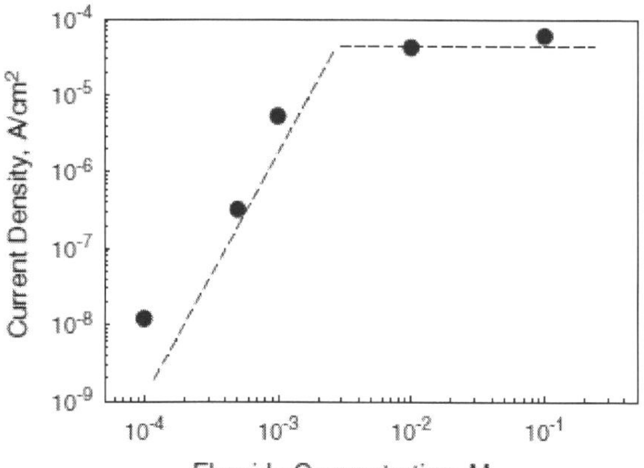

Figure 5.17 Effect of fluoride concentration on the dissolution rate of Ti Grade 7 after potentiostatic polarization at 0 V SCE for 453 h (deaerated, 1M NaCl at 95°C) [2].

Brossia and Cragnolino added Pd (0.155%) to Ti [2]. Pd additions increase the pitting and repassivation potentials in chloride solutions, but Pd did significantly affect the localized corrosion resistance of Ti. It was found that the Pd addition did not mitigate the deleterious effects of fluoride [2]. However, Nakagawa *et al.* [31] reported that the Pt or Pd introduced to the Ti-alloy improves the corrosion resistance of titanium in fluoride solutions. They measured corrosion properties of Ti alloys consisting 0.1–2 wt% Pt and Pd in artificial saliva containing 0.2% NaF (corresponding to 905 ppm F). The addition of over 0.5% Pt or Pd to Ti results in formation of a passive film on the Ti surface and hence high corrosion resistance in the Ti–Pt or Ti–Pd alloys [20].

During anodization of the Ti or its alloys, nanotubes (see Chapter 9) are often formed. The nanotubes formed in the anodization of Ti-based materials are a possible material for dental application and are investigated with respect to corrosion resistance too. For example, anodization of the Ti–Nb alloys results in the formation of oxide nanotubes with diameters range from 55 nm to 220 nm and lengths range from 730 nm to 2 μm [20]. These nanotubes show interesting corrosion properties (Fig. 5.18) [37]. The corrosion current density, for the non-anodized alloy (Fig. 5.18a) decreases with increasing Nb content. The lowest I_{corr} occur for the Ti–40Nb alloy. The increase of corrosion resistance with Nb content is attributed to rapid formation of a passive mixed TiO_2 and Nb_2O_5 film of a few nanometers thick on the Ti-Nb alloy surface. Anodic current rises in a transpassive region, indicates pitting corrosion, due to a breakdown of the TiO_2 and Nb_2O_5 film. Ti–xNb alloys with the oxide nanotubes on the surface (Fig. 5.18b) show lower corrosion resistance, than for the non-anodized alloys. It is seen that E_{corr} is lower and I_{corr} is higher for the Ti–Nb alloys with the surface nanotubes. The passive range is wider for the alloys with nanotubes, which means that the mixed TiO_2 and Nb_2O_5 film is stable.

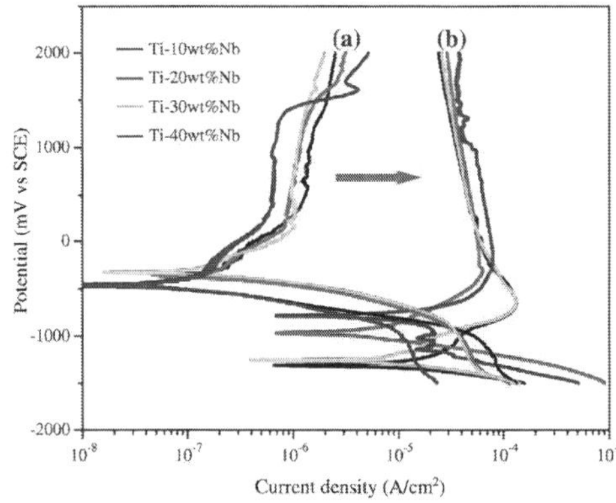

Figure 5.18 Potentiodynamic polarization curves of Ti–xNb alloys immersed in aqueous 0.9% NaCl solution at 36.5 ± 1°C: (a) non-anodized alloy and (b) anodized alloy [20]. See also Color Insert.

Raja *et al.* [37] in Ti anodization experiments found that addition of fluoride ions results in the corrosion potential to be more negative and the current density increases in the passive range. The critical passivation current density increases orders of magnitude with comparison to NaCl solution. Figure 5.19 shows sequence of oxide film formation during anodic polarization [37]. Dissolution of Ti metal occurs at free corrosion potential and a reprecipitated salt film is seen (a) till the critical passivation current density reaches (peak position). At this peak position, the Ti is maximally attacked and no external layer is formed (b). When the current decreases, a thin oxide layer covers the attacked grain facets (c) and further increase of potential results in nucleation of another layer of oxide over previous one is observed (d).

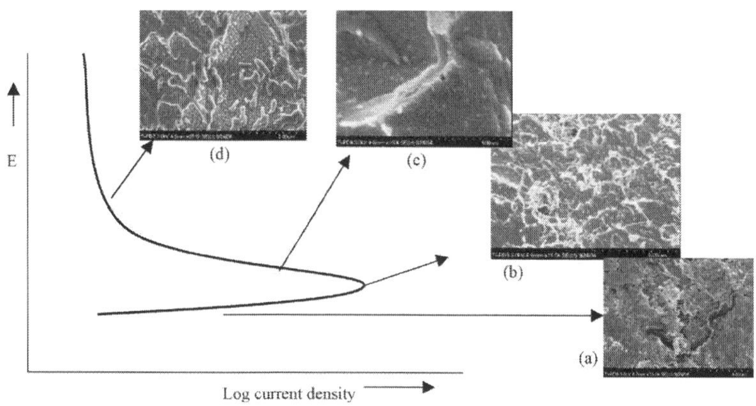

Figure 5.19 Stages of oxide film formation during polarization of Ti in acid fluoride solution. (a) A thick defective re-precipitated layer on the surface, (b) at the critical peak current density no visible film could be observed, (c) a thin film formed (inner-barrier layer) immediately after reaching critical current density for passivation, increases in potential resulting in nucleation of secondary (outer) oxide layer and (d) growth of outer oxide layer [49].

The surface treatment (not only anodic oxidation) is mainly responsible for the good corrosion resistance. For example, the effect of the laser surface remelting on corrosion resistance and pitting potential of pure titanium was investigated by Sun *et al.*

[49]. The laser surface remelting process results in suppressing of the pitting corrosion. The improved pitting corrosion resistance is related with microstructural changes caused by rapid solidification after laser remelting. Figure 5.20 shows an example of a single-track cross section of the laser remelted zone. Material after laser treatment consists of remelted zone (white), where remelting and resolidification occurs, and heat affected zone (light colored semi-circular) where heating and cooling process occurs without melting [49]. The width of the remelting zone depends from the laser scanning speed and was kept in the range of 1.7–2.0 mm. The Ti surface after laser remelting is shown in Fig. 5.21 [49]. The surface is relatively flat and the structure of the remelted zone is formed mainly by the martensitic phase. The corrosion test indicates an improvement of the corrosion resistance, especially pitting corrosion achieved by Nd-YAG laser Ti surface remelting.

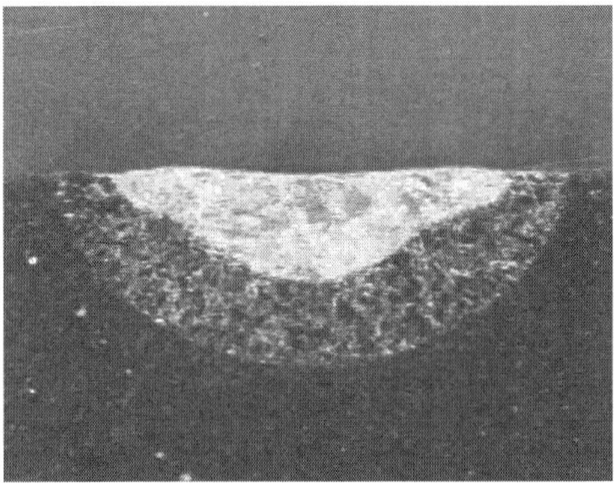

Figure 5.20 Cross section of a laser surface remelted zone [49].

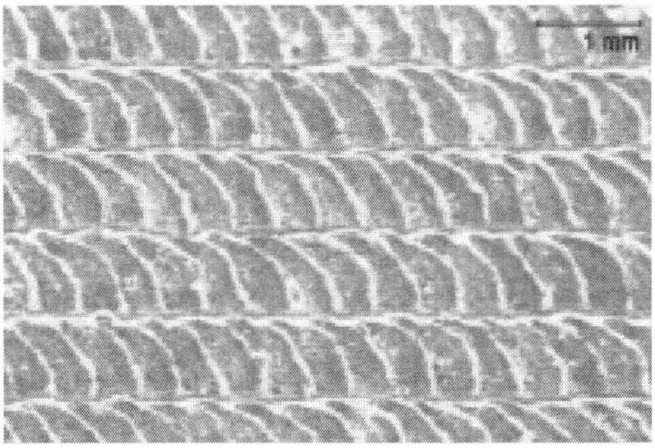

Figure 5.21 Surface after laser remelting [49].

Garbacz *et al.* [7] investigated the corrosion resistance of pure nano-Ti grade 2 after hydrostatic extrusion. They found that the grain size of the titanium substrate did not influence the thickness of oxide formed on the titanium. Passive oxide layers formed on the surface of coarse-grained Ti (CG Ti) and nano-grained Ti (NG Ti) had thickness of about 6 nm in both cases, and the same chemical composition. NG Ti shows slightly lower corrosion resistance in 0.9% NaCl solution with respect to CG Ti (Table 5.5).

Table 5.5 Specification of corrosion potentials values (E_{corr}) and corrosion currents density (I_{corr}) for CG and NC Ti [7]

Ni-type	E_{corr} (mV)	I_{corr} (A cm^{-2})
CG Ti	−152	18
NG Ti	−158	41

5.2 Corrosion of the Ti-Based Porous Materials

Porous metallic materials are attractive for biomedical applications, for example in dental or orthopedic hard tissue implants. The main problem of these materials is the relatively poor corrosion resistance, due to the rough surface. The corrosion resistance of porous titanium and its alloys in simulated body fluid is significantly lower than that of bulk counterparts [1, 8, 42], but the surface is attractive for osseointegration.

Jakubowicz investigated corrosion resistance of the oxidized porous Ti in Ringer's electrolyte, corresponding to human bio-environments (NaCl, 9g/L; KCl, 0.42 g/L; $CaCl_2$, 0.48 g/L; $NaHCO_3$, 0.2 g/L) [14]. Figure 5.22 presents potentiodynamic curves for pure Ti (a) and TiO_x (b), which characterize corrosion behavior of the materials with native and anodic oxides, respectively.

Polarization curves of both, pure Ti with native oxide (a) and Ti foil with porous oxide (b), prepared by electrochemical etching, shows good corrosion resistance, but it is evidence, that porous TiO_x layer results in significant increase of the corrosion resistance, with respect to pure Ti. The corrosion current for pure Ti is higher than for porous TiO_x. For porous TiO_x, a shift in the corrosion potential to the more negative values is observed . The lower corrosion current corresponds to better corrosion resistance of the TiO_x than pure

Ti, which is extremely important in biomaterials applications. The porous TiO_x has wide passive range (plateau on the polarization curve). This passive range suggests that the surface oxides are stable and no pitting occurs. The current density value in the passive range is almost the same in both cases. For pure Ti at higher positive potentials, a transpassivation and transformation of oxide into soluble salt is observed, which results in fast metal dissolution (a). This behavior was not observed for the porous TiO_x (b) and the surface layer protect material against dissolution.

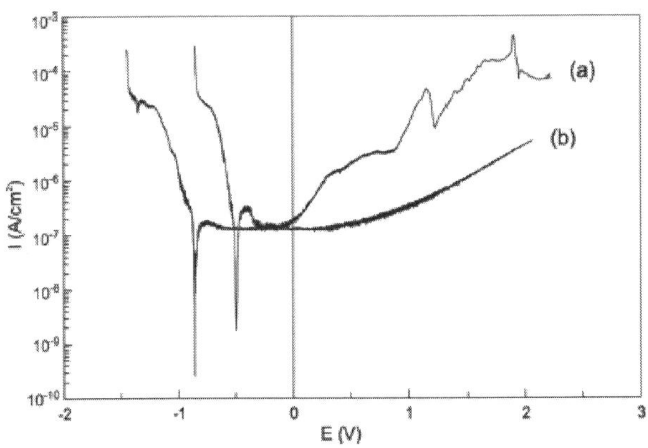

Figure 5.22 Potentiodynamic corrosion curves for pure Ti (a) and porous TiO_x layer (b) in Ringer's electrolyte [14].

The Ti-glass or Ti-HA bionanocomposites are a new class of biomaterials, with interesting mechanical and biomedical properties, prepared by Niespodziana *et al.* [33, 34]. The corrosion properties of the Ti-glass (45S5-Bioglass) were potentiodynamically investigated in Ringer's solution by Jakubowicz *et al.* [16] (Fig. 5.23). After electrochemical etching, the surface of Ti-glass nanocomposites has a slightly better corrosion resistance (b) than the parent not etched sample (a). The corrosion current density, estimated from the Tafel extrapolation, of the nanocomposite before and after electrochemical treatment is 5.1×10^{-6} A/cm^2 and 1.87×10^{-6} A/cm^2, respectively. The lower corrosion current density indicates better corrosion resistance and lower corrosion rate. The surface after etching is covered by thicker oxide, than samples not etched with native oxide. These anodic oxides provide better corrosion resistance of the electrochemically treated Ti-45S5 bionanocomposites [16].

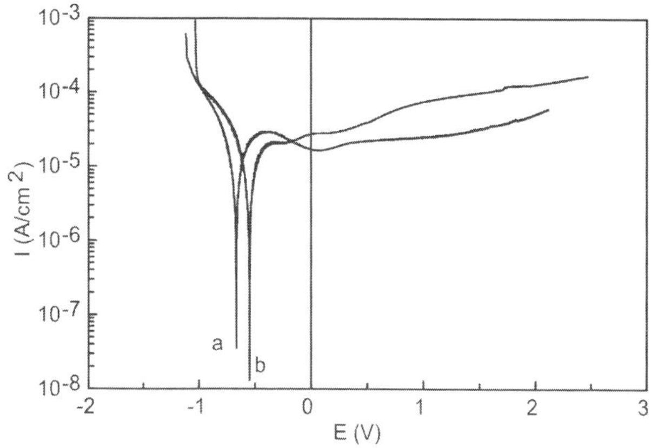

Figure 5.23 Potentiodynamic corrosion curves for flat Ti-45S5 (a) and porous Ti-45S5 (b) nanocomposites, etched in 1M H_3PO_4 + 2% HF at 10 V for 30 min [16].

Another example of the porous nanomaterial is Ti–6Al–4V alloy prepared by mechanical alloying and powder metallurgical processes [15]. Jakubowicz and Adamek investigated corrosion resistance of the nanocrystalline Ti–6Al–4V alloy before and after anodic oxidation [15]. They compared the results to microcrystalline counterpart. The corrosion resistance was investigated in simulated body fluids using potentiodynamic method (Fig. 5.24, Table 5.6). The results are surprising, because the authors expected improvement of the corrosion resistance due to the anodic oxidation, as was in the case of pure porous Ti [14]. The polarization curves clearly shows, that the lowest corrosion current has microcrystalline bulk alloy (curve a). Etching of this sample results in increasing the corrosion current (curve b). Corrosion resistance of the nanocrystalline alloy (curve c) is significantly lower than the microcrystalline bulk counterpart. After nanoparticles sintering, the sample has large volume of the grain boundaries and density of about 90% of the theoretical value, which means that the alloy has some pores inside, which is normal after sintering. This structure and larger volume of the grain boundaries are responsible for the total current flow during the corrosion test, different than for the bulk microcrystalline sample. After additional electrochemical treatment, the sample is etched through the grain boundaries, and anodic pores formation take place. The density decreases to 80% of the theoretical value, but surface of the grains/pores is covered by anodic oxide, resulting in corrosion current (curve d) almost of one order of magnitude smaller, than for the unetched sample (Table 5.6) [15].

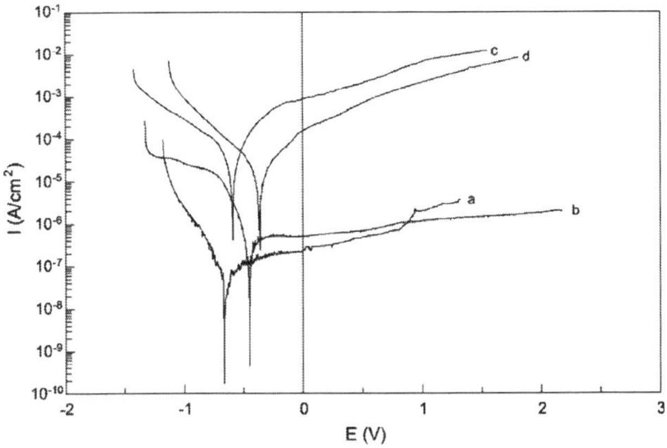

Figure 5.24 Corrosion polarization curves for Ti–6Al–4V: bulk microcrystalline (a), bulk microcrystalline electrochemically etched (b), nanocrystalline (c), nanocrystalline and electrochemically etched (d) [15].

Table 5.6 Corrosion current I_c and corrosion potential E_c for the Ti–6Al–4V [15]

Sample	I_c (A/cm^2)	E_c (V)
Ti-6Al-4V — bulk microcrystalline not etched	4.30×10^{-8}	-0.66
Ti-6Al-4V — bulk microcrystalline electrochemically etched	1.82×10^{-7}	-0.45
Ti-6Al-4V — nanocrystalline not etched	1.05×10^{-5}	-0.59
Ti-6Al-4V — nanocrystalline electrochemically etched	4.28×10^{-6}	-0.36

Due to toxicity of V and Al, new alloys without these elements are desirable, which provide properties accurate for hard tissue implants. The new V-free Ti–35Nb–5Ta–7Zr alloy anodized in H_3PO_4 + NaF which results in nanotubes and nanoporous oxide formation [40]. Saji *et al.* investigated theirs corrosion resistance (Fig. 5.25, Table 5.7) [40].

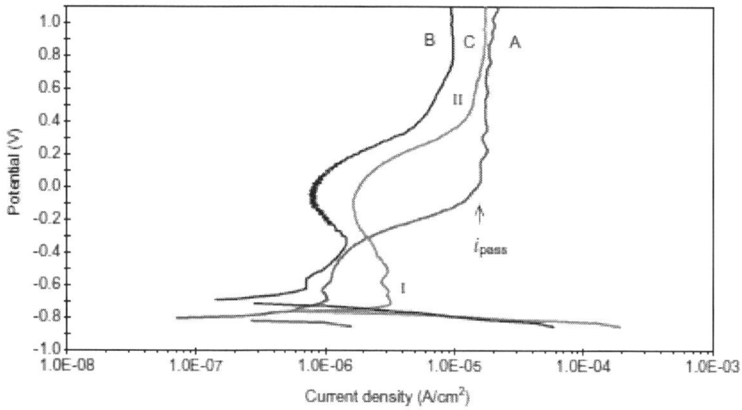

Figure 5.25 Potentiodynamic polarization curves of Ti–35Nb–5Ta–7Zr alloy in deaerated Ringer's solution at 37 ± 1°C: (A) bare, (B) nanoporous, and (C) nanotubular alloy [40]. See also Color Insert.

Table 5.7 Corrosion parameters from polarization plots [40]

Sample	E_{corr} (V)	I_{corr} (μA cm^{-2})	I_{pass} (μA cm^{-2})
Bare	-0.805	0.87	18.14
Nanoporous	-0.728	0.76	8.68
Nanotubular	-0.754	3.12	12.57

Nanoporous anodized layer significantly improved the corrosion resistance of the bare alloy. The nanotubular surface exhibits passivation behavior similar to the nanoporous surface. The passive region I extended over a wide potential range for both the nanoporous and nanotubular alloy (Fig. 5.25). The bare alloy shows a steady passivation (region II). The current density corresponding to the passivation region (I_{pass}) for both the nanoporous and nanotubular alloy was nobler than for the bare alloy, but the nanotubular alloy exhibited significantly higher corrosion current density (I_{corr}) values. The nanoporous surface consist perfectly passive pits due to the higher barrier oxide thickness and compact pore walls [40]. For the nanotubular surface, the tubes may act as the more effective channels for the electrolyte to reach the interface. The lower corrosion resistance of the nanotubular alloy surface may be associated with the concave shaped tube bottom and the distinctly separated tube bottom/barrier oxide interface [40].

5.3 Corrosion of the Hydroxyapatite Ti-Coated Compounds

Hydroxyapatite layers deposited on the Ti implant improve process of the osseointegration and should keep or improve high corrosion resistance of the applied metallic background.

Jakubowicz *et al.* investigated corrosion resistance of the hydroxyapatite (HA) electrochemically deposited on different Ti-based materials: micro-Ti [17], Ti-45S5 nanocomposites [19], and micro-/nanocrystalline Ti–6Al–4V alloys [18]. For the HA deposition process, they applied two different types of electrolytes. The source of Ca and P states: 0.1M HCl + 0.005M HA and 0.042M $Ca(NO_3)_2$ + 0.025M $(NH_4)_2HPO_4$ + 0.1M HCl electrolytes (see chapter 9). The corrosion resistance was investigated in Ringer's solution [17–19].

Corrosion potentiodynamic curves for Ti with HA deposits are shown in Fig. 5.26 and Fig. 5.27 and corrosion data are summarized in Table 5.8 [17].

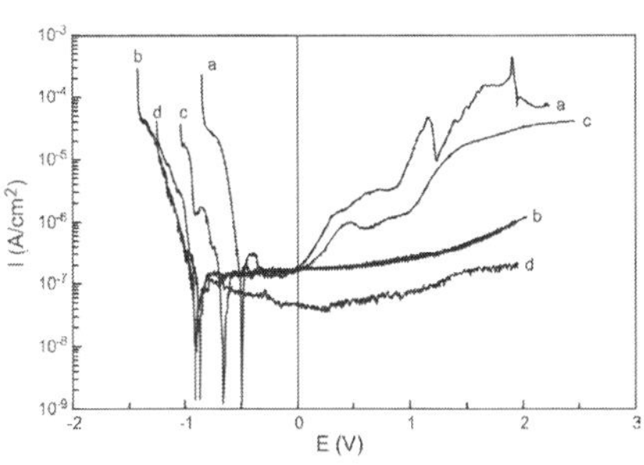

Figure 5.26 Potentiodynamic corrosion curves for flat Ti (a), porous Ti (b), flat Ti with calcium-phosphates (c), and porous Ti with calcium-phosphates (d); (HA deposited using HA + HCl electrolyte) [17].

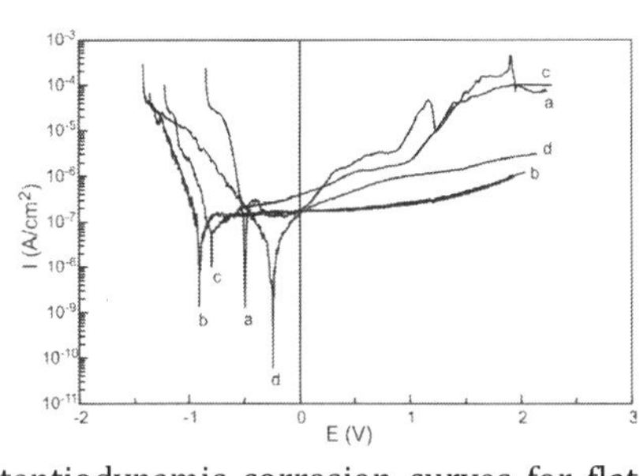

Figure 5.27 Potentiodynamic corrosion curves for flat Ti (a) porous Ti (b) flat Ti with calcium-phosphates (c) and porous Ti with calcium-phosphates (d); (HA deposited using $Ca(NO_3)_2$ + $(NH_4)_2HPO_4$ + HCl electrolyte) [17].

Table 5.8 Corrosion current I_c and corrosion potential E_c for the flat not etched Ti, porous electrochemically etched Ti, and Ti both flat and porous, with deposited calcium-phosphate (data taken from Figs. 5.26 and 5.27) [17]

Sample	I_c (A/cm^2)	E_c (V)	Figure
Ti — not etched	3.90×10^{-8}	−0.46	5.26, 5.27 (a)
Ti — not etched with deposited Ca-P (0.1M HCl + 0.005M HA)	1.57×10^{-8}	−0.66	5.26 (c)
Ti — not etched with deposited Ca-P (0.042M $Ca(NO_3)_2$ + 0.025M $(NH_4)_2HPO_4$ + 0.1M HCl)	3.92×10^{-8}	−0.77	5.27 (c)
Ti — etched	2.52×10^{-8}	−0.90	5.26, 5.27 (b)
Ti — etched with deposited Ca-P (0.1M HCl + 0.005M HA)	1.47×10^{-8}	−0.86	5.26 (d)
Ti — etched with deposited Ca-P (0.042M $Ca(NO_3)_2$ + 0.025M $(NH_4)_2HPO_4$ + 0.1M HCl)	1.073×10^{-8}	−0.25	5.27 (d)

Corrosion current I_c for flat and porous Ti is 3.9×10^{-8} A·cm^{-2} and 2.52×10^{-8} A · cm^{-2}, respectively. The anodization (pores formation) results in better corrosion resistance, due to the formed thick surface oxide. After HA deposition, the polarization curves significantly changes. The lowest corrosion current density 1.073×10^{-8} A·cm^{-2} (best corrosion resistance) has material anodically etched (porous) with deposited Ca-P using 0.042M $Ca(NO_3)_2$ + 0.025M $(NH_4)_2HPO_4$ + 0.1M HCl electrolyte. The anodically etched as well as anodically etched and HA deposited materials shows high resistance to pitting corrosion, too. These surfaces exhibit good passivation with extended passive range and lowest current in the passive range. Generally, the calcium-phosphate deposits enhance corrosion resistance, due to insulating behavior, with respect to surface without the Ca-P layer.

In the work [19] Jakubowicz *et al.* investigated the corrosion resistance of the flat, porous, and with Ca-P deposited [17–19] Ti-45S5 nanocomposites in Ringer's solution, and the results are shown in Fig. 5.28, Fig. 5.29 and Table 5.9.

The corrosion resistance increases (decreases the corrosion current) for the etched and with HA deposited nanocomposites, with respect to not etched materials. The best passivation (lowest current and wide plateau in the passive range) shows samples marked as c and d on both Figs. 5.28 and 5.29. These samples have flat and porous surface covered with Ca-P, respectively. After corrosion test, a large part of the particles deposited on flat surface was removed from the surface, which they were not observed on the porous nanocomposite [19].

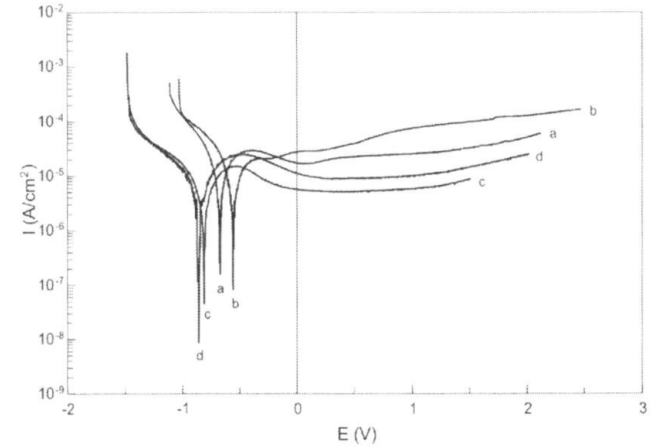

Figure 5.28 Potentiodynamic corrosion curves for Ti-glass nanocomposites: flat (a), porous (b), flat with HA (c), porous with HA (d); (HA deposited using HA + HCl electrolyte) [19].

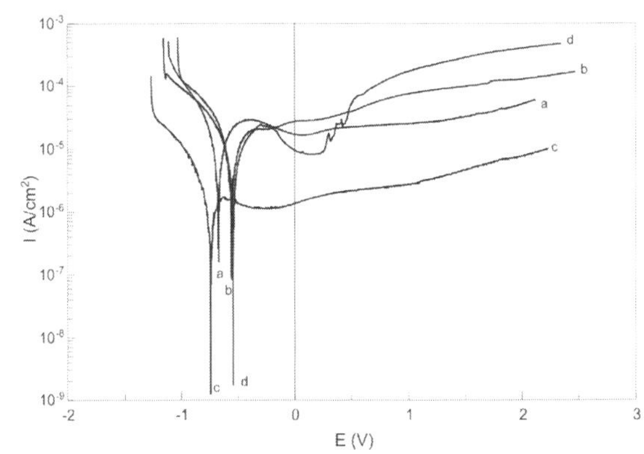

Figure 5.29 Potentiodynamic corrosion curves for Ti-glass nanocomposites: flat (a), porous (b), flat with HA (c), porous with HA (d); (HA deposited using $Ca(NO_3)_2$ + $(NH_4)_2HPO_4$ + HCl electrolyte) [19].

Table 5.9 Corrosion current I_c and corrosion potential E_c for the Ti-glass nanocomposites: flat not etched, porous electrochemically etched, and both flat and porous with deposited calcium-phosphate (data taken from Figs. 5.28 and 5.29) [19]

Sample	I_c (A/cm^2)	E_c (V)	Figure
Ti-glass — not etched	5.10×10^{-6}	−0.67	5.28a, 5.29a
Ti-glass — not etched with deposited HA (0.1M HCl + 0.005M HA)	7.32×10^{-7}	−0.8	5.28c
Ti-glass — not etched with deposited HA (0.042M $Ca(NO_3)_2$ + 0.025M $(NH_4)_2HPO_4$ + 0.1M HCl)	2.33×10^{-7}	−0.74	5.29c
Ti-glass — etched 30 min	1.86×10^{-6}	−0.55	5.28b, 5.29b
Ti-glass — etched 30 min with deposited HA (0.1M HCl + 0.005M HA)	7.46×10^{-7}	−0.86	5.28d
Ti-glass — etched 30 min with deposited HA (0.042M $Ca(NO_3)_2$ + 0.025M $(NH_4)_2HPO_4$ + 0.1M HCl)	1.87×10^{-6}	−0.54	5.29d

Jakubowicz *et al.* [18] also investigated the HA deposited on Ti–6Al–4V alloys. The corrosion properties of the flat (not etched), porous and with HA deposited Ti–6Al–4V samples investigated in Ringer's solution are shown in Fig. 5.30 and Table 5.10 [18]. Figure 5.30 shows examples polarization curves for micro and nanocrystalline materials with deposited HA layer (see chapter 9). The differences in corrosion properties are clearly visible (curves a, c with comparison to b, d). The microcrystalline materials have lower both corrosion current density and current density in the passive range, with comparison to nanocrystalline materials. Best corrosion resistance after electrochemical etching shows Ti (I_c = 2.52×10^{-8} A/cm^2).

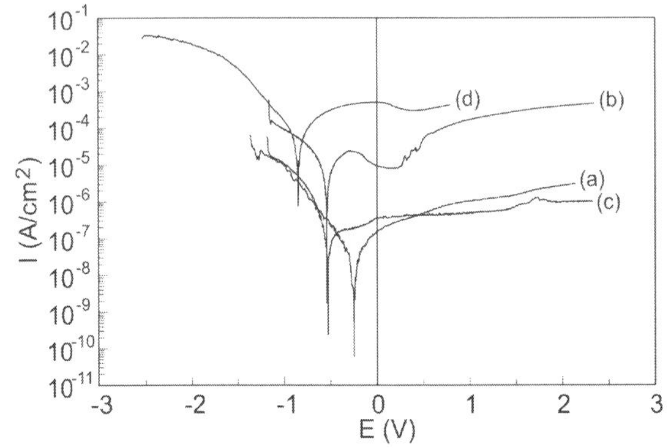

Figure 5.30 Polarization curves of microcrystalline Ti (a), Ti-45S5 nanocomposite (b), microcrystalline Ti–6Al–4V (c), nanocrystalline Ti–6Al–4V (d); all materials after electrochemical etching and HA deposition using 0.042M $Ca(NO_3)_2$ + 0.025M $(NH_4)_2HPO_4$ + 0.1M HCl electrolyte [18].

Table 5.10 Corrosion current I_c and corrosion potential E_c for the micro- and nanocrystalline Ti–6Al–4V: flat not etched, porous electrochemically etched, and both flat and porous with deposited hydroxyapatite (HA) [18]

Sample	I_c (A/cm^2)	E_c (V)
Ti–6Al–4V — microcrystalline not etched	4.30×10^{-8}	-0.66
Ti–6Al–4V — microcrystalline etched	1.82×10^{-7}	-0.45
Ti–6Al–4V — microcrystalline etched with deposited HA (HCl + HA)	1.50×10^{-9}	-0.48
Ti–6Al–4V — microcrystalline etched with deposited HA ($Ca(NO_3)_2$ + $(NH_4)_2HPO_4$ + M HCl)	3.83×10^{-8}	-0.53
Ti–6Al–4V — nanocrystalline not etched	1.05×10^{-5}	-0.59
Ti–6Al–4V — nanocrystalline etched	4.28×10^{-6}	-0.36
Ti–6Al–4V — nanocrystalline etched with deposited HA (HCl + HA)	1.13×10^{-5}	-0.51
Ti–6Al–4V — nanocrystalline etched with deposited HA ($Ca(NO_3)_2$ + $(NH_4)_2HPO_4$ + HCl)	2.47×10^{-5}	-0.85

After additional deposition of HA layer (HA + HCl electrolyte), best corrosion resistance shows Ti–6Al–4V ($I_c = 1.5 \times 10^{-9}$ A/cm^2) [18]. Nanocrystalline Ti–6Al–4V, although is easily etched during anodization resulting in largest pores, in corrosion tests shows lowest corrosion resistance (slightly improved by anodic oxidation with respect to the parent not etched nanocrystalline alloy). HA deposition on nanocrystalline Ti–6Al–4V does not improve corrosion resistance as was in the case of the rest materials [18]. Generally, worse corrosion resistance estimated by electrochemical method, shows nanocrystalline materials, especially Ti–6Al–4V. This behavior is related with large grain boundaries volume fraction. The grains boundaries state an easy way for the electrolyte penetration, which is useful during pores formation, but unfortunately highly unexpected during corrosion tests. The HA deposition improve the corrosion resistance for both types of electrolytes [18].

The corrosion resistance of the Ti–6Al–4V alloy after calcium ion implantation was examined by Krupa *et al.* [26]. The anodic polarization curves (in SBF) for the non-implanted and calcium ion implanted Ti–6Al–4V alloys are shown in Fig. 5.31. The increase of the anodic current density observed in the implanted samples when the potential exceeds about 1900 mV, suggests the pitting corrosion.

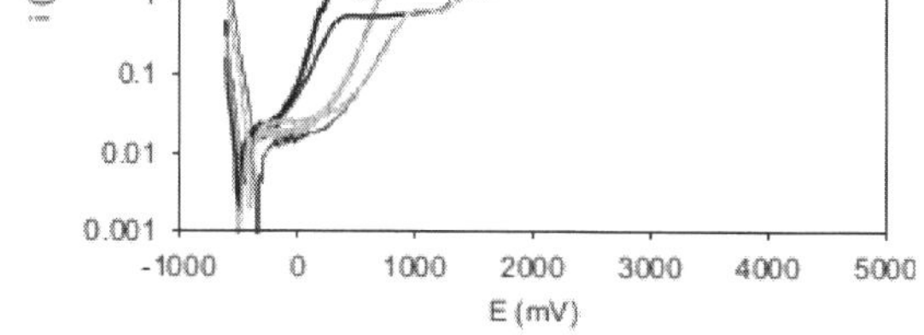

Figure 5.31 Anodic polarization curves for a non-implanted and calcium implanted Ti–6Al–4V alloy after various times of exposure in SBF: (1) non-implanted (13 h), (2) non-implanted (733 h), (3) calcium-implanted (13 h), (4) calcium-implanted (181 h) and (5) calcium-implanted (733 h) [26].

The results obtained for the Ti–6Al–4V alloy implanted to a level of 10^{17} Ca^+/cm^2 indicate that the calcium ion implantation affects the corrosion resistance of the alloy. The calcium ion implantation increases the corrosion resistance of the Ti–6Al–4V alloy, but unfortunately these implanted alloys underwent pitting corrosion [26]. This pitting corrosion is attributed to the reduced oxygen content at the interface between the passive layer and the implanted layer [45, 46]. The implanted ions act as a barrier to the diffusion of oxygen. The change of the corrosion resistance of the calcium-implanted Ti–6Al–4V alloy after long-term exposure into SBF is related with formation of calcium phosphates. The Ca-P precipitations are more numerous after longer exposures (Fig. 5.32), and the precipitated layer is continuous. The Ca/P ratio in the precipitations on the implanted surface after the exposure for 720 h was about 1.66, which suggests hydroxyapatite formation. Implantation of the calcium ions definitely enhances the calcium-phosphate precipitation.

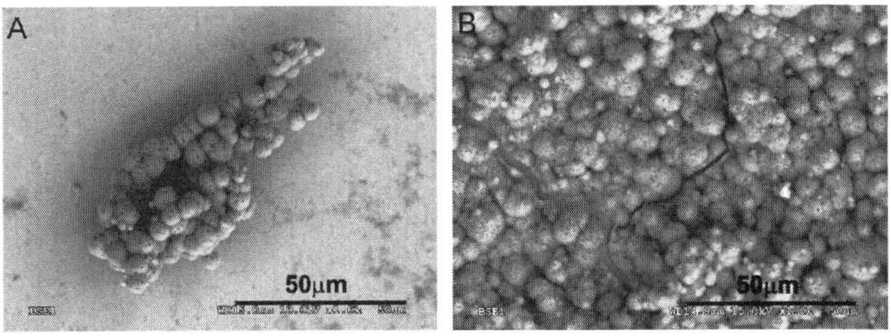

Figure 5.32 Surface of the Ti–6Al–4V alloy implanted with a 1×10^{17} Ca^+/cm^2 after exposure in SBF: (A) 168 h and (B) 720 h [26].

5.4 Corrosion of the Other Dental Materials

In medicine, the stainless steels which are typically used contain 17–20% Cr, 13–15% Ni, 2–3% Mo, and small amounts of other elements. Cr is the element responsible for the high passivation ability of these alloys. An increase in Cr and Mo content leads to an increase in the resistance against localized corrosion. Although the main problem is related with Ni, it is commonly added as the austenite forming element, but unfortunately, highly toxic for the human. The other Co-Cr-type alloys have high corrosion resistance [53]. The Co–Cr alloys are superior to stainless steel, both in fatigue and wear resistance. In all Cr-containing alloys, Cr_2O_3 forms on the surface acting as a corrosion protective layer.

Taher and Al Jabab [50] investigated the galvanic corrosion behavior of a different commercial dental alloys (Table 5.11) coupled with Ti implant. Galvanic corrosion is present and significant when the different alloys are in direct contact within the oral cavity or tissue environments. The combination of Ti together with metals with lower passivity can results in galvanic corrosion, whereas Ti combine together with higher resistant metal result in a passive coupling being formed. The current flow in the formed corrosion cell at the metals contact interface could lead to bone destruction. The coupled alloys should have the minimum potential differences (Fig. 5.1) which minimize the galvanic corrosion rate at the metal joint [21].

Table 5.11 Composition and manufacture of coupled alloy [50]

Alloy	Code	Trade name	Manufacturer	Composition (wt%)
Co–Cr	R2000	Remanium 2000	Dentaurum, Germany	61% Co, 25% Cr, 7% Mo, 5% W, 1.5% Si
Co–Cr	R800	Remanium 800	Dentaurum, Germany	63.5% Co, 30% Cr, 5% Mo, 1% Si
Ni–Cr	RCS	Remanium CS	Dentaurum, Germany	61% Ni, 26% Cr, 11% Mo, 1.5% Si, <1% Fe, Ce, Al
Ag–Pd	Jel	Jelstar	Jelenko, USA	60% Pd, 28% Ag, 6% Sn, 6% In
Au	Pont	Pontallor-4	Degussa, Germany	40% Au, 7.9% Pd, 35% Ag, 7% Cu, 2% Sn, 2% Ir, 5% In, 3.5% Zn
Ternary titanium	Ter Ti	Experimental		60% Ti, 10% Ag, 30% Cu
Endossseous Ti implant healing abutment (SCPB-010)	SSTi	Rematitan	Dentaurum, Germany	0.12% O, 0.05% N, 0.06% C, 0.013% H, rest Ti; (vol %)
Amalgam	Amalg	Dispersalloy	J&J, Ireland	34.8% Ag, 8.9% Sn, 5.9% Cu, 0.3% Zn, 50% Hg

The results obtained by Taher and Al Jabab (Table 5.12) indicates that Jelstar and Pontallor-4 alloys acts as the cathode in the coupling with Ti, which means that the Ti is anode and undergoes corrosion. For other materials, the Ti acts as the cathode [50]. The galvanic current measured between anode and cathode materials represents the ability to corrosion of the anode contacting with the cathode. The couples which generate the lowest current density value (Table 5.12) states the best materials for the implant joint. From the all investigated by them alloys, the Pontallor-4 (Au alloy) states an excellent couple with the Ti implant, whereas Jelstar (Pd-Ag alloy) shows acceptable corrosion behavior with Ti contact [50]. R800 and R2000 have acceptable galvanic corrosion with Ti contact. The RCS and SSTi shows unstable corrosion behavior, whereas Amalgam shows highest galvanic corrosion, when is coupled with Ti [50].

Table 5.12 The mean values of mixed potential, galvanic current density and integrated current density during the last 6 h for all Ti/alloy couples [50]

Couple	Count	Mixed potential (mV/SCE)	Galvanic current (μA/cm^2)	Current integration during the last 6 h (μA/cm^2)
Ti/Jelstar	3	66.8	0.00	34.4
Ti/Pontallor-4	3	1.7	0.00	-86.2
Ti/R2000	3	-107.1	-0.02	-95.5
Ti/R800	3	-27.4	-0.01	-205.4
Ti/RCS	3	-31	-0.02	-350.3
Ti/Ter Ti	3	-21.1	-0.01	-170.5
Ti/SSTi	3	-184.6	-0.04	-873.5
Ti/amalgam	3	-109.9	-0.08	-1127

To improve corrosion resistance of the stainless steel, the bioactive glass ($CaO–SiO_2–P_2O_5$) coating was successfully applied on the 316L substrate by sol–gel technique [5]. The size of the used bioactive glass powder was less than 100 nm [5], and finally deposited layer with 10–20 μm thickness is shown in Fig. 5.33. The coating strictly adheres to the surface. The corrosion parameters of the 316L steel, with and without coating are shown in Tables 5.13 and 5.14 in normal saline solution and Ringer's solution, respectively [5]. Bioactive glass coating improve the corrosion resistance of the

316L steel and additionally, the bioactive glass coated 316L show more pitting corrosion resistance in comparison with uncoated steel. The deposited coating is useful in improvement of biocompatibility and stronger bone bonding too.

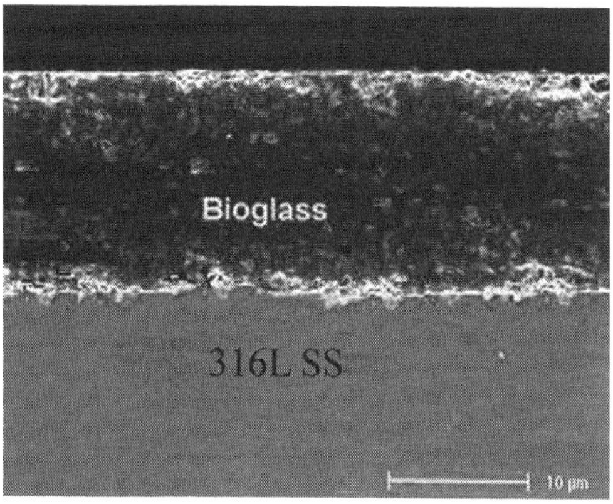

Figure 5.33 SEM image of a cross-section bioactive glass coating on the 316L substrate [5].

Table 5.13 Mean values (standard deviation) of corrosion current densities and corrosion potentials of the bioactive glass coated and uncoated 316L stainless steel in normal saline solution at 37°C [5]

		I_{corr} (nA/cm^2)	
Material	E_{corr} **(mV)**	**Tafel**	**Linear**
Pristine 316L SS	−174 ± 20	265 (16)	194 (16)
Bioactive glass coated 316L SS	−220 ± 20	85 (6)	50 (8)

Table 5.14 Mean values (standard deviation) of corrosion current densities and corrosion potentials of the bioactive glass coated and uncoated 316L stainless steel in Ringer's solution at 37°C [5].

		I_{corr} (nA/cm^2)	
Material	E_{corr} **(mV)**	**Tafel**	**Linear**
Pristine 316L SS	−195 ± 20	218 (17)	181 (20)
Bioactive glass coated 316L SS	−204 ± 20	78 (7)	59 (8)

The results presented in this chapter indicates, that nanocrystalline bulk materials have slightly lower corrosion resistance, due to the large volume of the grain boundaries, but surface with nanoroughness (nanopores) shows usually better corrosion resistance, passivation behavior and pitting resistance in comparison to microcrystalline counterparts.

References

1. Blackwood, D.J., Chua, A.W.C., Seah, K.W.H., Thampuran, R., and Teoh, S.H. (2000). Corrosion behaviour of porous titanium–graphite composites designed for surgical implants, *Corrosion Sci.*, **42**, pp. 481–503.
2. Brossia, C.S., and Cragnolino, G.A. (2004). Effect of palladium on the corrosion behavior of titanium, *Corrosion Sci.*, **46**, pp. 1683–1711.
3. Choubey, A., Balasubramaniam, R., and Basu, B. (2004). Effect of replacement of V by Nb and Fe on the electrochemical and corrosion behavior of Ti–6Al–4V in simulated physiological environment, *J. Alloys Comp.*, **381**, pp. 288–294.
4. Davidson, J.A., Mishra, A.K., Kovacs, P., and Poggie, R.A. (1994). New surface-hardened, low-modulus, corrosion-resistant Ti–13Nb–13Zr alloy for total hip arthroplasty, *Biomed. Mater. Eng.*, **4**, (1994), pp. 231–243.
5. Fathi, M.H., and Doostmohammadi, A. (2009). Bioactive glass nanopowder and bioglass coating for biocompatibility improvement of metallic implant, *J. Mater. Proc. Technol.*, **209**, pp. 1385–1391.
6. Fontana, M.G. (1987). *Corrosion Engineering*, 3rd ed., McGraw-Hill Book Company, New York.
7. Garbacz, H., Pisarek, M., and Kurzydłowski, K.J. (2007). Corrosion resistance of nanostructured titanium, *Biomol. Eng.*, **24**, pp. 559–563.
8. Gu, Y.W., Yong, M.S., Tay, B.Y., and Lim, C.S. (2009). Synthesis and bioactivity of porous Ti alloy prepared by foaming with TiH_2, *Mater. Sci. Eng. C*, **29**, pp. 1515–1520.
9. Güleryüz, H., and Cimenoglu, H. (2004). Effect of thermal oxidation on corrosion and corrosion-wear behaviour of a Ti–6Al–4V alloy, *Biomaterials*, **25**, pp. 3325–3333.
10. House, K., Sernetz, F., Dymock, D., Sandy, J.R., and Ireland, A.J. (2008). Corrosion of orthodontic appliances-should we care?, *Amer. J. Orthodont. Dentofac. Orthoped.*, **133**, pp. 584–592.
11. Hsu, R.W.-W., Yang, Ch.-Ch., Huang, Ch.-A., and Chen, Y.-S. (2004). Electrochemical corrosion properties of Ti–6Al–4V implant alloy in the biological environment, *Mater. Sci. Eng. A*, **380**, pp. 100–109.

12. Huang, H.-H. (2002). Effects of fluoride concentration and elastic tensile strain on the corrosion resistance of commercially pure titanium, *Biomaterials*, **23**, pp. 59–63.
13. Huang, H.-H. (2003). Effect of fluoride and albumin concentration on the corrosion behavior of Ti-6A1-4V alloy, *Biomaterials*, **24**, pp. 275–282.
14. Jakubowicz, J. (2008). Formation of porous TiO_x biomaterials in H_3PO_4 electrolytes, *Electrochem. Commun.*, **10**, pp. 735–739.
15. Jakubowicz, J., and Adamek, G. (2009). Preparation and properties of mechanically alloyed and electrochemically etched porous Ti-6Al-4V, *Electrochem. Commun.*, **11**, pp. 1772–1775.
16. Jakubowicz, J., Jurczyk, K., Niespodziana, K., and Jurczyk, M. (2009). Mechanoelectrochemical synthesis of porous Ti-based nanocomposite biomaterials *Electrochem. Commun.*, **11**, pp. 461–465.
17. Jakubowicz, J., and Adamek, G. (2010). Hydroxyapatite deposited on flat and porous Ti, *IEEE Xplore*, 10.1109/ICBBE.2010.5515482 (2010) 1-4, ISBN 978-1-4244-4712-1.
18. Jakubowicz, J., Adamek, G., and Jurczyk, M. (2010). Unpublished data.
19. Jakubowicz, J., Jurczyk, K., and Adamek, G. (2010). Unpublished data.
20. Jang, S.-H., Choe, H.-Ch., Ko, Y.-M., and Brantley, W.A. (2009). Electrochemical characteristics of nanotubes formed on Ti-Nb alloys, *Thin Solid Films*, **517**, pp. 5038–5043.
21. Jones, D.A. (1992). *Principles and Prevention of Corrosion*, Macmillan, New York.
22. Jones, D.A. (1996). *Principles and Prevention of Corrosion*, 2nd ed., Prentice Hall, New York.
23. Chadwick, D.J., Whelan, J. (eds)., and (1992). Neurotoxic effects of dietary aluminum, in *Aluminium in Biology and Medicine* (ed. Jope, R.S., and Johnson, G.V.W.), Wiley, Chichester, pp. 254–267.
24. Khan, M.A., Williams, R.L., and Williams, D.F. (1996). In vitro corrosion and wear of titanium alloys in the biologic environment, *Biomaterials*, **17**, pp. 2117–2126.
25. Khan, M.A., Williams, R.L., and Williams, D.F. (1999). Conjoint corrosion and wear in titanium alloys, *Biomaterials*, **20**, pp. 765–772.
26. Krupa, D., Baszkiewicz, J., Rajchel, B., Barcz, A., Sobczak, J.W., Bilinski, A., and Borowski, T. (2007). Effect of calcium-ion implantation on the corrosion resistance and bioactivity of the Ti-6Al-4V alloy, *Vacuum*, **81**, pp. 1310–1313.

27. Landsberg, J.P., McDonald, B., and Watt, F. (1992). Absence of aluminium in neuritic plaque cores in Alzheimer's disease, *Nature*, **360**, pp. 65–67.

28. Li, X., Wang, J., Han, E.-H., and Ke, W. (2007). Influence of fluoride and chloride on corrosion behavior of NiTi orthodontic wires, *Acta Biomater.*, **3**, pp. 807–815.

29. Lopez, M.F., Gutierrez, A., and Jimenez, J.A. (2002). In vitro corrosion behaviour of titanium alloys without vanadium, *Electrochim. Acta*, **47**, pp. 1359–1364.

30. Mabilleau, G., Bourdon, S., Joly-Guillou, M.L., Filmon, R., Baslé, M.F., and Chappard, D. (2006). Influence of fluoride, hydrogen peroxide and lactic acid on the corrosion resistance of commercially pure titanium, *Acta Biomater.*, **2**, pp. 121–129.

31. Nakagawa, M., Matono, Y., Matsuya, S., Udoh, K., and Ishikawa, K. (2005). The effect of Pt and Pd alloying additions on the corrosion behavior of titanium in fluoride containing environments, *Biomaterials*, **26**, pp. 2239–2246.

32. Narayanan, R., and Seshadri, S.K. (2008). Point defect model and corrosion of anodic oxide coatings on Ti–6Al–4V, *Corrosion Sci.*, **50**, pp. 1521–1529.

33. Niespodziana, K., Jurczyk, K., Jakubowicz, J., and Jurczyk M. (2010). Fabrication and properties of titanium — hydroxyapatite nanocomposites, *Mat. Chem. Phys.*, **123**, pp. 160–165.

34. Niespodziana, K., Jurczyk, K., Miklaszewski, A., and Jurczyk, M. (2010). Hybrid Ti-ceramic bionanomaterials for medical engineering, *Phys. Stat. Sol. C*, **7**, pp. 1363–1366.

35. Niinomi, M. (1998). Mechanical properties of biomedical titanium alloys, *Mater. Sci. Eng. A*, **243**, pp. 231–236.

36. Okazaki, Y., Rao, S., Ito, Y., and Tateishi, T. (1998). Corrosion resistance, mechanical properties, corrosion fatigue strength and cytocomp-atibility of new Ti alloys without Al and V, *Biomaterials*, **19**, pp. 1197–1215.

37. Raja, K.S., Misra, M., and Paramguru, K. (2005). Formation of Self-Ordered Nano-Tubular Structure of Anodic Oxide Layer on Titanium, *Electrochim. Acta*, **51**, pp. 154–165.

38. Rondelli, G., Vicentini, B. (1999). Localized corrosion behavior in simulated human body fluids of commercial Ni–Ti orthodontic wires, *Biomaterials*, **20**, pp. 785–792.

39. Ruzickova, M., Hildebrand, H., and Virtanen, S. (2005). On the stability of passivity of Ti-Al alloys in acidic environment, *Phys. Chem.*, **219**, pp. 1447–1459.

40. Saji, V.S., Choe, H.Ch., and Brantley, W.A. (2009). An electrochemical study on self-ordered nanoporous and nanotubular oxide on Ti–35Nb–5Ta–7Zr alloy for biomedical applications, *Acta Biomater.*, **5**, pp. 2303–2310.

41. Brunette, D.M, Tengvall, P., Textor, M., and Thomsen, P. (eds) (2001). The corrosion properties of titanium and titanium alloys, in., and *Titanium in Medicine: Material Science, Surface Science, Engineering, Biological Responses and Medical Applications* (ed. Schenk, R.), Springer-Verlag., Berlin, pp. 145–70.

42. Seah, K.H.W., Thampuran, R., Chen, X., and Teoh, S.H. (1995). A comparison between corrosion behaviour of sintered and unsintered porous titanium, *Corrosion Sci.*, **37**, pp. 1333–1340.

43. Semlitsch, M.F., Weber, H., Streicher, R.M., and Schon, R. (1992). Joint replacement components made of hot-forged and surface-treated Ti–6Al–7Nb alloy, *Biomaterials*, **13**, pp. 781–788.

44. Schiff, N., Grosgogeat, B., Lissac, M., and Dalard, F. (2002). Influence of fluoride content and pH on the corrosion resistance of titanium and its alloys, *Biomaterials*, **23**, pp. 1995–2002.

45. Schmidt, H., Konetschny, C., and Fink, U. (1998). Electrochemical behaviour of ion implanted Ti–6Al–4V in Ringer's solution, *Mater. Sci. Technol.*, **14**, pp. 592–598.

46. Schmidt, H., Stechemesser, G., Witte, J., and Soltani-Farshi, M. (1998). Depth distributions and anodic polarization behaviour of ion implanted Ti–5Al–3V, *Corrosion Sci.*, **40**, pp. 1533–1545.

47. Shi, P., Cheng, F.T., and Man, H.C. (2007). Improvement in corrosion resistance of NiTi by anodization in acetic acid, *Mater. Lett.*, **61**, pp. 2385–2388.

48. Shim, H.M., Oh, K.T., Woo, J.Y., Hwang, C.J., and Kim, K.N. (2005). Corrosion resistance of titanium-silver alloys in an artificial saliva containing fluoride ions., *J. Biomed. Mater. Res. B: Appl Biomater.*, **73**, pp. 252–259.

49. Sun, Z., Annergren, I., Pan, D., and Mai, T.A. (2003). Effect of laser surface remelting on the corrosion behavior of commercially pure titanium sheet, *Mater. Sci. Eng. A*, **345**, pp. 293–300.

50. Taher, N.M., and Al Jabab, A.S. (2003). Galvanic corrosion behavior of implant suprastructure dental alloys, *Dental Mater.*, **19**, pp. 54–59.

51. Upadhyay, D., Panchal, M.A., Dubey, R.S., and Srivastava, V.K. (2006). Corrosion of alloys used in dentistry: A review, Materials Science and Engineering, *Mater. Sci. Eng. A*, **432**, pp. 1–11.

52. Virtanen, S., Milosev, I., Gomez-Barrena, E., Trebse, R., Salo, J., and Konttinen, Y.T. (2008). Special modes of corrosion under physiological and simulated physiological conditions, *Acta Biomater.*, **4**, pp. 468–476.

53. Williams, D.F. (1981). The properties and clinical uses of cobalt-chromium alloys, in *Biocompatibility of Clinical Materials*, vol. I. (ed. Williams, D.F.), CRC Press, Boca Raton, FL, pp. 99–123.

54. Winship, K.A. (1992). Toxicity of aluminium: a historical review, part 1, *Adv. Drug React. Toxicol. Rev.*, **11**, pp. 123–141.

55. http://www.azom.com/article.aspx?ArticleID=1177

56. http://www.gamry.com/assets/Application-Notes/Getting-Started-with-Electrochemical-Corrosion-Measurement.pdf

57. Zhang, B.B., Zheng, Y.F., and Liu, Y. (2009). Effect of Ag on the corrosion behavior of Ti–Ag alloys in artificial saliva solutions, *Dental Mater.*, **25**, pp. 672–677.

Chapter 6

Nanostructured Stainless Steels

Maciej Tulinski
Institute of Materials Science and Engineering, Poznan University of Technology, Sklodowska-Curie 5 Sq., 60-965 Poznan, Poland
maciej.tulinski@put.poznan.pl

Metallic materials are widely used as biomaterials, replacing the structural components of the human body. This is because, in comparison to polymeric materials and ceramics, they have better mechanical properties such as tensile strength, fatigue strength, and resistance to cracking, characteristics that are crucial in biomedical applications [24].

316L stainless steel has inferior corrosion resistance comparing to cobalt-chromium alloys and titanium alloys [18]. For example, orthopedic implants made of 316L stainless steel corrode in the tissue environment, releasing iron, chromium, and nickel ions into the body.

Nickel ions are considered to be a very important factor in causing allergies and even cancer [32]. The number of women suffering from an allergy to nickel doubles every decade. Dermatologists estimate that about 20% of women and 4% of men on the earth have allergy to nickel. The main cause of allergy to nickel is wearing jewelry.

Bionanomaterials for Dental Applications
Edited by Mieczyslaw Jurczyk

ISBN 978-981-4303-83-5 (Hardcover), 978-981-4303-84-2 (eBook)
www.panstanford.com

World Health Organization concluded that metallic nickel and alloys containing nickel are carcinogenic. Nickel content of less than 0.2% is considered adequate for the human body. However, the reduction of nickel content below 0.2%, using conventional manufacturing methods, is associated with a significant increase in the cost of manufacture of the material.

In the following chapter, a concept of nickel-free austenitic stainless steels will be presented.

6.1 Nickel-Free Stainless Steels

From the chemical composition point of view, nitrogen is the most promising substitute for nickel. It is a gas at room temperature, stabilizes austenitic phase, increases the corrosion resistance, and prevents the formation of the sigma phase. Nitrogen dissolves interstitially in austenite structure in relatively large quantities, causing a large number of solution hardening. Drastic reduction in resistance to brittle fracture and corrosion resistance occurs with the presence of sigma phase with low nitrogen content and nitrides of chromium, such as Cr_2N with high nitrogen content. Therefore, the amount of nitrogen should be limited to obtain a homogeneous microstructure without inclusions (Fig. 6.1). Nitrides inclusions can lead to a local drop in chromium content, which in turn leads to local corrosion around the inclusions [14].

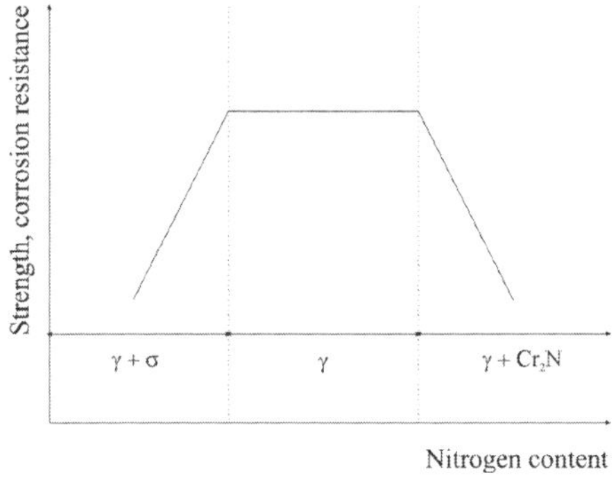

Figure 6.1 Schematic representation of strength and corrosion resistance as a function of nitrogen content in nickel-free austenitic stainless steel [14].

Alloying elements that increase the solubility of nitrogen are molybdenum, manganese, chromium, vanadium, niobium, and titanium. However, vanadium, niobium and titanium must not be used in large quantities due to their responsibility for the formation of nitrides. Elements that reduce the solubility of nitrogen are carbon, silicon, and nickel.

Austenitic steels with higher nitrogen content were produced for the first time in the forties of the 20th century. The main driving force for their creation was the need to replace the nickel with cheaper nitrogen. The resulting 18–8 steels with a nitrogen content of 0.3%, while retaining the 6% nickel, retained austenite structure and were characterized by both increased strength and yield point. The increase in these parameters was, however, accompanied with decreased ductility. As demonstrated by subsequent studies, the increase in strength without reducing ductility comes with the addition of at least 0.4% nitrogen.

The development of technology has enabled the introduction of more than 1% of nitrogen to the steel, which resulted in increase in strength by 134%, with decrease of elongation by only 8%.

Steels with high nitrogen content are examples of positive effects of nitrogen on the resistance to cracking. However, the addition of nitrogen to alloy does not guarantee resistance to cracking in the environment containing chloride ions. According to the graph shown in Fig. 6.2, it can be concluded that the addition of nitrogen to 18Cr–10Ni type steel results in reduced resistance to corrosion cracking, expressed in shortening the time to rupture. It is also the result of boiling in aqueous solution of $MgCl_2$. Nitrogen was added to stabilize the austenite, but in this case the presence of nitrogen does not eliminate the harmful effects of carbon.

Modern steels with carbon completely replaced by nitrogen and with very high concentration of the alloying elements show a much higher resistance to cracking. In Fig. 6.3, a measure of resistance to cracking is cracking initiation temperature — the higher the temperature, the less prone to cracking. The addition of nitrogen, nickel and molybdenum results in a high temperature to initiate cracks.

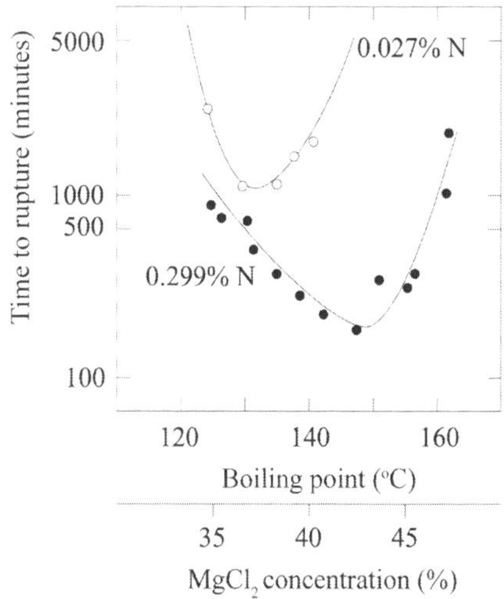

Figure 6.2 Influence of nitrogen content on the time to rupture of 18Cr–10Ni type steel.

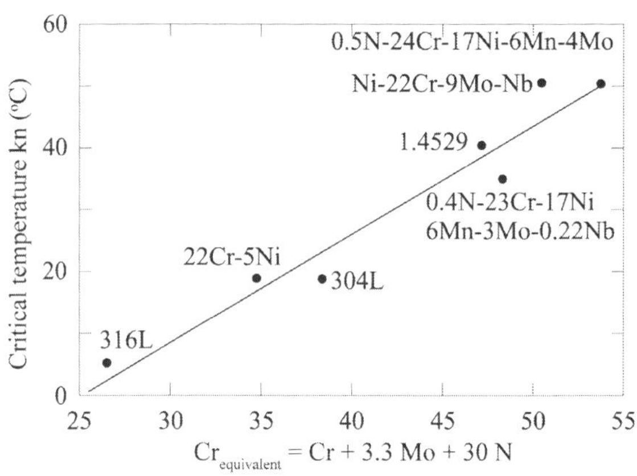

Figure 6.3 Relationship between corrosion equivalent of chromium and resistance to corrosion cracking in 6% $FeCl_3$.

Plastically deformed austenitic stainless steel exhibits magnetism. Such behavior occurs due to the stress caused by the transformation, where a martensitic phase is created. A characteristic feature of nitrogen-containing nickel-free austenitic stainless steels is the lack of ferromagnetism. This is due to the strong effect of stabilizing austenite by nitrogen, which prevents the formation of martensite, even in the most extreme conditions of cold-working or cryogenic cooling. These steels, even after a strong deformation, remain paramagnetic.

In a simulated biological environment, 316L steel shows pitting corrosion. Nickel-free austenitic stainless steel showed no corrosion of this type in all the electrolytes. Moreover, by preventing precipitation of $M_{23}C_6$ carbides, nitrogen reduces the risk of intergranular corrosion [4, 6, 16, 17].

Nickel-free austenitic stainless steels containing nitrogen are very promising metallic biomaterials [7]. They can be implemented as implants in the form of: stabilizing bone plates, screws or wires [9].

The most commonly used nickel-free stainless steels are shown in Table 6.1 [9, 14, 26, 32]. However, due to the large strengthening and low thermal conductivity of these steels, precise machining is obstructed. Therefore, the production of small, precise products, such as stents, from such steel is very expensive, and the range of possible dimensions is limited.

Table 6.1 Chemical compositions of commercially available nickel-free austenitic stainless steels

Chemical composition
Fe–(19–23)Cr–(10–12)Mn–(3–6)Mo–(0.85–1.1)N
Fe–15Cr–(10–15)Mn–4Mo–0.9N
Fe–18Cr–18Mn–2Mo–0.9N
Fe–(15–18)Cr–(10–12)Mn–(3–6)Mo–0.9N
Fe–23Cr–2Mo–1.5N

Efforts were taken to remedy this problem, and solution to this day is a new method of manufacturing. This method relies on a precise machining of small products made of nickel-free ferritic stainless steels, and then nitriding their surface at 1200°C where they become austenitic stainless steel with improved properties of strength and corrosion resistance [11].

Schematic representation of the process is shown in Fig. 6.4, and comparison of the properties of ferritic steels and nickel-free austenitic steels (after nitriding) is presented in Table 6.2.

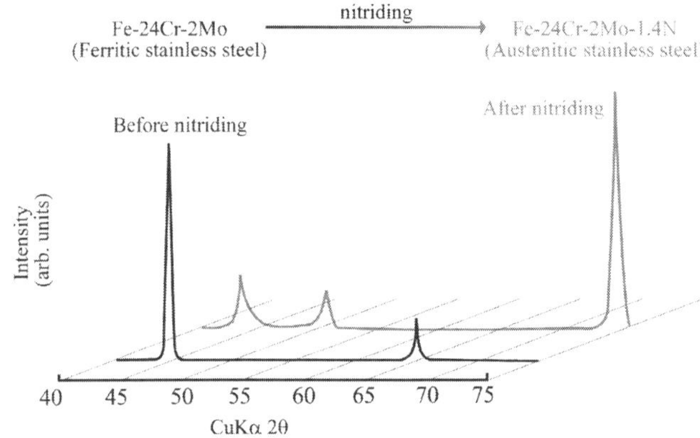

Figure 6.4 Schematic process of obtaining nickel-free stainless steels [11]. See also Color Insert.

Table 6.2 Comparison of properties of nickel-free ferritic and austenitic steels

Nickel-free ferritic steels	Nickel-free austenitic steels
Low biocompatibility	High biocompatibility
Good machinability	Poor machinability
Low corrosion resistance	High corrosion resistance
Low strength	High strength
Ferromagnetic	Paramagnetic

Previous methods of producing nickel-free stainless steels include: melting in induction furnaces, melting in the plasma furnaces, pressure electroslag remelting, and powder metallurgy combined with isostatic hot pressing (HIP) [1, 8, 10, 21]. The production by powder metallurgy uses the following techniques: gas atomization, nitriding in fluidized bed, and rotary furnaces.

Contemporary trends in the development of new materials are based on the obtaining of nanostructure, which guarantees the improvement of properties of the final product. When the grain size of austenitic stainless steel is constant, the strength increases with increasing nitrogen content. The following empirical equation, with the conventional yield strength 0.2% and nitrogen content N, concerns Fe–Cr–N, Fe–Cr–Mn–N and Fe–Cr–Ni–N austenitic stainless steels:

$$\sigma_{0.2} = 0.14 + 4.8\ N^{2/3}\ [GPa] \qquad (6.1)$$

When the nitrogen content is constant, the Hall-Petch equation involves grain size d and σ_y yield strength with the following dependence:

$$\sigma_y = K\, d^{-1/2} \qquad (6.2)$$

where K is constant. K value increases with increasing nitrogen content (as shown in Fig. 6.5) [25].

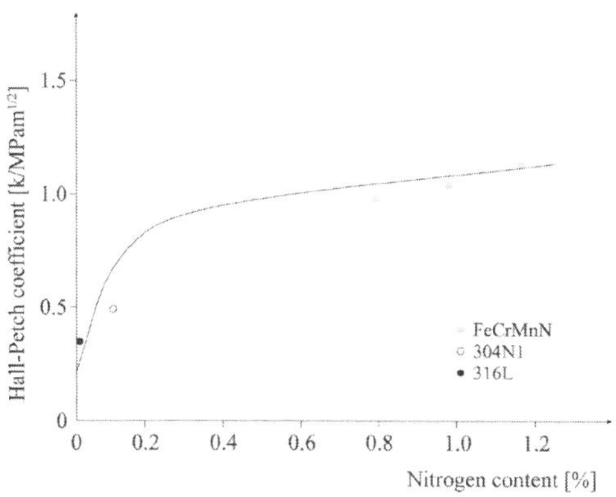

Figure 6.5 Dependence of Hall–Petch coefficient from nitrogen content [25].

6.2 Nanostructured Nickel-Free Stainless Steels

In recent years, nanostructured materials — also called nanomaterials — has become a subject of research in academic centers worldwide. Nanostructure can be obtained in many ways. Among other processes, nickel-free austenitic stainless steel with nanostructure can be obtained by mechanical alloying. Mechanical alloying (MA) was developed in the 1970s at the International Nickel Co. as a technique for dispersing nanosized inclusions into nickel-based alloys [3]. During the last years, the MA process has been successfully used to prepare a variety of alloy powders including powders exhibiting supersaturated solid solutions, quasicrystals, amorphous phases, and nano-intermetallic compounds. MA technique has been proven as a novel and promising method for alloy formation.

The effect of MA processing was studied by X-ray diffraction and microstructural investigations. Ni-free austenitic stainless steels with nanostructure were synthesized by mechanical alloying (MA), heat treatment, and nitriding of elemental microcrystalline Fe, Cr, Mn and Mo powders. After 48 h of MA, the alloy had decomposed into an amorphous phase and nanocrystalline α-Fe [29]. Heat treatment performed after MA process results in crystallization into ferritic phase. Then, compacted material was nitrided at 1210°C, which resulted in phase transformation from ferritic to fully austenitic (Fig. 6.6). Average crystallite size of so produced material increased during processing from 16 to 27 nm.

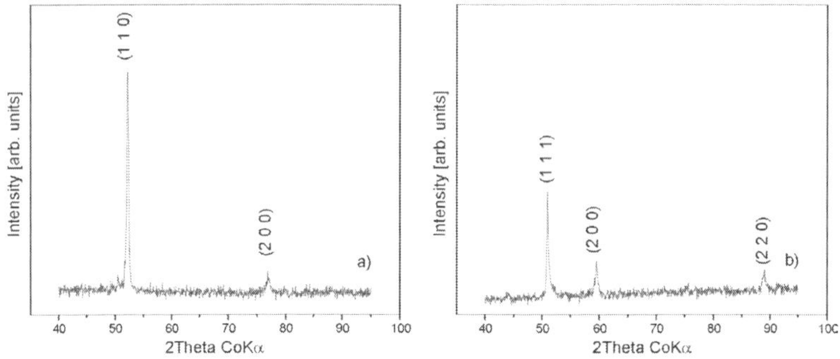

Figure 6.6 Phase transformation from ferritic (a) to austenitic (b) in $Fe_{74}Cr_{24}Mo_2N$ stainless steel [29].

Using confocal laser scanning microscope, it is possible to reveal the surface of nanocrystalline nickel-free austenitic stainless steels. The surface of $Fe_{54}Cr_{24}Mn_{21}Mo_1N$ sample shown in Fig. 6.7 reveals pores. These pores have irregular shapes and dimensions.

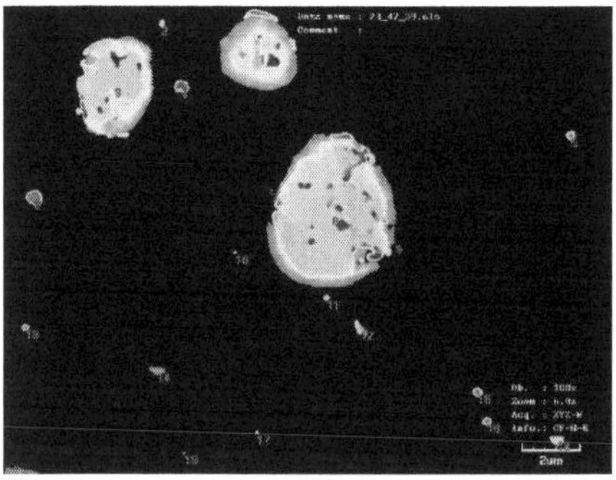

Figure 6.7 Surface of $Fe_{54}Cr_{24}Mn_{21}Mo_1N$ sample under confocal laser scanning microscope [30].

From an analysis of the surface profile it can be derived that pores penetrate to a depth of 152 to 460 nm. On the other hand, their percentage of the entire surface of the material is 9.5% and the largest pore diameter up to 4 μm.

Compared with the widely used medicine 316L stainless steel (248 HV0.2), microhardness of sintered nanocrystalline austenitic nickel-free nitrogen containing stainless steels obtained by mechanical alloying is significantly higher (378 to 520 HV0.2, see Fig. 6.8). The result is two times greater than in austenitic steel obtained by conventional methods. This effect is directly connected with structure refinement and obtaining of nanostructure as well as the introduction of nitrogen. Nitrogen dissolved in austenitic stainless steel increases its strength, which is caused by large amount of solution hardening. The grain size hardening in N-alloyed austenitic stainless steels is based on the grain size dependence of the yield strength described by the Hall–Petch equation. The effect of N content on grain boundary hardening increases proportionally as the N content of the steel increases. Grain boundary hardening therefore increases with increasing N content of the steel and is related to the strong affinity between Cr, Mo and N atoms.

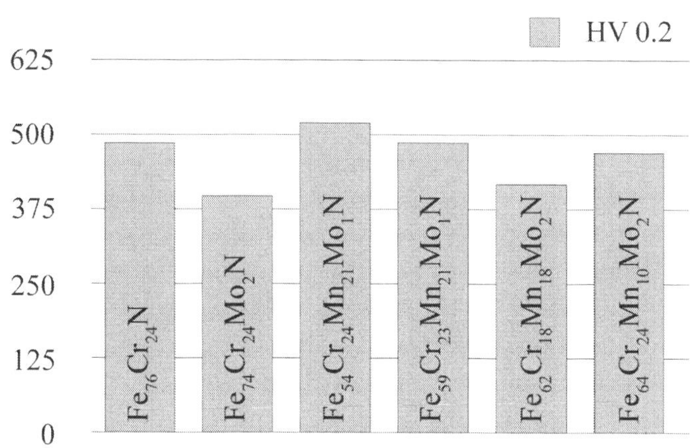

Figure 6.8 Microhardness of FeCrMnMoN materials [29].

As one can see from nanoindentation test (see Table 6.3), microhardness remains on the level of 500 HV in case of Ni-free stainless steels with nanostructure and 276 HV for 316L stainless steel.

Table 6.3 Results of microhardness and Young's modulus tests for selected samples of Ni-free austenitic stainless steels and 316L stainless steel

Sample	HV	H [MPa]	E [GPa]
316L	276	2977	172
$Fe_{74}Cr_{24}Mo_2N$	650	7014	201
$Fe_{54}Cr_{24}Mn_{21}Mo_1N$	542	5846	210
$Fe_{59}Cr_{23}Mn_{12}Mo_6N$	525	5663	213
$Fe_{64}Cr_{24}Mn_{10}Mo_2N$	469	5027	199

Young's modulus of obtained steels is about 210 GPa, which is slightly higher than in conventional stainless steel 316L (~170 GPa) and comparable with cobalt-chromium alloys (220–240 GPa) that makes this material applicable for production of e.g. stents.

As one may expect, materials with high value of microhardness (~500 HV) could show embrittlement. To exclude or endorse that unwanted phenomenon, stress intensity factor — K_{IC} — should be determined. It could be done by calculating K_{IC} from hardness, diameter of indentation and length of Palmquist's cracks measured directly in Vicker's method. Niihara's equation applies here [19].

Imprint of Vicker's indenter on polished surface of Ni-free stainless steel with nanostructure ($Fe_{59}Cr_{23}Mn_{12}Mo_6N$) is shown on the Fig 6.9. Load of 200 g, 500 g, 1 kg, 2 kg, 5 kg, 10 kg, and even 30 kg did not reveal any marks of Palmquist's cracks. It can be concluded, that in case of Ni-free stainless steels obtained by mechanical alloying high microhardness did not imply embrittlement of the samples.

Figure 6.9 Imprint of Vicker's indenter (30 kg load) for $Fe_{59}Cr_{23}Mn_{12}Mo_6N$ sample [29].

Corrosion tests of Ni-free stainless steels with nanostructure were performed in H_2SO_4 and Ringer's solution. In H_2SO_4 the corrosion current, I_{corr}, in the case of $Fe_{59}Cr_{23}Mn_{12}Mo_6N$ is 1.3×10^{-5} A/cm^2. The corresponding values of $Fe_{54}Cr_{24}Mn_{21}Mo_1N$ and $Fe_{64}Cr_{24}Mn_{10}Mo_2N$ are 1.6×10^{-4} and 5.1×10^{-4} A/cm^2, respectively. I_{corr} values indicate that addition of Mo and reduction of Mn resulted in appreciable decrease in the corrosion current density — two orders of magnitudes and corrosion rate. Measured values are considerably improved compared to widely used in medicine 316L stainless steel.

In Ringer's solution tests, the corrosion current, I_{corr}, in the case of $Fe_{59}Cr_{23}Mn_{12}Mo_6N$ is 2.6×10^{-5} A/cm^2. The corresponding values for $Fe_{54}Cr_{24}Mn_{21}Mo_1N$ and $Fe_{64}Cr_{24}Mn_{10}Mo_2N$ are 6.9×10^{-6}, and 4.9×10^{-6} A/cm^2, respectively. Calculated corrosion rates for $Fe_{54}Cr_{24}Mn_{21}Mo_1N$ and $Fe_{64}Cr_{24}Mn_{10}Mo_2N$ steels, 0.073, and 0.052 mm/year, respectively, show that corrosion rate is four times lower than in 316L stainless steel.

Table 6.4 Results of corrosion test in H_2SO_4 and Ringer's solution [28]

	0.1 M H_2SO_4		**Ringer's solution**	
Sample	**I_{corr} [A/cm^2]**	**Corrosion rate [mm/year]**	**I_{corr} [A/cm^2]**	**Corrosion rate [mm/year]**
316L	3.6×10^{-3}	1.0	9×10^{-5}	0.30
$Fe_{54}Cr_{24}Mn_{21}Mo_1N$	1.6×10^{-4}	1.2	6.9×10^{-6}	0.073
$Fe_{59}Cr_{23}Mn_{12}Mo_6N$	1.3×10^{-5}	0.18	2.6×10^{-5}	0.27
$Fe_{64}Cr_{24}Mn_{10}Mo_2N$	5.1×10^{-4}	2.1	4.9×10^{-6}	0.052

The comparison of results obtained for samples in H_2SO_4 and Ringer's solution (Table 6.4) show that synthesized stainless steels manage much better in chloride ions environment. Chemical composition of steels plays a bigger role in case of H_2SO_4, while in Ringer's solution, differences are much smaller.

These results indicate that FeCrMnMoN stainless steels have superior corrosion resistance in H_2SO_4 and Ringer's solution environments than the conventional metallic biomaterial, 316L. The addition of nitrogen to a stainless steel influences corrosion resistance of the steel. The nitrogen adsorption treatment of FeCrMnMoN results in the change of its microstructure from ferrite to austenite without forming CrN or Cr_2N that leads to higher Vickers microhardness. These facts suggest that austenitization by adsorbed nitrogen contribute to higher corrosion resistance, which may be a key factor in providing higher cytocompatibility of produced materials.

Both mechanical alloying and nitriding of Ni-free steels are very effective ways to improve the corrosion resistance and microhardness. Nitrogen absorption treatment contributes to the higher corrosion resistance and also in the presence of wear. With regard to austenitic stainless steels, it could help obtain better biomedical implants (e.g. stents) with better mechanical properties, corrosion resistance, and biocompatibility.

6.3 Nickel-Free Stainless Steels/ Hydroxyapatite Nanocomposites

Hydroxyapatite (HA) ceramics have been intensively studied for bone repair and replacement applications. That is because of their high biocompatibility and ability to bond to bone [2]. However, compared to the mechanical properties of natural bone, HA ceramics need improvement [13, 20]. Recently, various HA based composites have been investigated [23]. Many reinforcements have been proposed, e.g. zirconia [12] and ceramic fibers [22]. Results obtained for FeCr alloy reinforcement [5] and HA-316L stainless steel composites [15] are very promising. Thus, since nickel-free austenitic stainless steels seem to have better mechanical properties, corrosion resistance, and biocompatibility compared to 316L stainless steels [27], it is possible that composite made of this steel and HA could have improved properties.

Nickel-free austenitic stainless steel/hydroxyapatite composites can be produced by means of mechanical alloying and nitrogen absorption treatment, which leads to metallic-ceramic nanocomposite. Obtaining nanostructure via mechanical alloying process as well as high biocompatibility of hydroxyapatite and good mechanical properties of stainless steel seem to be excellent combination of complementary parameters that could results in new promising biomaterial.

The microhardness of the final bulk material was studied using Vickers method and the results are presented in Table 6.5. Compared with widely used medicine 316L microcrystalline stainless steel (250 ± 10 HV0.2), microhardness of sintered nanocrystalline austenitic nickel-free nitrogen containing stainless steel $Fe_{54}Cr_{24}Mn_{21}Mo_1N$ obtained by mechanical alloying is significantly higher (520 ± 10 HV0.2). The result is two times greater than in austenitic steel obtained by conventional methods. This effect is directly connected with structure refinement and obtaining of nanostructure as well as introduction of nitrogen. Nitrogen dissolved in austenitic stainless steel increases its strength, which is caused by large amount of solution hardening. The grain size hardening in N-alloyed austenitic stainless steels is based on the grain size dependence of the yield strength as described by the Hall–Petch equation. The effect of N content on grain boundary hardening increases proportionally as the N content of the steel increases. Grain boundary hardening increases therefore with increasing N content of the steel and is related to the strong affinity between Cr, Mo, and N atoms.

Table 6.5 Results of microhardness and corrosion tests in Ringer's solution [31]

Sample	HV0.2	I_{corr} [A/cm²]	E_{corr} [mV]	Corrosion rate [mm/year]	R_p [Ohm/cm²]
$Fe_{65}Cr_{18}Ni_{12}Mo_2Mn_2$ (316L)	250	9.0×10^{-5}	-220	0.300	450
$Fe_{54}Cr_{24}Mn_{21}Mo_1N$	520	6.9×10^{-6}	-263	0.073	3747
$Fe_{54}Cr_{24}Mn_{21}Mo_1N$ + 5% HA	510	2.8×10^{-6}	-414	0.030	14385
$Fe_{54}Cr_{24}Mn_{21}Mo_1N$ + 10% HA	580	9.2×10^{-7}	-322	0.009	28143
$Fe_{54}Cr_{24}Mn_{21}Mo_1$ + 5% HA	170	1.3×10^{-4}	-1021	1.420	22

Addition of 5% of HA slightly decreases microhardness (510 ± 10 HV0.2) while 10% of HA results in growth of microhardness to 580 ± 10 HV0.2. It is worth to mention that microhardness of the same material before nitrogen absorption treatment is three times lower (170 ± 10 HV0.2). It can be concluded that strengthening is due to nitrogen absorption process. Strengthening due to HA addition need more detailed studies.

Table 6.5 summarizes also results of corrosion tests in Ringer's solution performed at temperature 37°C. In Ringer's solution tests, the corrosion current, I_{corr}, in the case of starting composition: $Fe_{54}Cr_{24}Mn_{21}Mo_1N$ is 6.9 × 10^{-6} A/cm^2 and is associated with the corrosion potential of −263 mV. The comparison of the values E_{corr} and I_{corr} indicates that addition of HA resulted in the shift of the corrosion potential to a more negative value and appreciable decrease in the corrosion current density. When the addition of 5% of HA results in small decrease of the corrosion current to 2.8 × 10^{-6} A/cm^2, the corresponding value in case of 10% addition of HA is appreciable lower — one order of magnitude. Measured values are considerably improved compared to widely used in medicine 316L microcrystalline stainless steel. Comparison of the corrosion current of the same sample ($Fe_{54}Cr_{24}Mn_{21}Mo_1$ + 5% HA) before and after nitrogen absorption treatment emphasizes the influence of this process on the corrosion resistance — the difference is two orders of magnitude. The polarization resistance values, R_p, calculated from the potentiodynamic curve were 14385 and 28143 Ohm/cm^2 for nanocomposites with 5% and 10% of HA, respectively, which compared to 3747 Ohm/cm^2 for nickel-free austenitic stainless steel also indicates huge impact of HA on corrosion resistance of obtained nanocomposites. Calculated corrosion rates for $Fe_{54}Cr_{24}Mn_{21}Mo_1N$/HA composites show that corrosion rate is two ($Fe_{54}Cr_{24}Mn_{21}Mo_1N$ + 5% HA) to eight ($Fe_{54}Cr_{24}Mn_{21}Mo_1N$ + 10% HA) times lower than in alloy without HA content. Composite without nitrogen content ($Fe_{54}Cr_{24}Mn_{21}$ Mo_1 + 5% HA) even with the addition of 5% HA shows unacceptable corrosion rate of 1.420 mm/year, which confirms necessity of nitrogen absorption treatment. From further research point of view, $Fe_{54}Cr_{24}Mn_{21}Mo_1N$ + 10% HA composite have the most promising results of corrosion resistance (lowest corrosion rate combined with lowest corrosion current).

Both nitrogen absorption treatment and the presence of HA contribute to the higher corrosion resistance in Ringer's solution. While pure nickel-free stainless steel with nanostructure has significantly better corrosion properties and microhardness compared to conventional 316L stainless steel, introduction of hydroxyapatite and obtaining a nanocomposite enhances these parameters even further. The improved properties combined with excellent biocompatibility of HA could help to obtain better biomedical implants with better mechanical properties, corrosion resistance and biocompatibility.

Obtained nickel-free austenitic stainless steels and their nanocomposites with hydroxyapatite have superior cytocompatibility to the conventional metallic biomaterial, 316L stainless steel. Additionally, the nitrogen adsorption treatment increases corrosion resistance with the existence of wear. The addition of nitrogen to a stainless steel influences the corrosion resistance of the steel.

Nanocrystalline nickel-free stainless steels can be used in variety of applications where corrosion resistance and biocompatibility have to be significantly higher than in case of conventional microcrystalline materials. As shown in Fig. 6.10, using powder metallurgy process (mechanical alloying), heat treatment, and nitriding, bulk homogenous samples can be prepared. Thus biomedical implants with better mechanical properties, corrosion resistance, and biocompatibility could be obtained.

Figure 6.10 Bulk nanocrystalline nickel-free stainless steels samples.

References

1. Alvarez, K., Hyun, S., Tsuchiya, H., Fujimoto, S., and Nakajima, H. (2008). Corrosion behaviour of lotus-type porous high nitrogen nickel-free stainless steels, *Corr. Sci.*, **50**, pp. 183–193.

2. Bagambisa, F.B., Ulrich, J., and Schilli, W. (1993). Mechanisms and structure of the bond between bone and hydroxyapatite ceramics, *J. Biomed. Mater. Res.*, **27**, pp. 1047–1055.
3. Benjamin, J.S. (1976). Mechanical alloying, *Sci. Am.*, **234**, pp. 40–57.
4. Briant, C.L., Mulford, R.A., and Hall, E.L. (1982). Sensitization of austenitic stainless steels, I. Controlled purity alloys, *Corrosion*, **38**, pp. 468–477.
5. De With, G., and Corbijn, A.J. (1998). Metal fibre reinforced hydroxyapatite ceramics, *J. Mater. Sci.*, **24**, pp. 3411–3415.
6. Eckenrod, J.J., and Kovach, C.W. (1979). Effect of nitrogen on the sensitization, corrosion and mechanical properties of 18Cr–8Ni stainless steels, *ASTM STP*, **679**, pp. 17–41.
7. Fini, M., Nicoli Aldini, N., Torricelli, P., Giavaresi, G., Bosari, V., Lenger, H., Bernauer, J., Giardino, R., Chiesa, R., and Cigada, A., (2003). A new austenitic stainless steel with negligible nickel content: an *in vitro* and *in vivo* comparative investigation, *Biomaterials*, **24**, pp. 4929–4939.
8. Foct, J., and Hendry, A. (1989). High nitrogen steels, *Proceedings of the Conference on High Nitrogen Steels, Lille, France, The Institute of Metals, London*, 461 p.
9. Gebeau, R.C., and Brown, R.S. (2001). Tech spotlight: Biomedical implant alloy, *Adv. Mater. Proc.*, **159**, pp. 46–48.
10. Kikuchi, M., and Mishima, Y. (1996). Nitrogen content of 316L weld metal and its FINE Particle by means of high-pressure MIG arc welding, *ISIJ International*, **36**, pp. 927–931.
11. Kuroda, D., Hanawa, T., Hibaru, T., Kuroda, S., Kobayashi, M., and Kobayashi, T. (2003). New manufacturing process of nickel-free austenitic stainless steel with nitrogen absorption treatment, *Mater. Trans.*, **44**, pp. 414–420.
12. Li, J., Liao, H., and Hermansson, L. (1996). Sintering of partially stabilized zirconia and partially stabilized zirconia-hydroxyapatite composites by hot isostatic pressing and pressureless sintering, *Biomaterials*, **17**, pp. 1787–1790.
13. Lopes, M.A., Monteiro, F.J., and Santos, J.D. (1999). Glass reinforced hydroxyapatite composites: fracture toughness and hardness dependance on microstructural characteristics, *Biomaterials*, **20**, pp. 2085–2090.
14. Menzel, J., Kirschner, W., and Stein, G. (1996). High nitrogen containing Ni-free austenitic steels for medical applications, *ISIJ International*, **36**, pp. 893–900.

15. Miao, X. (2003). Observation of microcracks formed in HA-316L composites, *Mater. Let.*, **57**, pp. 1848–1853.
16. Mozhi, T.A., Betrabet, H.S., Jagannathan, V., Wilde, B.E., and Clark, W.A.T. (1986). Thermodynamic modeling of sensitization of AISI 304 stainless steels containing nitrogen, *Scripta Metall.*, **20**, pp. 723–728.
17. Mozhi, T.A., Clark, W.A.T., Nishimoto, K., Johnson, W.B., and MacDonald, D.D. (1985). The effect of nitrogen on the sensitization of AISI 304 stainless steel. *Corrosion*, **35**, pp. 555–559.
18. Nakayama, Y., Yamamura, T., Kotoura, Y., and Oka, M. (1989). *In vivo* measurement of orthopedic implant alloys: comparative study of *in vivo* and *in vitro* experiments, *Biomaterials*, **10**, pp. 420–424.
19. Niihara, K., Morena, R., and Hasselmann, D.P.H. (1982). Evaluation of K_{IC} of brittle solids by the indentation method with low crack-to-indent ratios, *J. Mater. Sci. Lett.*, **1**, pp. 13–16.
20. Ramachandra Rao, R., Roopa, H.N., and Kannan, T.S. (1997). Solid state synthesis and thermal stability of HAP and HAP — β-TCP composite ceramic powders, *J. Mater. Sci.: Mater. Med.*, **8**, pp. 511–518.
21. Ren, Y., Yang, K., and Zhang, B. (2005). *In vitro* study of platelet adhesion on medical nickel-free stainless steel surface, *Mater. Lett.*, **59**, pp. 1785–1789.
22. Slosarczyk, A., Klisch, M., Blazewicz, M., Piekarczyk, J., Stobierski, L., and Rapacz-Kmita, A. (2000). Hot pressed hydroxyapatite-carbon fibre composites, *J. Eur. Ceram. Soc.*, **20**, pp. 1397–1402.
23. Suchanek, W., and Yoshimura, M. (1998). Processing and properties of hydroxyapatite-based biomaterials for use as hard tissue replacement implants, *J. Mater. Res.*, **13**, pp. 94–117.
24. Sumita, M., Hanawa, T., and Teoh, S.H. (2004). Development of nitrogen-containing nickel-free austenitic stainless steels for metallic biomaterials — review, *Mater. Sci. Eng. C*, **24**, pp. 753–760.
25. Takaki, S. (2010). Effect of carbon and nitrogen on the Hall–Petch coefficient of ferritic iron (review on the Hall–Petch relation in ferritic iron), *Mater. Sci. Forum*, **638–642**, pp. 168–173.
26. Thomann, I., and Uggowitzer, J. (2000). Wear-corrosion behavior of biocompatible austenitic stainless steels, *Wear*, **239**, pp. 48–58.
27. Tulinski, M., Jurczyk, K., and Jurczyk, M. (2008). Nanoscale nickel-free austenitic stainless steel, *Solid State Phenom.*, **140**, pp. 179–184.
28. Tulinski, M., Jurczyk, K., and Jurczyk, M. (2008). Nickel-free nanocrystalline austenitic stainless steels, *Mater. Sci. Poland*, **26**, pp. 381–388.

29. Tulinski, M., and Jurczyk, M. (2008). Mechanical and corrosion properties of Ni-free austenitic stainless steels, *Arch. Met. Mater.*, **53**, pp. 955–959.

30. Tulinski, M., and Jurczyk, M. (2010). Mechanical properties and corrosion resistance of nickel-free nanocrystalline austenitic steel, *Adv. Man. Sci. Tech.*, **34**, pp. 73–79.

31. Tulinski, M., and Jurczyk, M. (2010). Nickel-free austenitic stainless steels and their nanocomposites with hydroxyapatite, *Phys. Stat. Sol. C*, **7**, pp. 1359–1362.

32. Uggowitzer, P.J., Magdowski, R., and Speidel, M. (1996). Nickel-Free High Nitrogen Austenitic Steels, *ISIJ Int.*, **36**, pp. 901–908.

Chapter 7

Ti-Based Ceramic Nanocomposites

Katarzyna Niespodziana* **and Mieczyslaw Jurczyk**
Institute of Materials Science and Engineering, Poznan University of Technology, Sklodowska-Curie 5 Sq., 60-965 Poznan, Poland
*katarzyna.niespodziana@put.poznan.pl

7.1 Introduction

There is a high demand for biomaterials to assist the replacement of organs and their functions. Interaction between biomaterials and natural tissues is a significant subject for biomedical science and is essential to aid the design and fabrication of new biocompatible and bioactive materials. For this reason, researchers search for new biomaterials with advanced mechanical and biological properties and develop new technologies for the enhancement of those properties. Over the past years, nanoscale biomaterials have become very popular in medical applications [30, 31].

Titanium and titanium alloys are preferred materials in the production of implants. These materials possess favorable properties, such as relatively low modulus, low density, and high strength.

Bionanomaterials for Dental Applications
Edited by Mieczyslaw Jurczyk

ISBN 978-981-4303-83-5 (Hardcover), 978-981-4303-84-2 (eBook)
www.panstanford.com

Titanium materials are resistant to corrosion because of the stable passivity of the surface oxide film. Apart from that, titanium and titanium alloys are generally regarded to have good biocompatibility, although there are reports that show the accumulation of titanium in tissues adjacent to the implant, signifying metal release and corrosion *in vivo* [2,17]. In addition, these metal implants may loosen and even separate from the surrounding tissues during implantation [16, 18, 29]. Titanium and titanium-based alloys have relatively poor tribological properties because of their low hardness [17].

One of the methods that allow the change of biological properties of Ti alloys is to produce a composite, which will exhibit the favorable mechanical properties of titanium, excellent biocompatibility and bioactivity of ceramic. The main ceramics, used in medicine are hydroxyapatite, silica, or bioglass [3]. Hydroxyapatite (HA, $Ca_{10}(PO_4)_6(OH)_2$) shows good biocompatibility because of its chemical and crystallographic structure being similar to that of living bone. HA has porous nature and is bioactive, which means that after some time it is partially resorbed and replaced by natural bone [32]. Besides, HA has the ability to form strong chemical bonds with natural bone. Unfortunately, the HA cannot be used for load bearing applications, due to its poor mechanical properties with regard to natural bone. The ceramic coating on the titanium, improves the surface bioactivity, but often flakes off as a result of poor ceramic/metal interface bonding, which may cause the surgery to fail [1, 33]. For this reason, the nanocomposite materials containing titanium and ceramic as a reinforced phase are promising alternatives to conventional materials, because they can potentially be designed to match the properties of bone tissue in order to enhance patients' quantity of life [4–7, 27].

Earlier, microcrystalline Ti-20 vol% HA composite with a relative density of 97.86% was fabricated by a hot pressing technique [27]. The phase constitution of Ti-20 vol% HA composite is similar to that of HA-based composite with Ti and HA as the predominant phases. Elastic modulus and Vicker's hardness of Ti-20 vol% HA composite are 102.6 and 3.41 GPa, respectively. Additionally, the osteointegration ability of the composite is better than that of pure titanium, especially in the early stage after the implantation, which may be due to the presence of HA ceramic in the Ti-matrix composite [27].

7.2 Ti-HA Nanocomposites

Hydroxyapatite is considered as the most promising biomaterial for clinical use. However, it is well known that its poor mechanical properties compared to those of living bone are one of the most serious obstacles for wider applications, especially for load bearing implants. Intensive research is ongoing to produce HA composites with improved mechanical properties [4–7, 27]. Thus, it is necessary to introduce some matrix materials — in our case, titanium. Current research on the synthesis of nanoscale metallic and composite biomaterials, shows that Ti-bioceramic nanocomposites posses better mechanical and corrosion properties than microcrystalline titanium [8–15, 19–24].

7.2.1 Microstructure and Phase Constitution

Titanium-hydroxyapatite nanocomposites can be produced by mechanical alloying method [19]. The typical XRD patterns of titanium (ICDD: 5-682) and hydroxyapatite (ICDD: 9-432) before mechanical alloying are shown in Figs. 7.1a,b, respectively. During MA process, the original sharp diffraction lines of the Ti and HA gradually become broader and their intensity decreases with milling time (not shown). The peak broadening represents a reduction in the crystallite size and increase in the internal strain in the mechanically alloyed materials. After 44 h of MA, the amorphous phase forms directly from the starting mixture, without the formation of other phases (Fig. 7.1c). But differentiation between a "truly" amorphous, extremely fine grained material and a material in which very small crystals are embedded in an amorphous matrix in produced materials has not been easy on the basis of diffraction basis.

During the mechanical alloying process, the powder Ti and HA particles are periodically trapped between colliding balls and are plastically deformed. Such a feature occurs by the generation of a wide number of dislocations as well as other lattice defects. Furthermore, the ball collisions cause fracturing and cold welding of the elementary particles, forming clean interfaces at the atomic scale. Further milling leads to an increase of the interface number and the sizes of the elementary component area decrease from millimeter to submicrometer lengths. Concurrent to this decrease of the elementary distribution, some nanocrystalline intermediate phases are produced inside the particles or on their surfaces. As the milling duration increases, the content fraction of such intermediate compounds increases, leading to a final product whose properties are the function of the milling conditions.

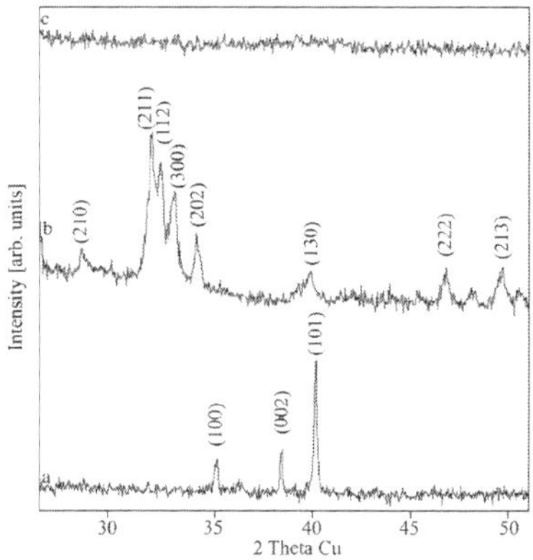

Figure 7.1 XRD spectra of Ti and HA (10 vol%) powders mechanically alloyed: (a) Ti — 0 h, (b) HA — 0 h, (c) Ti-HA — 44 h MA [23].

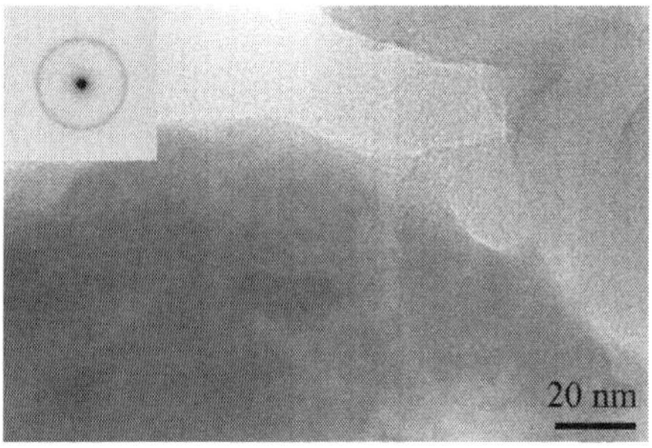

Figure 7.2 TEM micrographs and electron diffraction patterns (insets) of the milled Ti-10 vol% HA sample for 44 h — typical amorphous fragment [23].

The microstructure of milled titanium and hydroxyapatite powder was also studied by TEM. The sample milled for 44 h was mostly amorphous (broad rings in the SEAD pattern) as appears from high-resolution image (Fig. 7.2). Apart from prevailing amorphous phase, the milled powders contained a small amount of fine-crystalline and crystalline phases. Lack of any sharp reflections in the XRD pattern (Fig. 7.1c) suggests that the amount of the crystalline phase is very low and/or it forms during TEM observation. During TEM studies, it has been found that the amorphous powders was unstable upon exposure to electron beam and underwent some crystallization.

The SEM morphologies of the fabricated Ti-10 vol% HA powder mixtures, at the various stages during milling, are shown in Fig. 7.3. It can be seen that after 10 h of milling, the powder shows inhomogeneous size distribution (Fig. 7.3d). With the increase of milling time, the size of the mixed powders decreases gradually and the microstructure is more homogenous (Fig. 7.3e,f).

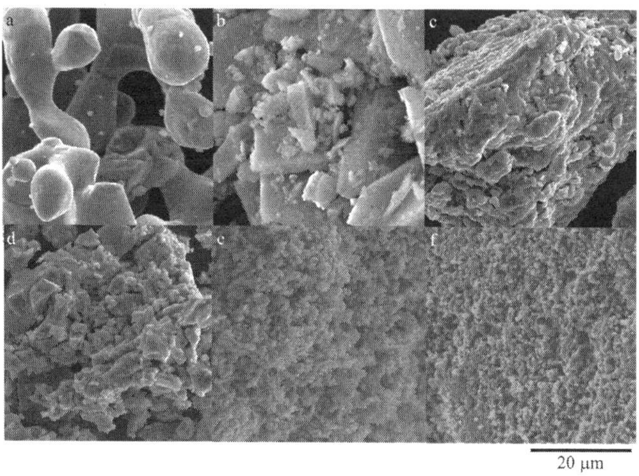

Figure 7.3 SEM images show morphologies of powder mixtures of Ti-10 vol% HA composite at the various stages during milling: (a) Ti — 0 h, (b) HA — 0 h (c) Ti-HA — 3 h, (d) Ti-HA — 10 h, (e) Ti-HA — 20 h, (f) Ti-HA — 44 h [23].

The formation of the bulk nanocomposites was achieved by cold uniaxial pressing and sintering of the amorphous materials. In all cases, XRD analysis of Ti-HA nanocomposites showed the presence of α-Ti (hexagonal-type structure), but hydroxyapatite no longer exists (Fig. 7.4). When 10 or 20 vol% of hydroxyapatite is added to titanium, the lattice constants of Ti increases, as manifested by a shift of the diffraction peaks of the (1 0 0), (0 0 2), and (1 0 1) crystal planes of titanium towards smaller angles in comparison with pure microcrystalline titanium. Calcium has higher atomic diameter in comparison to titanium, so solid solution of Ca in Ti was forming. It is important to note, that mechanical alloying technique allows alloying of elements that are difficult or impossible to combine by conventional melting methods. In general, the process can be viewed as a means of assembling alloys constituents with a controlled microstructure. At 1300°C, calcium is soluble in titanium to the extent of at least 0.13% [23]. According to the Scherrer method of XRD profiles, the mean crystallite size of heat treated Ti-HA nanocomposites was 5–10 nm (Table 7.1).

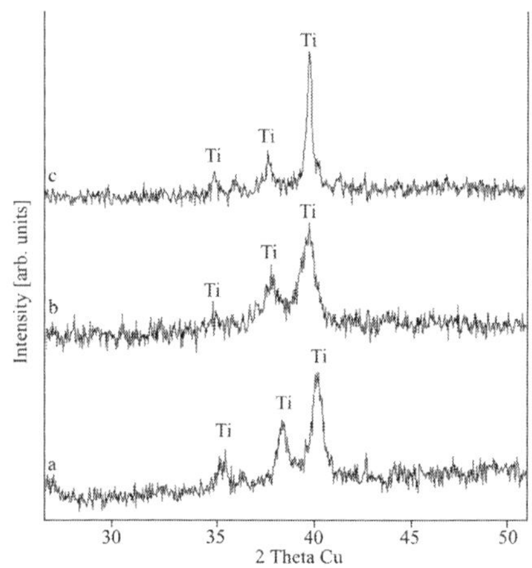

Figure 7.4 XRD spectra of Ti and HA powders mechanically alloyed for 44 h and heat treated at 1150°C for 2 h under an argon atmosphere: (a) Ti-3 vol% HA, (b) Ti-10 vol% HA and (c) Ti-20 vol% HA [23].

Table 7.1 Structural parameters and grain sizes for Ti-HA nanocomposites mechanically alloyed for 44 h and heat treated at 1150°C for 2 h in comparison with microcrystalline titanium and hydroxyapatite

Sample	*a* [Å]	*c* [Å]	*V* [Å^3]	*d*
Ti-3 vol% HA	2.95	4.68	35.2	5 nm
Ti-10 vol% HA	2.98	4.76	36.6	6 nm
Ti-20 vol% HA	2.97	4.78	36.5	10 nm
Ti (microcrystalline)	2.95	4.68	35.3	45 μm
HA (microcrystalline)	9.42	6.87	527.9	10 μm

The effect of the initial composition on the phase composition of microcrystalline Ti-HA composites was studied previously by

Ning and Zhou [25, 26]. Neither HA and Ti is found in the 30Ti-70HA composite, which has the lowest titanium content. In addition to $CaTiO_3$ (ICDD: 22-0153) and CaO (ICDD: 4-777), some Ti_xP_y compounds were also detected. With the increase in titanium content, Ti_2O (ICDD: 11-218) and α-Ti appeared in addition to the above phases. In the 70Ti-30HA composite, the main crystal phase was only α-Ti. This result is in good agreement with our current studies (see Fig. 7.4). The HA phase is unstable at high temperatures, which is correlative with the features of its crystal structure [26]. Additionally, the decomposition of HA phase plays an important role on the final properties of the sintered Ti-HA nanocomposites [25].

7.2.2 Mechanical Properties

The change of processing parameters, like mechanical alloying and annealing times, as well as the chemical compositions of Ti-HA nanocomposites leads to a distribution of the properties. For example, the Vickers hardness of bulk samples prepared by application of mechanically alloyed for 10 h Ti-10 vol% HA powder is low. On the other hand, the increase of milling time to 44 h leads to a strong increase in hardness. For Ti-10 vol% HA and Ti-20 vol% HA nanocomposites Vickers hardness are higher than of pure microcrystalline Ti metal (250 HV) and reaches 1300 HV and 1050 HV, respectively. Additionally, the Vickers hardness of the samples mechanically alloyed for 44 h and sintered at 1150°C depends on annealing time. Hardness increases additionally with increasing of the annealing time from 1300 HV to 1500 HV for 0.5 h and 2 h of heat treatment time, respectively (for a sample containing 10 vol% HA) [23]. This effect is directly connected with structure refinement and obtaining of nanostructure.

The research conducted by Popa *et al.* showed that application of higher pressure causes more swelling of samples [28]. Swelling is due to the diffusion of calcium/phosphorus in the titanium matrix. The phenomenon was proven by subsequent EDX analysis. The higher loading of the grains in the metal matrix during the compaction stage, the diffusion coefficient of elements from HA to the Ti increases. This is caused by the more numerous lattice defects induced by a more severe plastic deformation.

If the material is hot pressed, the plastic deformation and flow of Ti matrix and HA ceramic under the pressure can reduce the cracks deriving from the decomposition reaction of HA phases and the difference of sintering shrinkages between Ti and HA, which can promote the sintering densification and improve the mechanical properties of Ti-HA composites [7]. Table 7.2 compares the mechanical properties of Ti-HA nanocomposites and microcomposites with different HA contents, produced by various methods.

Table 7.2 Mechanical properties of Ti-HA nano- and microcomposites

Materials	ρ_{rel} (%)	P (%)	HV (GPa)	E (GPa)	σ_{bs} (MPa)	K_{IC} (MPa·m$^{1/2}$)	Ref.
Ti-3% HA nano	78.5	22	4.7	—	—	—	[23]
Ti-10% HA nano	91.6	8	14.7	127	—	1.64	[23]
Ti-20% HA nano	91.3	9	10.9	110	—	—	[23]
Ti-50% HA nano	92.8	7	8.8	124	—	—	[22]
Ti-20% HA micro	97.9	2.1	3.4	103	170	3.57	[7]
Ti-60% HA micro	93.3	6.7	2.9	80	92	2.69	[6]
Ti-80% HA micro	90.2	9.8	3.1	76	79	0.99	[6]

Note: ρ_{rel}, relative density; P, porosity; HV, Vickers hardness; E, Young's modulus; σ_{bs}, bending strength; K_{IC}, fracture toughness.

It could be found that the relative density of Ti-10 vol% HA is 91.6%, while the one of pure Ti can reach 98.6%. The Vickers microhardness of the Ti-HA nanocomposites exhibits various distributions corresponding to constitutional change and increase with the rise of HA contents (Table 7.2). Reinforced by HA particles, the Vickers hardness of Ti-10 vol% HA composite is greatly higher than that of Ti-3 vol% HA composite and about six times that of pure Ti. It should be pointed out that Young's modulus of Ti-HA nanocomposities can be significantly reduced by introduction a porous structure. Parameter E obtained for Ti-HA nanocomposites decreases from 153 GPa (microcrystalline Ti) to 110 GPa (Ti-20 vol% HA nanocomposite). A certain degree of porosity is essential for this type of materials because, it has been documented that, the porous structure provides better cell attachment, differentiation and ingrowth osteoblasts and vascularization [30–32].

Table 7.2 clearly shows that the Ti-HA nanocomposites possess much higher hardness compared to Ti-HA microcomposites. The change of the chemical compositions in Ti-HA composites leads to a continuous distribution of the properties (Fig. 7.5) [23]. The value of such properties ranks among the values obtained for pure titanium and hydroxyapatite.

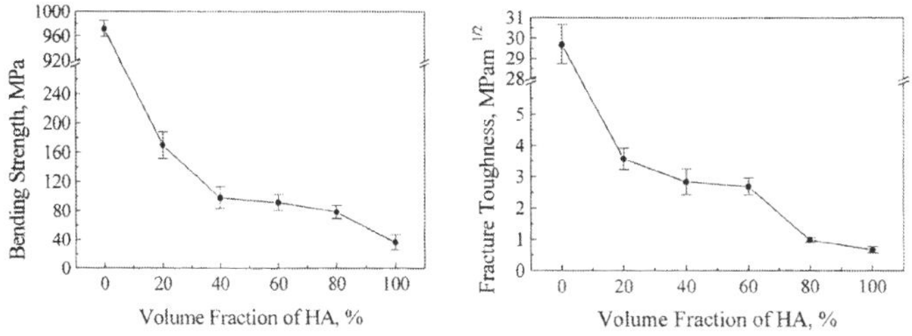

Figure 7.5 Relationship between bending strength (left) or fracture toughness (right) and volume fraction of HA in HA-Ti microcrystalline material [4].

For both nanocomposite and microcomposite materials, Young's modulus changed corresponding to the relative density. Both bending strength and fracture toughness are sensitive to the microstructure of the materials. They are associated with the configuration and distribution of the constitutional phases and the pores in the materials. The pores reduce the effective area bearing the load and can result in the stress concentration in the composites. Consequently, bending strength and fracture toughness of the composites decrease sharply according to the approximately exponential relation with the increase of the porosity.

The bending strength and fracture toughness of pure HA (37 MPa and 0.66 MPa·m$^{1/2}$) are far lower than those of pure Ti (972 MPa and 30 MPa·m$^{1/2}$). For this reason, the effect of HA phase on the bending strength and fracture toughness of Ti-HA composites is similar to the one of pores in Ti-matrix of the composite. Thus the sharp decrease of bending strength and fracture toughness of Ti-20 vol% HA composite to only about 170 MPa and 3.57 MPa·m$^{1/2}$, respectively, and can be contributed to the existence of 20% HA phase and the porosity of 2.1%. It should be mentioned that bending strength and fracture toughness of compact human bond for load-bearing applications can reach about 130 MPa and 2 MPa·m$^{1/2}$, respectively. Thus, Ti-20 vol% HA composite, with higher bending strength and fracture toughness than the compact human bone, is suitable for load-bearing applications from the point of view of mechanical properties [7].

Both nano- and microcomposites have a lower Young's modulus as compared to titanium. From the perspective of future application of this type of composites as biomaterials for heavy load-bearing implants, the decrease of modulus of elasticity is very favorable. Materials for such implants should have a modulus of elasticity as close as possible to the modulus of elasticity of bone.

Fracture behaviors vary with the rise of HA content (Fig. 7.6) [4]. In the HA rich composite, HA matrix presents intergranular fracture without macroscopic plastic deformation. Most Ti particles protruding from the fracture surface are intact and HA matrix has numerous cracks and large pore left behind due to the pull-out of titanium particles. This confirms that the HA matrix is very weak and the pull-out of titanium particles is result of crack propagation along the interface, which indicates the interface bonding appears to be weak. In contrast, in the Ti-rich composite, the fracture behavior is controlled by Ti matrix and many pores among the Ti particles can also be observed.

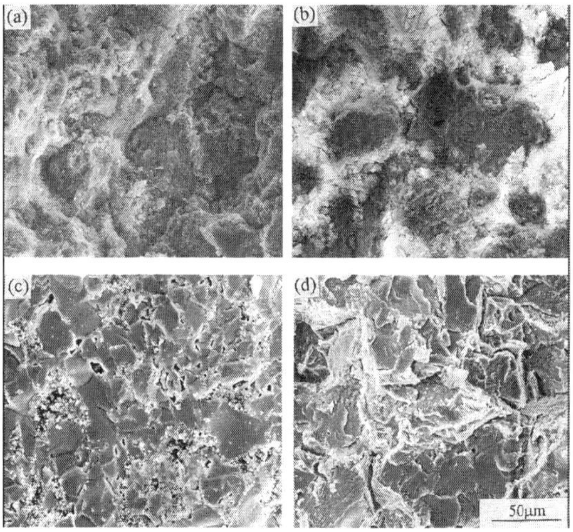

Figure 7.6 SEM fractograp of different composition regions in HA-Ti microcrystalline material with: (a) 20, (b) 40, (c) 80, (d) 100 vol% Ti [4].

7.2.3 Corrosion Properties

One of the main factors for determining the suitability of a biomaterial is corrosion resistance, which is very closely linked to biocompatibility. Corrosion of implants may lead to pathological changes in the body, caused by changing the chemical composition of tissues. Initially, the corrosion resistance is determined mainly in comparative studies, which leads to the accepted methods of research in the laboratory, and approximately simulating real biological environment. These methods allow the determination of such values as corrosion potential, corrosion current density, and polarization.

The polarization data obtained for Ti-HA nanocomposites and microcrystalline titanium, including corrosion potentials (E_C), corrosion current densities (i_C), and corrosion rate (C_R) values are summarized in Table 7.3. From this table it is possible to observe that hydroxyapatite doped to titanium had a positive effect on corrosion resistance of Ti. The corrosion test results indicate that the microcrystalline titanium possesses lower corrosion resistance and thus higher corrosion current density in Ringer's solutions. The result indicated that there was no significant difference in corrosion resistance among Ti-3 vol% HA ($i_C = 9.06 \times 10^{-8}$ A/cm^2, $E_C = -0.34$ V vs. SCE) and Ti-20 vol% HA ($i_C = 8.5 \times 10^{-8}$, A/cm^2, $E_C = -0.55$ V vs. SCE) although there was a significant difference in porosity. The polarization curves (Fig. 7.7) of these nanocomposites had a wider passive region, which can provide an excellent corrosion resistance, even if the applied potential rises to 850 mV vs. SCE for Ti-20 vol% HA and to 2500 mV vs. SCE or higher for Ti-HA (3, 10 vol%) composites. For Ti-20 vol% HA, the current density in the passive range is 2 and 1 orders of magnitude smaller than for Ti-10 wt% HA and Ti-3 vol% HA, respectively. Besides, the hydroxyapatite reinforced with titanium decreased about hundredfold the corrosion rate of Ti.

Corrosion tests show that the modification of microstructure of titanium by hydroxyapatite significantly improves the corrosion resistance. The rate of corrosion is affected, inter alia, by the density of the material. Although the obtained nanocomposites possess certain porosity, corrosion rate decreases.

Table 7.3 Mean values of corrosion current densities, corrosion potentials and corrosion rate of Ti-HA nanocomposites and microcrystalline titanium (grade 4) in Ringer's solutions

Materials	i_C (A/cm^2)	E_C vs SCE (V)	C_R (mm/year)
Ti-3 vol% HA	9.06×10^{-8}	−0.34	0.003
Ti-10 vol% HA	1.19×10^{-7}	−0.41	0.004
Ti-20 vol% HA	8.5×10^{-8}	−0.55	0.003
Ti (microcrystalline)	1.31×10^{-5}	−0.36	0.363

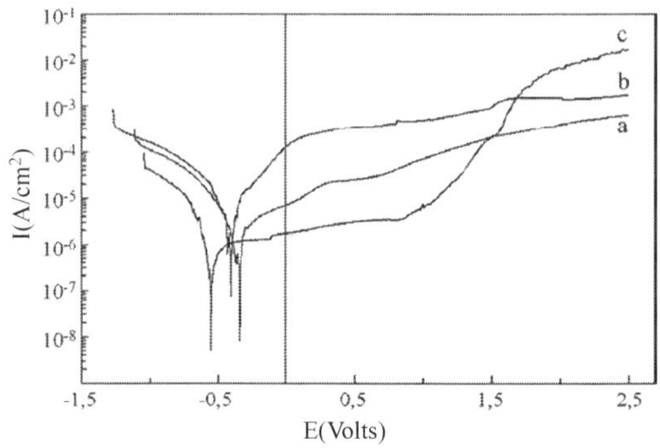

Figure 7.7 Potentiodynamic polarization curves of: (a) Ti-3 vol% HA, (b) Ti-10 vol% HA, (c) Ti-20 vol% HA nanocomposites in Ringer's solution [23].

7.3 Ti-SiO_2 Nanocomposites

Silica is chemically inert. It has good biocompatibility in the environment of the living organism and fairly good strength properties, particularly compressive strength. However, the toughness of silica is low. The manufacturing of silica-titanium composites can significantly improve its properties. Such nanocomposites can be produced by mechanical alloying and powder metallurgical method [21].

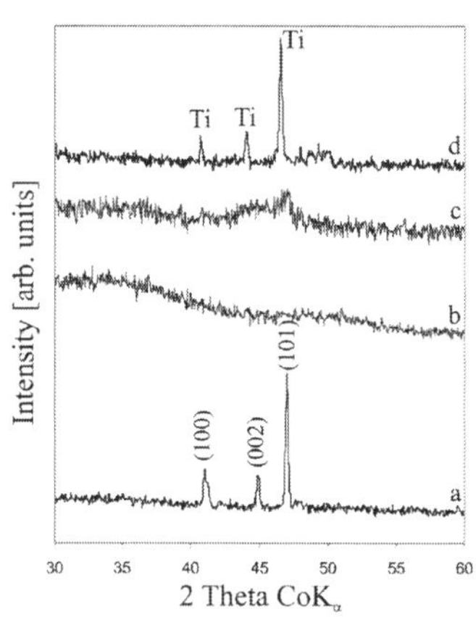

Figure 7.8 XRD spectra of Ti and SiO_2 (10 vol%) powders mechanically alloyed for different times: (a) Ti — 0 h, (b) SiO_2 — 0 h, (c) 20 h, (d) after annealing at 1150°C for 2 h [21].

X-ray diffraction was employed to study the effect of mechanical alloying on Ti-SiO_2 composites. Fig. 7.8 a, b shows the XRD patters of the starting titanium and amorphous silica powders. During MA process, the intensity of diffraction line of titanium decreases with milling time and after 20 h of milling has transformed completely to an amorphous phase, without the formation of any other phases (Fig. 7.8c).

TEM results shows that the powder milled for 20 h was mostly amorphous (Fig. 7.9a). SEAD pattern contains broad rings at position expected for Ti with hexagonal structure. Fig. 7.9b shows the high-resolution image of grain containing nanoparticles of titanium with visible lattice planes. The interplanar distance was 0.234 and 0.224 nm, which corresponds to the crystallographic planes of titanium. Apart from grains with nano Ti particles, the milled powders contained small amount of crystals of Ti. The same structure was observed for Ti-3 vol% SiO_2 composite. Lack of any sharp reflections in the XRD pattern suggests that the amount of the crystalline phase is very low and/or it forms during TEM observation. During TEM studies, it has been found that the amorphous powders was unstable upon exposure to electron beam and underwent some crystallization.

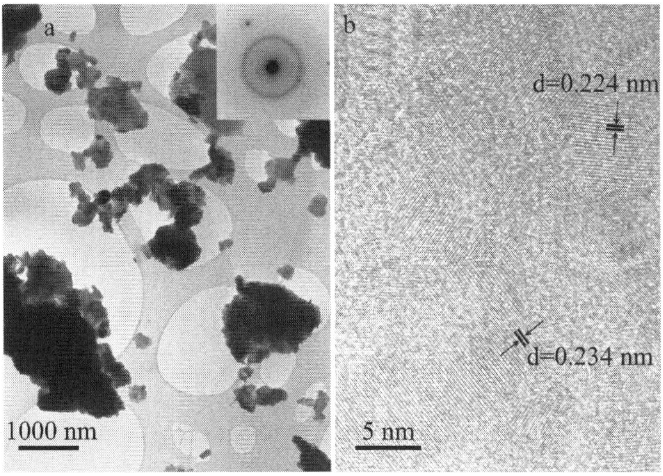

Figure 7.9 TEM micrographs of the milled Ti-10 vol% SiO_2 sample for 20 h: (a) typical amorphous parts with corresponding SAED pattern, (b) grain with nanoparticles of titanium [24].

The formation of the bulk nanocomposites were achieved by annealing of the amorphous materials in high purity gas atmosphere composed of 95% Ar and 5% H_2 at 1150°C for 2 h (Fig. 7.8d). XRD analysis of Ti-10 vol% SiO_2 showed the presence of α-Ti type structure with cell parameters a = 2.972 Å, c = 4.774 Å. The formation of crystalline SiO_2 phase was not observed. According to the Scherrer method of XRD profiles, the average size of heat treated Ti-SiO_2 nanocomposites is about 40–50 nm.

EDS results indicate that the predominant phase in Ti-SiO_2 composites is titanium with content of silica or silicon particles. The presence of some amount of iron atoms in the sintered nanocomposites could be explained by Fe impurities trapped in the MA powders from erosion of the milling media.

The Vickers microhardness of the sintered nanocomposites exhibit various distribution corresponding to constitutional change and increased with the rise of ceramic contents. The Vickers hardness for Ti-3 vol% SiO_2, and Ti-10 vol% SiO_2 nanocomposites reaches 550 HV0.2 and 670 HV0.2, respectively, and are two times higher than that of pure microcrystalline Ti metal (225 HV0.2); see Fig. 7.10.

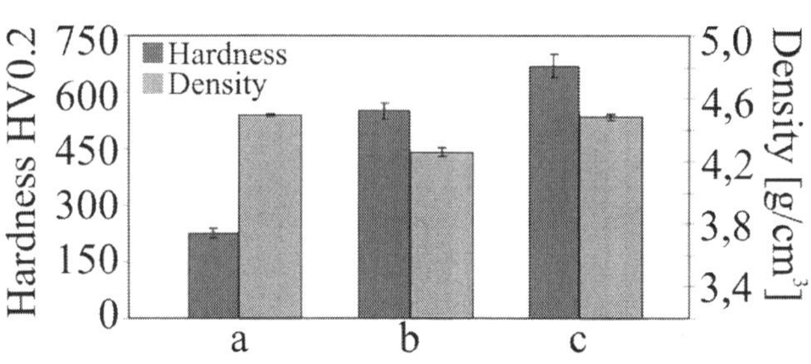

Figure 7.10 Vickers microhardness of studied nanocomposites; (a) Ti, (b) Ti-3 vol% SiO_2, (c) Ti-10 vol% SiO_2 [21].

Table 7.4 shows the polarization data obtained for sintered composites and microcrystalline titanium, including corrosion potentials (E_C), corrosion current densities (i_C), and corrosion rate (C_R) values. According to Table 7.4, it is possible to observe that ceramic doped to titanium had a positive effect on corrosion resistance of Ti. Titanium composites with 10 vol% of silica have better corrosion resistance than that with a small amount of silica. Besides, the ceramic reinforced with titanium decreased the corrosion rate of Ti.

Table 7.4 Vickers microhardness, mean values of corrosion current densities, corrosion potentials and corrosion rate at Ringers solution of Ti-ceramic nanocomposites and microcrystalline titanium (grade 4)

	Materials	HV	i_C [A/cm^2]	E_C [V]	C_R [mm/year]
Microcrystalline titanium		250	1.31×10^{-5}	-0.36	0.363
Nanocomposites	Ti-3 wt% 45S5	500	8.12×10^{-7}	-0.31	—
	Ti-10 wt% 45S5	620	1.20×10^{-7}	-0.42	0.004
	Ti-3 wt% SiO_2	550	1.91×10^{-6}	-0.37	0.055
	Ti-10 wt% SiO_2	670	4.60×10^{-7}	-0.43	0.006
	Ti-3 wt% Al_2O_3	600	8.65×10^{-7}	-0.29	—
	Ti-10 wt% Al_2O_3	700	1.51×10^{-6}	-0.63	0.046

7.4 Ti-45S5 Bioglass Nanocomposites

Bioglass is a ceramic material that is characterized by some degree of surface reactivity. These materials are specially selected chemical composition, which allows the physiological environment selective chemical reaction with certain ingredients of the bioglass. This leads to a lasting bond with living tissue and biomaterial surface.

After 20 h of mechanical alloying process, the Ti-45S5 Bioglass (3, 10 vol%) mixtures have transformed completely to an amorphous phases without the formation of on other phase [24]. Nanocrystalline Ti-45S5 Bioglass composites were achieved by annealing of the amorphous material. According to the Scherrer method or XRD profiles, the average crystallite size of titanium-45S5 Bioglass nanocomposites mechanically alloyed and heat treated was about 40 nm.

Combined with analysis of EDS, it can be confirmed that the Ti-45S5 Bioglass nanocomposites mainly consist of titanium matrix with nanoparticles of Si, P, Ca, and Na (Fig. 7.11).

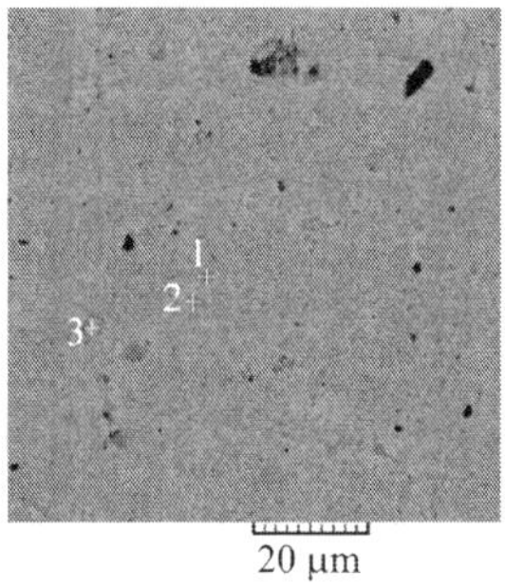

1

Element	Wt %
Ti	98.6
P	0.2
Na	0.2
Ca	0.2
Si	0.5
Al	0.3
Total	100.0

2

Element	Wt %
Ti	96.2
P	0.2
Na	0.8
Ca	1.7
Si	1.0
Al	0.2
Total	100.0

3

Element	Wt %
Ti	97.7
P	0.2
Na	0.3
Ca	0.7
Si	0.8
Al	0.3
Total	100.0

Figure 7.11 EDS spectra of surface of Ti-10 vol% 45S5 Bioglass mechanically alloyed for 20 h and heat treated at 1150°C for 2 h [24].

As shown in Table 7.4 the Vickers microhardness of Ti-3 wt% 45S5 Bioglass and Ti-10 wt% 45S5 Bioglass nanocomposites are 500 HV and 620 HV, respectively and are higher than that of pure microcrystalline titanium. The addition of silica to titanium has a positive effect on the corrosion resistance of titanium. The corrosion resistance increases with the rise of ceramic contents. The Ti-10 wt% 45S5 Bioglass nanocomposite possesses higher corrosion resistance and thus higher corrosion current densities than microcrystalline titanium.

7.5 Ti-Al_2O_3 Nanocomposites

Alumina ceramics have good biocompatibility and good mechanical properties (particularly the parameters of friction and wear). It is biologically inert, which means that it practically does not emit any components to the biological environment. Due to it high resistance to wear by friction, this can significantly improve poor tribological properties of titanium.

Amorphous material is form after 20 h of mechanical alloying process [24]. The formation of the bulk nanocomposites was achieved by annealing of the amorphous materials. XRD and EDS analysis of Ti-10 vol% Al_2O_3 showed the presence of α-Ti type structure with nanograins of aluminum or aluminum oxide (Fig. 7.12). The average crystallite size of heat treated Ti-Al_2O_3 nanocomposites is about 35 nm.

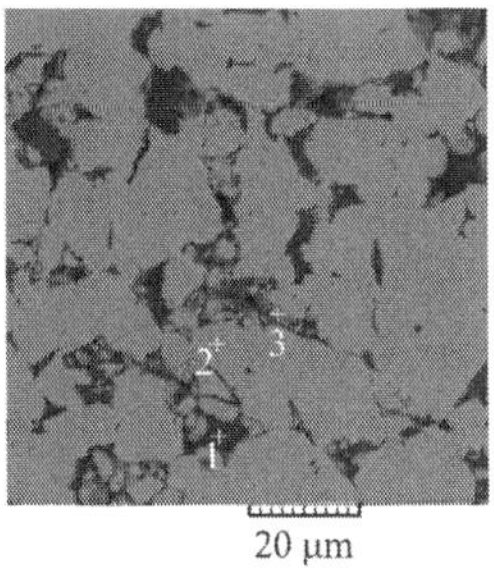

1

Element	Wt %
Ti	13.0
O	18.6
Al	64.7
Si	0.1
Ca	0.6
S	0.8
Cl	1.0
Fe	1.5
Total	100.0

2

Element	Wt %
Ti	89.8
O	1.5
Al	8.8
Si	0.1
P	0.2
Total	100.0

3

Element	Wt %
Ti	79.4
O	6.0
Al	12.5
Si	2.0
P	0.2
Total	100.0

Figure 7.12 EDS spectra of surface of Ti-10 vol% Al_2O_3 mechanically alloyed for 20 h and heat treated at 1150°C for 2 h [24].

The Vickers microhardness of titanium-alumina nanocomposites increases about threefold in comparison with microcrystalline titanium (Table 7.4). Titanium composites with 3 vol% of Al_2O_3 have better corrosion resistance than microcrystalline titanium [23].

References

1. Brossa, F., Cigada, A., Chiesa, R., Paracchini, L., and Consonni, C. (1994). Post-deposition treatment effects on hydroxyapatite vacuum plasma spray coatings, *J. Mater. Sci. Mater. Med.*, **5**, pp. 855–857.
2. Browne, M., and Gregson, P.J. (2000). Effect of mechanical surface pretreatment on metal ion release. *Biomaterials*, **21**, pp. 385–392.
3. Cao, W.P., and Hench, L. (1996). Bioactive materials, *Ceramics Int.*, **22**, pp. 493–507.
4. Chenglin, C., Jingchuan, Z., Zhongda, Y., and Shidong, W. (1999). Hydroxyapatite-Ti functionally graded biomaterial fabricated by powder metallurgy, *Mater. Sci. Eng. A*, **271**, pp. 95–100.
5. Chu, C., Lin, P., Dong, Y., Xue, X., Zhu, J., and Yin, Z. (2002). Fabrication and characterization of hydroxyapatite reinforced with 20 vol% Ti particles for use as hard tissue replacement, *J. Mater. Sci. Mater. Med.*, **13**, pp. 985–992.
6. Chu, C., Xue, X., Zhu, J., and Yin, Z. (2004). Mechanical and biological properties of hydroxyapatite reinforced with 40 vol.% titanium particles for use as hard tissue replacement, *J. Mater. Sci. Mater. Med.*, **15**, pp. 665–670.
7. Chu, C., Xue, X., Zhu, J., and Yin, Z. (2006). Fabrication and characterization of titanium-matrix composite with 20 vol% hydroxyapatite

for use as hard load-bearing hard tissue replacement, *J. Mater. Sci. Mater. Med.*, **17**, pp. 245–251.

8. Jakubowicz, J. (2008). Formation of porous TiO_x biomaterials in H_3PO_4 electrolytes, *Electrochem. Commun.*, **10**, pp. 735–739.
9. Jakubowicz, J., Jurczyk, K., Niespodziana, K., and Jurczyk, M. (2009). Mechanoelectrochemical synthesis of porous Ti-based nanocomposite biomaterial, *Electrochem. Commun.*, **11**, pp. 461–465.
10. Jakubowicz, J., Jurczyk, K., and Jurczyk, M. (2010). Electrochemical formation and corrosion properties of porous TiO_x biomaterials, *Mat. Sci. Forum*, **636–637**, pp. 15–21.
11. Jurczyk, K., Niespodziana, K., and Jurczyk, M. (2006). Preparation and characterization of nanocomposite Ti-hydroxyapatite materials, *Eur. J. Med. Res.*, **11**, Suppl. II, p. 133.
12. Jurczyk, K., Niespodziana, K., Stopa, J., and Jurczyk, M. (2007). Composite Ti-hydroxyapatite bionanomaterials for application in modern dentistry, *Polish J. Environ. Studies*, **16**, no 2CII, pp. 323–327.
13. Jurczyk, K., Niespodziana, K., Stopa, J., and Jurczyk, M. (2009). Titanium-ceramic dental bionanomaterials, *J. Clin. Periodontol.*, **36**, Suppl. 9, pp. 73–74.
14. Jurczyk, K., Niespodziana, K., Stopa, J., and Jurczyk, M. (2009). Nanotechnology in perspective of applications in modern dentistry, *Polish J. Environ. Stud.*, **18**, no 1A, pp. 21–27.
15. Jurczyk, M., Jurczyk, K., Niespodziana, K., and Miklaszewski, A. (2010). Hybrid Ti-ceramic bionanomaterials for dental engineering, *Eur. Cells Mater.*, **19**, Suppl. 1, p. 1.
16. Khan, M.A., Williams, R.L., and Williams, D.F. (1999). Conjoint corrosion and wear in titanium alloys. *Biomaterials*, **20**, pp. 765–772.
17. Liu, X.Y., Chu, P.K., and Ding, C.X. (2004). Surface modification of titanium, titanium alloys, and related materials for biomedical applications. *Mat. Sci. Eng. R*, **47**, pp. 49–121.
18. Long, M., and Rack, H.J. (1998). Titanium alloys in total joint replacement — A materials science perspective. *Biomaterials*, **19**, pp. 1621–1639.
19. Niespodziana, K., Jurczyk, K., and Jurczyk M. (2006). The manufacturing of Ti-hydroxyapatite nanocomposites for bone implant applications, *Nanopages*, **1**, pp. 219–229.
20. Niespodziana, K., Jurczyk, K., and Jurczyk, M. (2007). Titanium-ceramic nanocomposite materials, *Adv. Mat. Sci.*, **7**, pp. 103–107.

21. Niespodziana, K., Jurczyk, K., and Jurczyk, M. (2008). Titanium-silica nanocomposities: preparation and chatracterization. *Arch. Metallurgy Mater.*, **53**, pp. 875–880.

22. Niespodziana, K., Jurczyk, K., and Jurczyk, M. (2009). Mechanical and corrosion properties of titanium hydroxyapatite nanocomposites, *Solid State Phenomena*, **151**, pp. 217–221.

23. Niespodziana, K., Jurczyk, K., Jakubowicz, J., and Jurczyk, M. (2010). Fabrication and properties of titanium–hydroxyapatite nanocomposites, *Mater. Chem. Phys.*, **123**, pp. 160–165.

24. Niespodziana, K., Jurczyk, K., Miklaszewski, A., and Jurczyk, M. (2010). Hybrid Ti-ceramic bionanomaterials for medical engineering, *Phys. Status Solidi C*, **7**, pp. 1363–1366.

25. Ning, C.Q., and Zhou, Y (2002). *In vivo* bioactivity of a biocomposite fabricated from HA and Ti powders by powder metallurgy method, *Biomaterials*, **23**, pp. 2909–2916.

26. Ning, C.Q., and Zhou, Y. (2004). On the microstructure of biocomposites sintered from Ti, HA and bioactive glass, *Biomaterials*, **25**, pp. 3379–3387.

27. Papargyri, S.A., Tsipas, D., Stergioudis, G., and Chlopek, J. (2005). Production of titanium and hydroxyapatite composite biomaterial for use as biomedical implant by mechanical alloying process. *Eng. Biomater.*, **46**, pp. 27–30.

28. Popa, C., Simon, V., Vida-Simiti, I. Batin, G., Candea, V., and Simon, S. (2005). Titanium-hydroxyapatite porous structures for endosseous applications, *J. Mater. Sci., Mater. Med.*, **16**, pp. 1165–1171.

29. Takeshita, F., Ayukawa, Y., Iyama, S., Murai, K., and Suetsugu, T. (1997). Long-term evaluation of bone-titanium interface in rat tibiae using light microscopy, transmission electron microscopy, and image processing. *J. Biomed. Mater. Res.*, **37**, pp. 235–242.

30. Webster, T.J., (2001). Nanophase ceramics: The future orthopedic and dental implant material, In *Advances in Chemical Engineering. Nanostructured Materials*, (ed. Ying, J.Y.), Academic Press, San Diego, pp. 125–166.

31. Webster, T.J. (2007). Nanotechnology for the improvement of all implants, *Mater. Sci. Forum*, **539–543**, pp. 511–516.

32. Williams, D.F. (1987). Review: tissue-biomaterial interactions, *J. Mater. Sci.*, **22**, pp. 3421–3445.

33. Yang, C.Y., Wang, B.C., Chang, E., and Wu, B.C. (1995). Bond degradation at the plasma-sprayed HA coating/Ti–6Al–4V alloy interface: an *in vitro* study, *J. Mater. Sci. Mater. Med.* **6**, pp. 258–265.

Chapter 8

Shape Memory NiTi Materials

Andrzej Miklaszewski
Poznan University of Technology,
Institute of Materials Science and Engineering,
Sklodowska-Curie 5 Sq., 60-965 Poznan, Poland
andrzej.miklaszewski@put.poznan.pl

8.1 Background

The main goal of this short background is to introduce readers to shape memory NiTi materials so that the attributes of the entire family to which they belong can be clearly recognized. We are thinking about the group of the materials whose properties dominate their appearing terminology, for instance, multifunctional material, intelligence, and, finally, smart. All those synonyms are related to the possibility of changing properties through sensing the surrounding environment, actuating by process function of sensor information and in effect, transform one kind of signal to another with multiple amplitude.

Bionanomaterials for Dental Applications
Edited by Mieczyslaw Jurczyk

ISBN 978-981-4303-83-5 (Hardcover), 978-981-4303-84-2 (eBook)
www.panstanford.com

At the turn of the 70s and 80s in the last century, scientists tried to define intelligent structures/systems, and the discussions finally came up with the following four qualifying features: actuation, sensing, control, and time [35]. The created definition [1] of smart structures finally sounds like this:

"A system or material which has built-in or intrinsic sensor(s), actuator(s) and control mechanism(s) whereby it is capable of ***sensing*** *a stimulus,* ***responding*** *to it a predetermined manner and extent, in a* ***short/appropriate*** *time and reverting to its originally state as soon as it stimulus is removed."*

Changing the properties in context of sensing environment through implemented reaction mechanism seems to be the feature of intelligence being. When these variants are assigned to the systems, our attention is guided to the variable operating levels of data exchange. When we think about materials, it is hard to imagine sensing coupling with actuating and managing by supervision function, as illustrated in Fig. 8.1, which presents the classification of smart structures by Rogers.

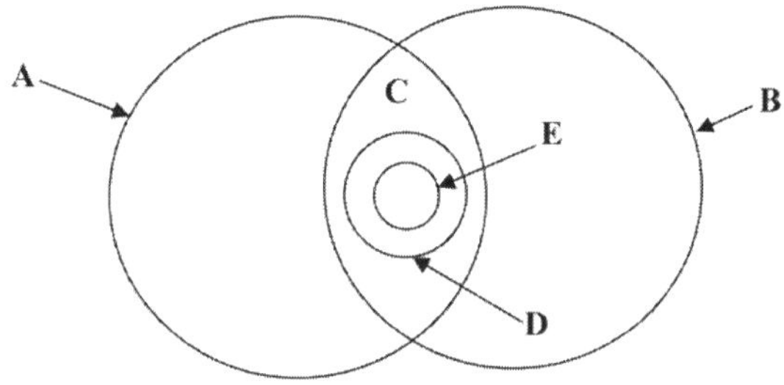

A: Sensory structures; **B:** Adaptive structures; **C:** Controlled structures; **D:** Active structures; **E:** Intelligent structures.

Figure 8.1 Classification of smart structures [36].

During the US Army Research Office Workshop, Rogers and co-workers [37], defined smart materials as *materials, which possess the ability to change the physical properties in a specific manner in response to specific stimulus input.*

Following this expression, we may say that smart materials are a new class of materials, possessing adaptation features on external stimuli with their innate intelligence.

The ability to change physical properties in a proper manner happens through external stimuli, which may include pressure, temperature, pH, radiation, magnetic, or electrical field. Associated variable physical properties may be viscosity, stiffness, or shape. Intelligence of a system manifests in designed chemical composition, appropriate manufacture process, and intended defects initiation or microstructure modification in such way that it may easily adapt to a different levels of stimuli in a controlled manner. One signal transforms to another in a controlled transformation mechanism, and the level of conversion is always higher than for conventional materials.

In this specific and interesting group, the following materials are identified:

- Chromic materials
- Thickness-changing fluids such as magneto and electro-rheological fluids
- Self-repairing materials
- Self-organized materials
- Thermoelectric materials
- Light-emitting materials
- Shape- and size-changing materials

Materials that possess shape memory abilities are classified as active structure and can be used to generate motion or force. They can also accumulate energy, and most important, they exhibit shape recovery effect.

The materials that possess actuation properties are useful in many interesting applications, depending on many factors. The most important factor that engineers have to pay attention to during designing is the actuation energy density — available work per unit volume and actuation frequency. The best situation for an active material perhaps is to have both of them, as high as possible [22].

The actuation energy density is shown in Fig. 8.2 by dotted lines, which is a result of the relation of actuation stress to actuation strain, assuming that the material is operating at a constant stress level. For a better understanding of signaled relations, we make a comparison: Imagine a 1 cm^3 of shape memory alloy that can exert enough force to move an object weight 4650 kg.

Specific value of actuation energy density expressed by work output per unit mass can be easy calculated for the material by dividing the actuation energy density by the mass density, as shown in Fig. 8.3.

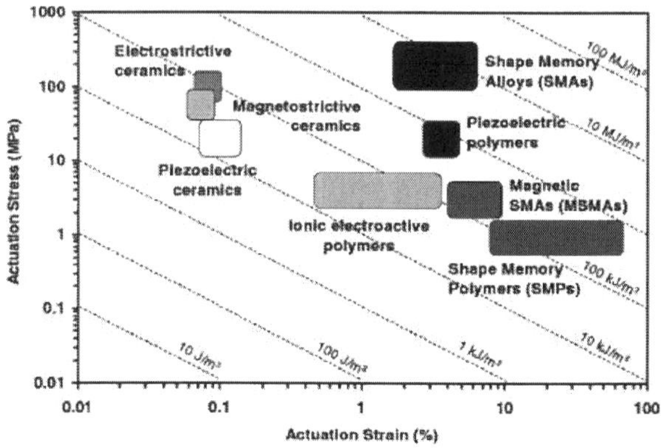

Figure 8.2 Actuation energy density diagram that presents actuation stress to actuation strain relation with typical ranges for active materials [22].

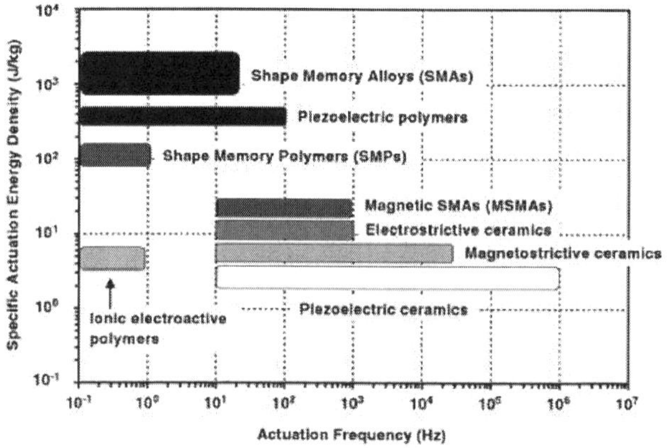

Figure 8.3 Comparison of various active materials in relation to actuation frequency and specific actuation energy density [22].

The comparison in Fig. 8.3 gives readers an outlook of the range of applications of smart materials and could be the next step for further consideration. This short background introduction should realize the existence of broad range of smart materials, including not only shape memory alloys (SMAs) but also magnetic shape memory materials (MSMMs) and shape memory polymers (SMPs). These will be discussed briefly in the following sections.

8.2 Overview of Shape Memory Materials

As we can see from the above considerations, shape memory behavior can occur in different types of materials, for a different kind of external stimuli influence. Materials from the group of shape memory alloys may retain their original shape depending on the prevailing temperature and pressure circumstances. This gives rise to the possibility of using them as a source of displacement or force. Temperature is used as the external stimulus for the shape memory effect control.

These crucial facts led to the idea of looking for more precise systems, and the result was magnetic shape memory alloys. Different expression brings a novel class of the smart materials — shape memory polymers, which featured highly recoverable strains (near 400%), ease of processing, and low density and costs [21].

For a better understanding, magnetic shape memory alloys (MSMA) and shape memory polymers (SMP) will be discussed briefly now.

8.2.1 MSMAs — Magnetic Shape Memory Alloys

The unique ability of magnetic shape memory materials is that they can recover from up to 6%, and for a single crystal up to 10%, of strains in a moderate magnetic field [16]. These materials can work at higher frequency range, which is not attainable by conventional SMAs. Also, transformation control mechanism is more precise when we use magnetic field. Ullakko [47] was the first to observe the reversible transformation due to magnetic field influence. In general assumption the material should possess shape memory effect and strong magnetocrystalline anisotropy. These two issues allow the control of reorientation structure processes.

To further explain the existing mechanism, we will use the most examined example of MSM: NiMnGa alloy. We also simplify the structure representation. Figure 8.4 shows a non-deformed NiMnGa austenite body center cubic cell with lattice parameter of a_0 and non-deformed, possible tetragonal martensite cell with lattice parameters a and c, in which we can distinguish the local magnetization vector in each ferromagnetic variant, which is oriented in one crystallographic direction called easy axis of magnetization.

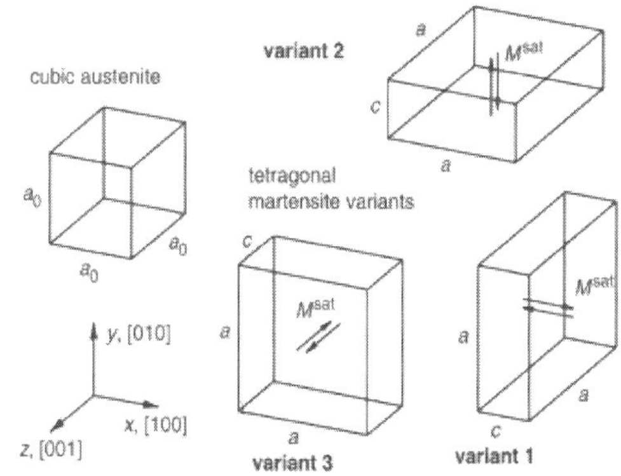

Figure 8.4 High-temperature austenite structure of NiMnGa and its low-temperature tetragonal variants of martensite with visible easy magnetization axis [16].

The influence of the external magnetic field on a structure forces the material to behave similarly as a ferromagnetic, and it responds by the mechanism of magnetic domain wall motion and magnetization vector rotation. The uniqueness of MSMA mechanism is the result of the magnetic field and the martensite variant reorientation behavior.

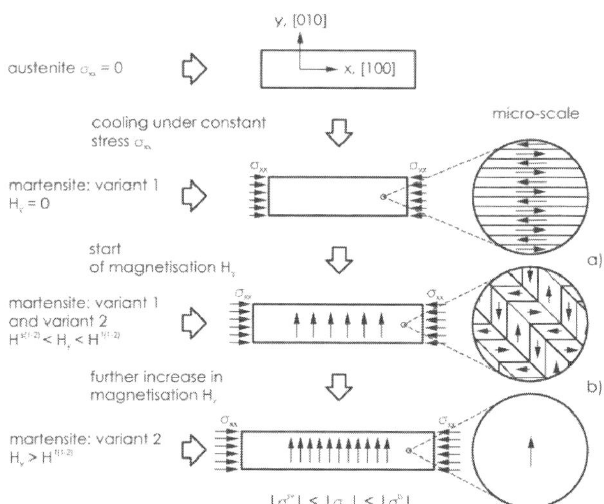

Figure 8.5 Magnetization process of NiMnGa sample with visible microscale wall domain reorientation for circumstances review in text [16].

In the case shown in Fig. 8.5, the sample was cut along the [100] direction and in the same direction, stress σ_{xx} was applied during cooling from the austenite to the martensite phase. For the phenomenon, occurring σ_{xx} stress magnitude has to be larger than σ^{sv}, which is the minimum value of stress necessary to obtain single martensite variant, and smaller than σ^{b} blocking stress level that is a relatively low 6–10 MPa, which suppressed the reorientation of martensite variants.

In the first stage where the stress is applied, we observe a small shortening of the sample, as shown in Fig. 8.5, resulted in single martensite variant 1, in the whole sample. At the microscale, we can also distinguish several magnetic domains resulting from the variant structure — the existing easy axis of magnetization, which in the discussed case, lies along the shorter edge of the tetragonal structure.

Furthermore, external magnetic field H_y is applied orthogonal to the direction of the applied stress and also in direction of the easy magnetization axis of martensite variant 1. The subsequent magnetization of the sample begins nucleation and growth of martensite variant 2, whose easy magnetization axis is parallel to it. This process is realized through the magnetic domain wall motion, reorientation, and migration of the twins. The growth of martensite variant 2 goes on at the cost of martensite variant 1 and results in strain reorientation of the sample. Finish stadium of magnetization, where the value of external field accedes $H_y > H^{f(1-2)}$, results in microscale by obtaining single martensite variant 2, with preferable direction of magnetic domains by the influence of external field, together with elongation of sample.

When the value of acting compressive stress is higher than σ^{b}, the process of martensite variant reorientation (Fig. 8.6) cannot be realized as it was mentioned earlier. The polarization of single martensite variant by external magnetic field is possible, but can be done only at the same axis with working stress, because the small amount of energy is necessary for reversible rotation of magnetic domain walls.

For the orthogonal direction of applied stress to external magnetic field, the rotation of magnetic domain walls is not available because it requires work against the magnetocrystaline anisotropy. Therefore, the domain walls stay perpendicular to the direction of the easy axis of magnetization.

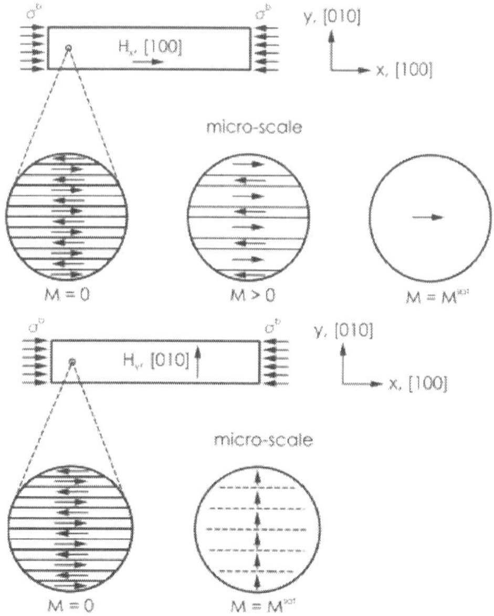

Figure 8.6 Two situations exhibit possible polarization of a single martensite variant, along easy and hard magnetization axis [16].

The brief discussion of mechanisms should give the basic outlook for knowledge in this area and may also be the next step for further consideration on properties, problems, receiving techniques, and applications of MSMAs.

8.2.2 SMPs — Shape Memory Polymers

For SMP, the shape memory effect cannot be considered as intrinsic property like in MSMAs or SMAs. It results from a specific polymer architecture and processing conditions. Specific architecture of polymer is seen as a network consisting of molecular switches that are sensitive to specific external stimuli and net points that can be realized chemically (by covalent bonds) or physically (by intermolecular interactions). Memorizing the shape involves processes such as cross-linking, crystallization, entanglement of the polymer chains, and formation of the domain structure.

Most SMPs demonstrate morphology that consist of at least two segregated domains as for blocky copolymers, where we can distinguish higher temperature (hard segment) domain that acts as a net point and the second one with T_{trans} temperature that acts as molecular switch linked with chain segments in domains (switching segments), as shown in Fig. 8.7.

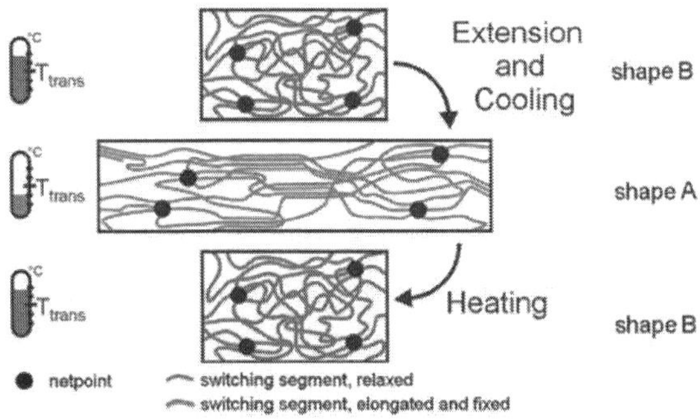

Figure 8.7 Shape memory effect induced thermally, with characteristic polymer architecture; switching molecule segments are related to the thermal transition temperature T_{trans} [19].

Switching or transition in shape memory polymer can be stimulated by heat or magnetism like in SMAs or MSMAs, and also by light, water immersion, or even pH value. This gives a broad range of possible applications that can be found in smart fabrics (for example, in Fig. 8.8.), self-deployable sun sails for the spacecraft, intelligence medical devices, or implants for minimally invasive surgery.

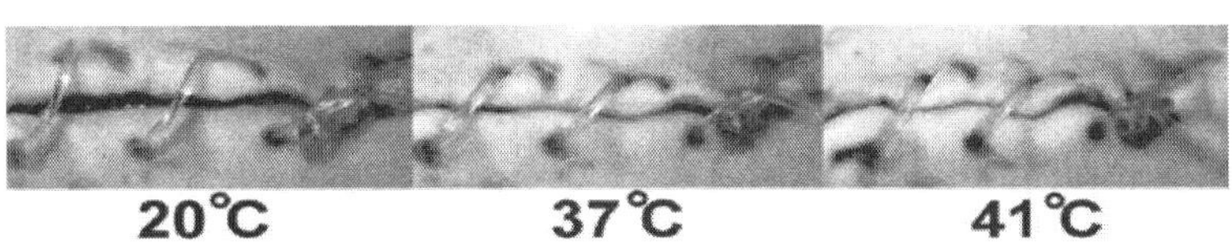

Figure 8.8 Degradable shape memory suture for wound closure, animal experiment shows the shrinkage of fiber with temperature increase from left to right [19].

For brief considerations, we will focus only on the mechanism of the shape memory effect and on the example of the thermo-mechanical cycle of thermo-responsive SMPs.

In typically three-dimensional stress–strain–temperature diagrams, we can observe controlled programming procedure with the recovery of the test specimen. It is also important to mention that such procedure, called training in SMAs, can take much longer time, and deformation rates from the initial to the deformed shape assume a much lower value. Also for other types of external stimulus, understandable as different molecular switch reaction, polymer will behave similarly as shown in Fig. 8.9.

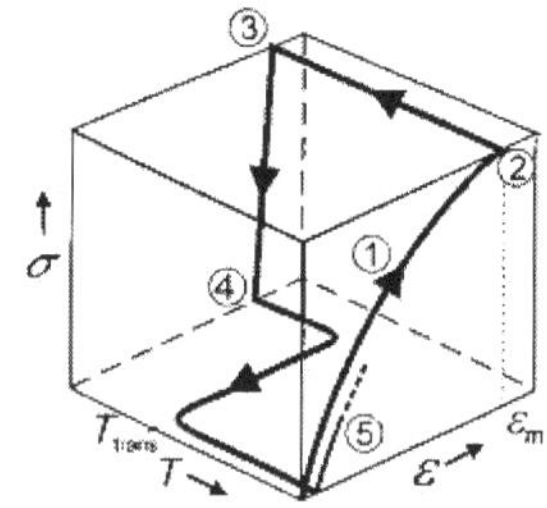

Figure 8.9 Schematic representation of cyclic thermomechanical behavior [20].

The first step brings an elongation ε_m of the specimen that is realized in the temperature range above T_{trans}, the molecular switch is open and the stress-induced reaction of material is not blocked by its bonds. After that, the strain is maintained to allow the relaxation of the polymer chains.

When the sample is cooled below T_{trans} in the second step, the molecular switches closes while the temporary shape that was introduced in the earlier stage is retained (shape A in Fig. 8.7). The third step reduces the applied stress, but the attained strain remains, because of closed switches are blocked and elastically retained in that state. The fourth stage brings the contraction and resumption of the permanent shape of the specimen, with the temperature rise above T_{trans} (shape B on a Fig. 8.7).

This short overview of shape memory materials should provide a good introduction to the existing mechanism in the different types of materials and its possible applications. It also brings a broader vision of discussed behaviors and its similarities for the upcoming section of this chapter.

8.3 Brief History of SMAs

Since the discovery of martensite in steels by Adolf Martens around 120 years ago, scientists have wondered about the nature of mechanism transformation and got involved in exploring its applications mostly from metallurgical point of view. They assume the nonreversible transformation of martensite in steels.

In 1932, Swedish physicist Arne Ölander [32] observed that the applied stress to gold–cadmium alloy could be taken off easily (material returned to its original shape) by heating. In 1938, scientific research on CuZn revealed the temperature dependence of nucleation and disappearance of martensite phase.

For the first time, the concept of thermoelastic martensitic transformation was introduced by Kurdjumov and Khandros in 1949 [17]. Research on the thermally reversible martensitic transformation of CuAl and CuZn alloys gives the reason to think that this phenomenon could occur in a broader manner.

In 1951, Chang and Read used x-ray analysis [5]. They assumed some rules of transformation mechanism in AuCd. They demonstrate that the systems possessing shape memory effect could perform work.

The first significant research started in 1961 over the shape memory alloys by the group of William Beuehler from Naval Ordinance Laboratory, brought a breakthrough for the engineering application. Beuehler's earlier research on intermetallic compounds (for the nose cone Navy's missile SUBROC), which are supposed to possess high melting point and impact resistance, showed that the NiTi alloy reveal the best of it, with the satisfactory properties of elasticity, malleability, and fatigue. The alloy was called NITINOL in memory of the place where it was discovered (the acronym of Naval Ordinance Laboratory was added to the NiTi intermetallic compound's name). During the research on the NiTi alloy, Beuehler made an observation related to the acoustic effect during cooling. In 1963, he confirmed the occurrence of the shape memory effect [2], and for the first time, this phrase was used to determine material property [3].

Scientific research carried out by Wang [48] shows that the addition of the third alloying element, Fe or Co, to the main composition NiTi had a strong influence on the temperature of transformation.

The first application of shape memory alloys was introduced by Raychem Corporation for pipe couplings in F14 jet fighters [7, 38]. The temperature of transformation was so low that before assembly, couplings were transported in liquid nitrogen, which undoubtedly from here called Cryofit (Fig. 8.10). None of the over million manufactured couplings used in jet fighters and pipe couplings reported any damage.

In the 1970s, scientists elaborated high-temperature shape memory alloys through the addition of Pd, Pt, and Au [6]. The temperature of transformation was increased, and in 1978, the addition of Cu by Melton and Marcier resulted in slight temperature decrease with visible constraint of stress hysteresis [24].

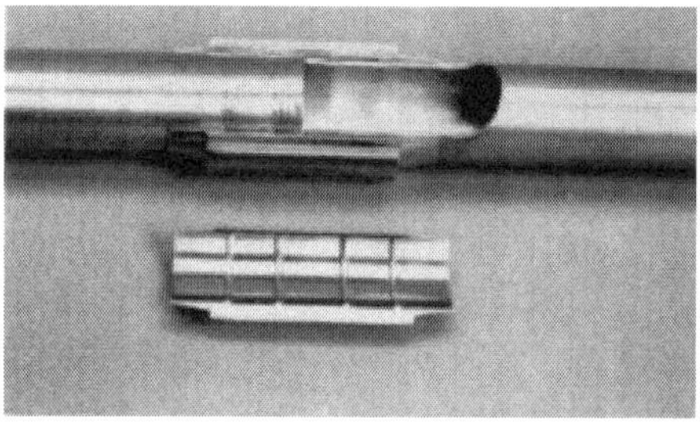

Figure 8.10 Cryofit hydraulic coupling (by permission of Raychem Corporation).

Continued research aiming to improve the parameters of the work temperatures was the main issue of that time. Finally, the year 1989 witnessed the development of the NiTiNb alloy that was easier to handle due to its broader range of temperature hysteresis, which found a wider range of applications [49].

Further, Myazaki and colleagues demonstrated an improvement in the fatigue behavior through the addition of Cu [26].

8.4 SMAs — Introduction to Properties

Materials with shape memory effect derive their unique properties from the fact that they can be exposed to a higher load in the elastic area and easily deformed up to 8% strain, which is more or less 20 times bigger than that for steels but still preserve the ability to recover their original shape.

SMAs can exist in two temperature-dependent phases: (1) basic and higher temperature phase austenite and (2) lower temperature martensite. Both phases differ from each other significantly, and getting involved mechanical, thermal, electrical, acoustic, and optic properties results from differences in crystallographic structure

[9]. For example, martensite is softer and less stiff, while its elastic modulus reaches 31–35 GPa compared with austenite's 84–98 GPa [4, 18].

The austenite phase, also called the mother phase, reveals the B2 type CsCl structure with a_0 = 3.015 Å. The martensite phase appears to be the most common in monoclinic structure where a = 2.889 Å, b = 4.120 Å, c = 4.622 Å and γ = 96.8° [21].

From the crystallographic point of view, it is important to establish the vision of structure deformation that shows diffusionless mechanism driving shape memory effect. Transformation from the austenite to the martensite phase in 3-D crystallographic structure is shown in Fig. 8.11. We may distinguish 24 possible martensite variants that may appear as an effect of twinning — the mirror symmetry of atoms displacements along a specific plane. This specific material situation results from the capability of twin boundaries that can be easily moved in the structure, without the creation of defects such as dislocation, contrary to most metals, in which deformation directly uses the slip or the dislocation movement [41].

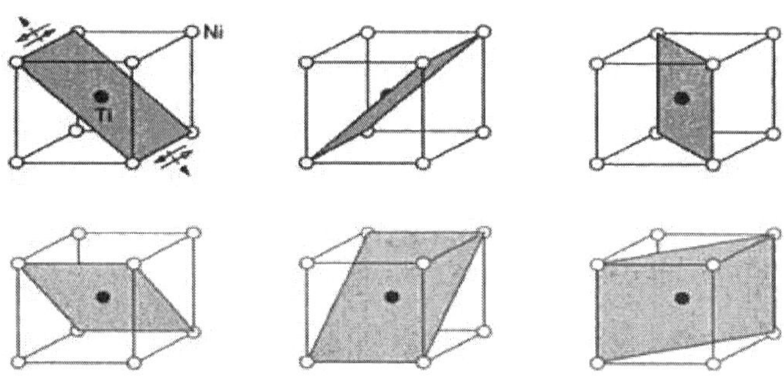

Figure 8.11 Lattice cell of NiTi illustrating the six-face diagonal planes; each of them can shift in two directions or shear in two directions, as shown by the arrows, to produce a total of 24 martensite variants [41].

For the plane consideration of uniaxial loading, we may distinguish only two variants of deformation, which in detwinned form of martensite enter the positive or the negative strain to the structure as shown in Fig. 8.12.

The thermodynamic stability behavior of austenitic and martensitic phases in SMA provides the ability to "remember" different kinds of shape designed in the austenite phase.

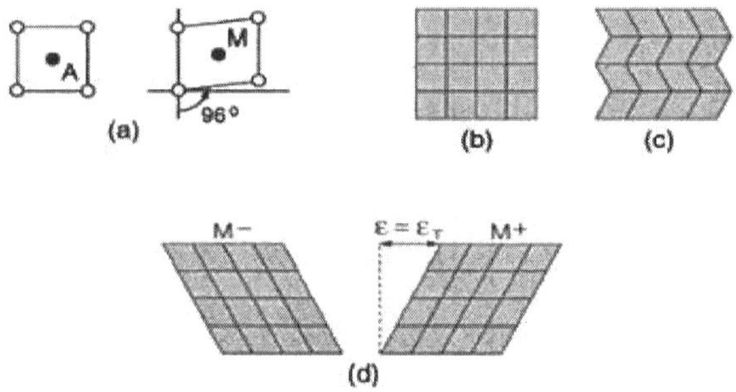

Figure 8.12 Different states of structure in SMA: (a) austenite to martensite transformation, (b) austenite phase, (c) twinned (self-accommodated) martensite, and (d) detwinned martensite possible configuration [41].

8.4.1 Thermally Induced Transformation

In the situation without external stress applied, the shape memory material reacts with temperature decrease by a phase transition from austenite to martensite and reversely with temperature increase. No macroscopic shape change associated with that transformation occurs; the atoms remain on their lattice position in the absence of diffusion process. From the energetic point of view in the situation of temperature reduction and lack of stress, the formation of self-accommodated twins in structure during austenite to martensite transformation, where microscopic strain is negligible, is favorable.

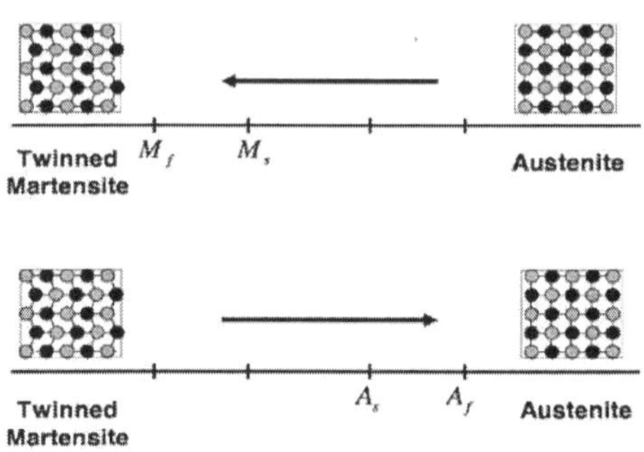

Figure 8.13 Thermally induced phase transformation in state of absence of external load [22].

The characteristic temperature at which martensite transformation begins is marked by M_S, and the temperature at which the transformation ends is denoted by M_f. Subsequent heating starts reversible transformation from martensite to original symmetric crystallographic state of austenite phase, and similarly A_s and A_f are characteristic temperatures (Fig. 8.13).

8.4.2 Mechanically Induced Transformation

Apart from temperature dependence, phase transformation in material can be induced by the applied loads as it has been mentioned earlier.

At the applied mechanical load, the transformation of the austenite structure generates only the type of martensitic structure variants that reduce stress. When the loads are applied to the martensitic grain structure, most favorable variants are generated from the existing structure of different variants that lead to the situation mentioned earlier and thus create the detwinned form of martensite. This phenomenon is called reorientation. Also essential in this point of discussion is to introduce the characteristic M_d temperature, defined as the value above the martensitic mechanically induced transformation cannot occur; it is usually 25–50°C higher than A_f.

For the temperature range of thermodynamic stability of the austenite phase, above A_f and beneath M_d during loading, nearly linear relation between stress and strain occurs — the variants of martensite that reduce stress are generated. When the level of σ_{Ms} is exceeded, the transformation of austenite to martensite begins. The structure starts to accumulate energy in elongation form, which manifests as a macroscopic change of microstructure. It undergoes further transformation — reorientation into the detwinned form of martensite M+ (Fig 8.14).

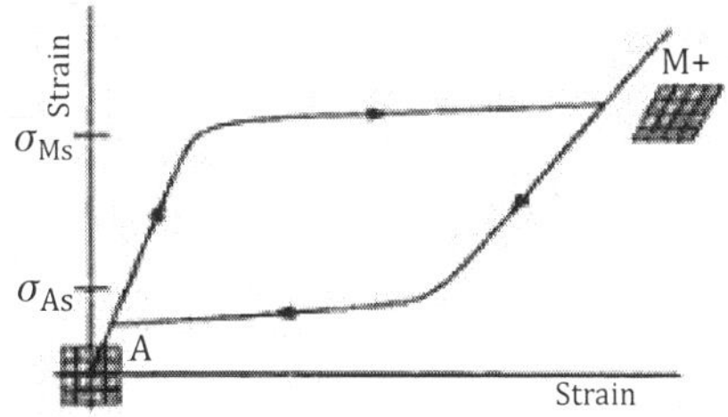

Figure 8.14 Stress to strain relation during loading for SMA at thermodynamic stable austenite temperature range [41].

Similar reaction occurs during unloading. Material tends to reveal almost linear relation of stress to strain when the level of decreasing stress exceeded σ_{As}, and it starts a reversible transformation from martensite to austenite. It is critical to have a temperature range, which has to be considered for fully reversible transformation.

In a high-temperature regime, where the temperature of material is maintained above A_f, fully shape recovery is observed and during unloading, this phenomenon is call pseudoelasticity or transformational superelasticity.

During the full cycle of pseudoelastic transformation, comprehended as austenite to martensite and reverse martensite to austenite, we deal with generating in stress to strain system — hysteresis curve, which is a result of the energy dissipation during process. The curve's shape and values of stress levels strongly depend on the material and measurement conditions.

For the temperature range of thermodynamic stability of martensite phase, beneath M_f during loading, macroscopic change of shape occurs as a result of detwinning mechanism. During unloading, deform detwinned structure of martensite phase retains a residual strain in the form of elongation ε_r as shown in Fig 8.15.

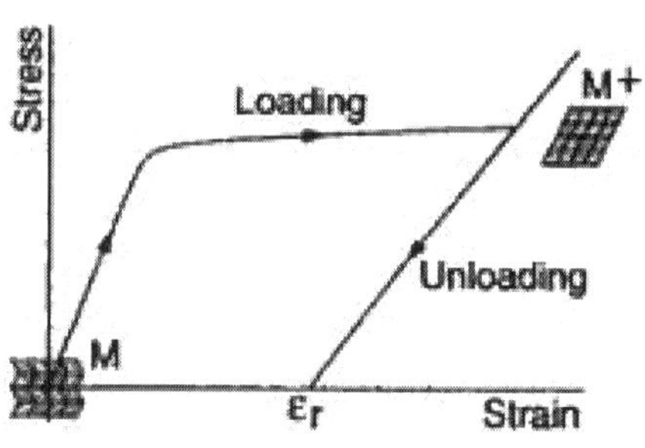

Figure 8.15 Stress to strain relation during loading for SMA at thermodynamic stable martensite temperature range [41].

This low-temperature behavior is call **quasiplasticity** as opposed to plastic deformation, characterized by permanent deformation of structure. After heating the material in temperature range higher than A_f, the accumulate strain is removed from the structure and the reversible transformation to austenite is achieved as shown in Fig. 8.16. This behavior is called shape memory effect (SME) and will be discussed now.

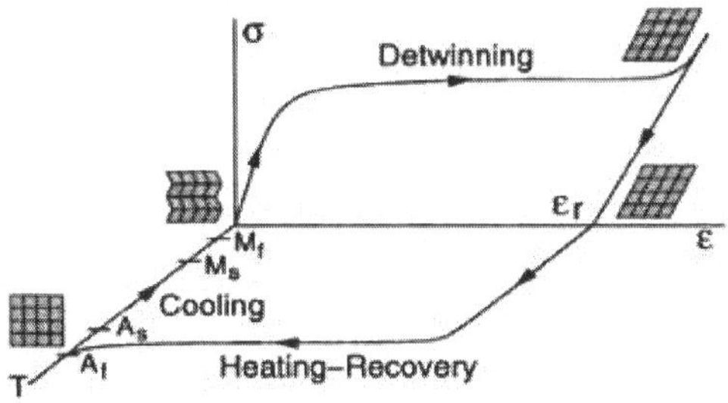

Figure 8.16 Shape memory effect (SME) [41].

8.4.3 SME — Shape Memory Effect

The effect of shape memory requires a more precise analysis and getting involved in a broad range of application of this phenomenon. When we look at a three-dimensional diagram of dependency of stress, strain, and temperature (Fig 8.17), we can notice the relations that have been discussed earlier.

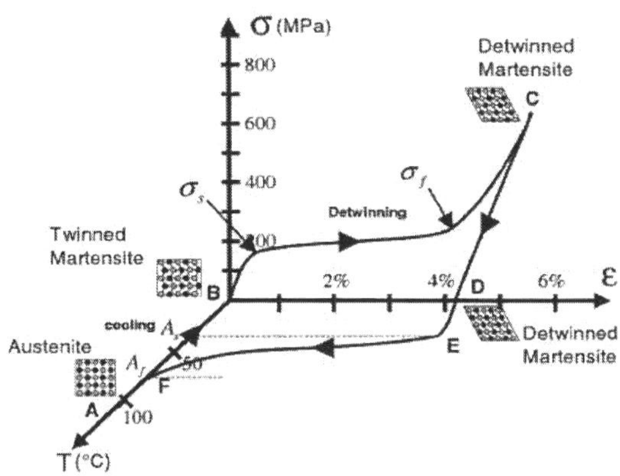

Figure 8.17 Three dimensional diagram that presents a full cycle of transformation in SMA [22].

The temperature decrease from point A, which demonstrates fully stress-free austenite structure, to point B results in self-accommodated transformation into the twinned form of martensite structure. From point B the load is applied and its increase in linearly stress to strain relation up to σ_S value.

From that point, detwinning process of martensite structure begins, which was mentioned earlier. Energy is accommodated in the form of macroscopic strain, and apparently, no stress increase is observed and the relation receives a plateau character. When the stress level achieves σ_f, the process of detwinning is complete and we can consider it, as the end of plateau. The structure is orienting twinned planes in one preferred manner by mechanical load direction. Further stress increase causes elastic deformation of detwinned structure and the maximum elongation is revealed. If the stress increase were sustained (to amount dependent from material composition and receiving process — for the bulk tests 8–10% of elongation), it could generate dislocation movement that plastically deformed the material, and it would not be recovered in the characteristic temperature range.

Next, the material is unloaded from point C to D where we can observe small amount of elastic spring-back; it also now retains the deform state. During the temperature increase from point D, the process of the release of accumulated strains starts, and at point E, detwinned martensite begins to transform into austenite, and at point F, reversible transformation ends with fully austenite structure. For a better understanding, the sequence of transformation is shown in Fig. 8.18., with macro and atomic scale of behavior.

The maximum amount of elongation regain strongly depends on the system; for the polycrystalline structure, it is usually 1–7%.

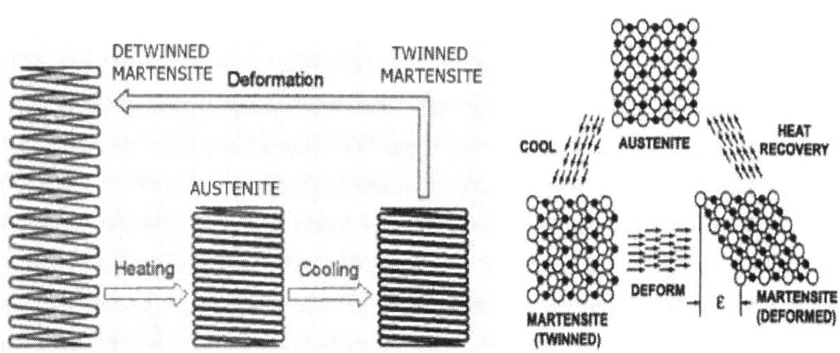

Figure 8.18 Sequence of transformation in SMA [13, 42].

8.4.4 TWSME — Two-Way Shape Memory Effect

The behavior discussed earlier is the result of the reversible transformation occurring in the material. To obtain the two-way shape memory effect, it is necessary to expose the material to a cyclic thermomechanical loading according to a determined way, also called "training."

This can be done in two different ways:

(1) Cyclic loading of the material in the temperature range of thermodynamic stabile austenite phase above A_f beneath M_d,

(2) Cyclic loading of the material in the temperature range of thermodynamic stabile martensite phase beneath M_f and next heating above A_f.

Repeat path of material load in considerable amount of cycles leads to the emergence of a small value of the microstructure change, as a consequence of partial relaxation of accumulated stress during transformation ε^p.

As indicated earlier, the shape of hysteresis σ–ε strongly depends on the material and the loading path. Here an increase in the values of stress results in an increase in characteristic temperature as shown in Fig. 8.19.

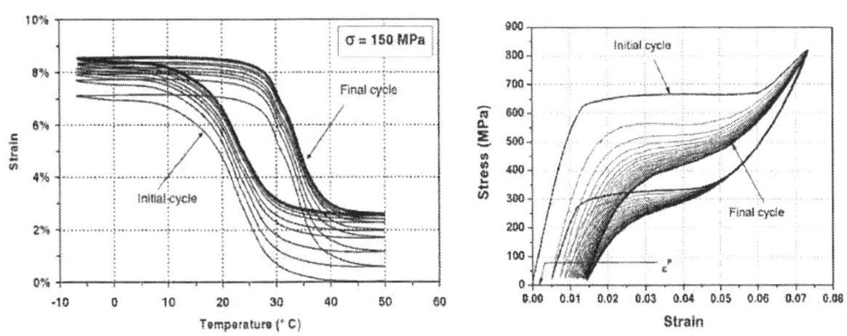

Figure 8.19 From left, hysteresis shape behavior during thermal cycling loading, wire form of NiTi under constant load of 150 MPa and from right, pseudoelastic response of an as receive NiTi wire with A_f = 65°C tested at temperature of 70°C [25].

Cyclic repeating of thermomechanical loading path of the material should remain so long as the saturation effect of nonelastic stress occurs. Permanent strain associated with each cycle gradually decreases until there is no more accumulation. The two-way effect finds practical application, and the material after training can transform into two ways with the shape maintained for austenite and martensite phase as shown in Fig. 8.20.

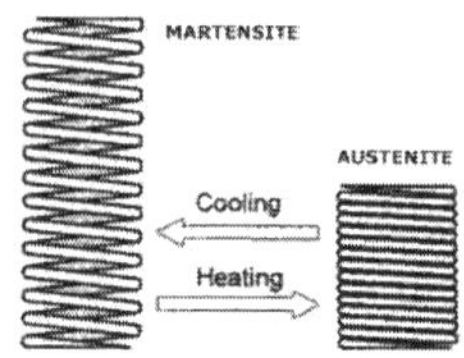

Figure 8.20 Sequence of transformation after training in SMA [13].

8.5 Receiving of Nickel–Titanium Shape Memory Alloy

For the receiving of NiTi is crucial to maintain high-purity requirements. Slight fluctuation of composition causes a strong deviation in temperatures of the transformation. For instance, 1% shift in the amount of composition can cause a 100°C change, as shown in Fig. 8.21.

It is very important to control a manufacturing process using proper methods and protective atmospheres. Usual melting methods employed to obtain pure titanium such as electron beam melting (EBM), vacuum induction melting (VIM), and vacuum arc remelting (VAR), are used to mix well the composition and achieve ingot homogeneity and uniformity [31].

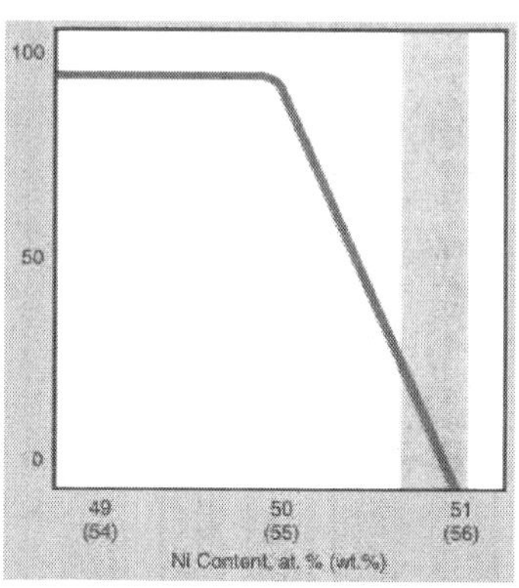

Figure 8.21 Schematic of the effect of the element content on the alloy phase transformation temperature [33].

The impurities mainly result from elemental raw materials, atmosphere, or crucible materials. Oxygen content in the alloy causes a decrease in M_S temperature. It also reduces the grain grown through precipitation formation, similar reaction causes nitrogen (see Table 8.1.) [30]. Crucial to present research is the influence of both elements that is additive.

Table 8.1 Schematic illustrating the effects of various melting impurities on the resulting ingot properties [33]

Element	O	N	H	C	Cu	Cr	Co	Fe	V	Nb
Temperature	↓	↓	↓	↓	→	↓↓	↓↓	↓↓	↓	→
Strength	↑	↑	↑	↑	→	↑	↑	↑	↑	↑
Ductility	↓	↓	↓	↓	→	↓	↓	↑	↓	→

The substitution of Fe, Co, Cr, Al, and V strongly influences the decrease of the M_S temperature, and this behavior is additive also. The main function of the mentioned additions is to increase the stability and strength of the alloy [9]. Substitution of Pd or Pt in small amount causes a decrease in M_S temperature, but with an increase in their contribution, the temperature begins to rise and can reach 350°C [8]. The substitution of Cu or Nb gives the possibility of controlling the hysteresis shape and also strengthening the martensite phase [7].

The next crucial fact that the reader must be aware of, is a strong relationship of the alloy properties with the temperature, stress, and history. A range of characteristic temperatures strongly depends on the chemical composition and also on the applied loads that we mentioned earlier while discussing the hysteresis shape.

Melting and fabrication of Nitinol sometimes become a challenge, considering the material's strong sensitivity toward the chemistry and processing conditions. Forming techniques in some cases are not appropriate when considering a spring-back effect or thermal recovery of a shape memory alloy. Shape change by the reduction of thickness, for instance, during cold rolling of sheet, is lost when the material needs to be heat treated. Therefore, it is so important to understand the underlying mechanism.

For the primary ingot methods, the reduction of dimensions by a hot working process is usually employed. It is breaking up as cast microstructure that exhibits poorly mechanical properties. Finally, the shape of the product depends on the forming technique where for the hot working, this could be forging, rotary forging, bar rolling, rod rolling, sheet rolling, extrusion, and swaging. Typically, the temperature range for the process for the NiTi system is 600°C to 900°C approximately (0.55–0.75) T_m.

For the final product shape with the finished surface, understood as a refined microstructure with optimum mechanical properties, a combination of cold working and annealing is required. For Nitinol wires in the as-drawn conditions, material microstructure consists of deformed martensite with the high density of dislocation.

As shown in Fig. 8.22, to achieve the best stress–strain behavior — maximum superelasticity — the material should be heat treated in the temperature range of 400–500°C. To obtain the best shape settings, a combination of temperature, time, and strain has to be considered during the designing and manufacturing process selection (see Fig. 8.23).

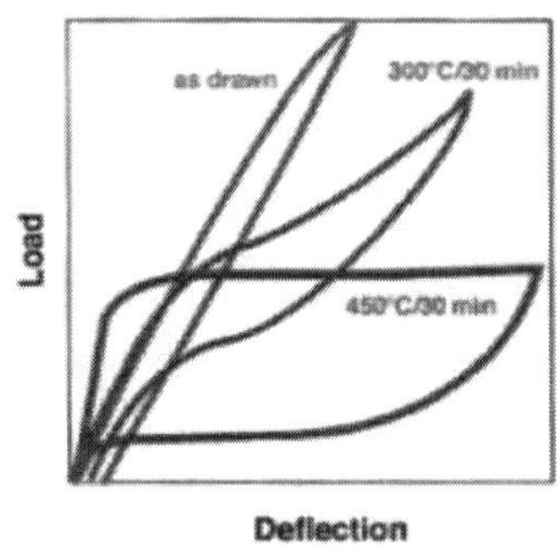

Figure 8.22 Stress–strain behavior of NiTi alloy in various treatment conditions [40].

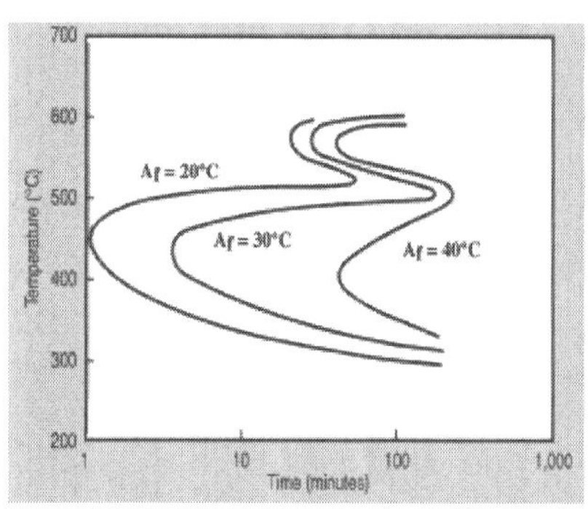

Figure 8.23 Temperature time transformation diagram for $Ni_{50.8}Ti_{49.2}$ wire form [31].

The difficulties discussed above seem to be very simple to overcome, but in the light of the fact that the Ni-rich precipitation forms compounds such as Ti_3Ni_4, Ti_2Ni_3, and $TiNi_3$ during ageing process, it turns out to be more complicated [31].

Precipitations evoke the occurrence of the multi-stage transformation behavior that introduces the R-phase orthogonal structure between austenite and martensite and changes characteristic transformation temperatures as shown in Fig. 8.24.

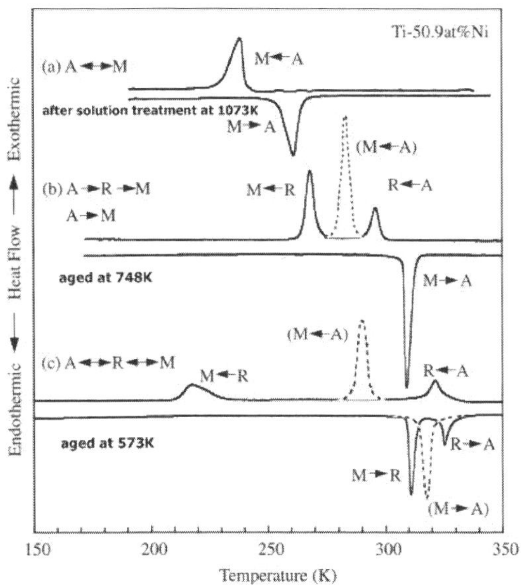

Figure 8.24 Thermal transformation with possible occurring sequence behaviors of Ti-50.9 at% Ni wire form, after temperature heat treatment [14].

The characterization of the mechanical behavior of a few common NiTi alloys prepared by Myazaki and Otsuka gives certain specific ageing conditions used widely as reference for material processing in research [27–28, 31].

Machining process such as drilling, milling, and turning are possible, but they cause difficulties such as excessive tool wear. Abrasive processes, laser machining, electro discharge machining, and photochemical etching are more appropriate for this demanding material. Most important, a cost–efficiency calculation needs to be considered in the designing stage.

The incentive to use different methods for achieving consistent product came from the possibility of cost reduction, improvement in properties, and elimination of the problems that appear in conventional methods. Powder metallurgy gives a precise control of the composition and the achievement of a complex shape without machining it also eliminates the castings problems such as segregation or extensive grain growth. A great challenge for these methods is to produce SMA with properties comparable or better with those of the cast alloys.

The main disadvantage of PM processes is porosity and the manner of its control that affect the mechanical properties of manufactured products.

Near-net shape part methods that employ powder metallurgy use both pre-alloyed and elemental powders.

For pre-alloyed powders, we can distinguish a few fabrication methods such as gas atomization, where molten alloy composition is blasted through inert gas under high pressure. The melt spinning process produces a ribbon from the rapid solidification of liquid composition, which should be hydrated and pulverized for conversion to the powder form next. The process of hydrating uses the strong solubility of hydrogen in NiTi, with rapid diffusion at 40 at.% causing the full embrittlement of the material. Checked by our team for nanocrystalline electrode purposes, the mechanical alloying process, which relies on mechanically induced synthesis in the solid state, could start independently from prealloyed atomized powders or pure powders to form the amorphous form of the material [10]. The nanocrystalline NiTi alloy was synthesized by mechanical alloying followed by annealing [15].

The powder mixture that was milled for more than 5 h transformed completely to the amorphous phase, without formation of the other phase. The formation of the nanocrystalline alloy NiTi was achieved by annealing the amorphous material in high-purity argon atmosphere at 700°C for 0.5 h. All diffraction peaks were assigned to those of CsCl-type structure with cell parameter a = 3.018 Å.

Microstructure and possible local ordering in the NiTi samples was studied by TEM. The microstructure of the annealed sample is shown in Fig. 8.25. The analysis of the high-resolution images (Fig. 8.25a, b) revealed the presence of well-developed crystallites with broad range of sizes from 4 to more than 30 nm. SAED pattern obtained from a large area (200 μm) (Fig. 8.25c) contains sharp rings corresponding to the NiTi alloy with CsCl structure. The method approves the possibility of nanocrystalline NiTi material receiving [29], although no further research was done to characterize the shape memory properties in obtained alloy.

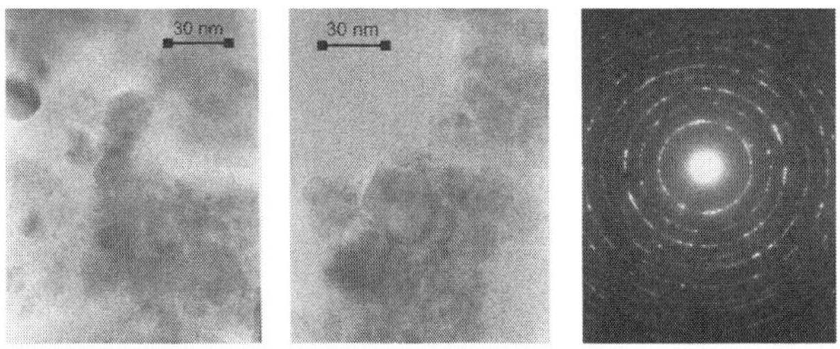

a) b) c)

Figure 8.25 TEM micrographs (a,b) and electron diffraction patterns (c) of the annealed NiTi sample; nanocrystallites of an alloy are clearly visible in (a) and (b) [15].

The powders obtained from the above methods of fabrication should be compacted and sintered. The successfully applied methods include cold and hot isostatic pressing, hot and cold uniaxial die compaction, direct powder rolling, hot extrusion, and consolidation by atmospheric pressure.

For the elemental Ti and Ni powders, we focus on blending, pressing, and sintering, where additional problem arises during thermal treatment. Strong exothermic reaction for Ni + Ti = NiTi results in a much bigger diffusion factor of Ti to Ni, causing the Kirkendall porosity effect. The produced heat can be used to synthesize material like in the combustion synthesis method as shown in Fig. 8.26. Self-propagating high-temperature synthesis, understood as a local heating of cold compacted material above ignition temperature, leads to Ni–Ti formation, which retaliates heat propagation from the reaction with the surrounding material, affect the creation of the synthesis front throughout the entire capacity.

Another approach presents the method of thermal explosion where the entire sample of material is heated up until it reaches ignition temperature. This leads to fusion energy release and generates enough heat to sinter the material. If the heat exceeds the melting point, a cast structure with no visible prior powder boundaries will be formed. To break down the received microstructure, the alloy should be hot rolled or extruded. Methods that use elemental powders generally fabricate highly porous materials and may contain the not so-desired intermetallic phases; remelting or extrusion seem to be necessary for the microstructure improvement.

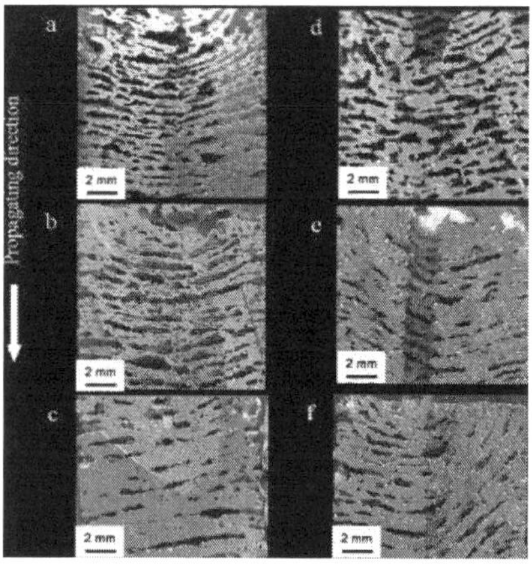

Figure 8.26 Composition with 50 at.% of Ni, SHS longitudinal views of synthesized parts at different preheating temperatures and compaction pressures (a) 200°C, 70 MPa, (b) 200°C, 140 MPa, (c) 200°C, 210 MPa, (d) 300°C, 70 MPa, (e) 300°C, 140 MPa, (f) 300°C, 210 MPa [44].

Cold pressing and sintering for elemental powders do not guarantee high-quality product. The shape memory effect remains, but obtained mechanical properties are low, and compact densities exceed the level of 80% of theoretical volume. Furthermore, sintering can reduce density. This is referred to as the Kirkendall effect, which creates large pores or even craters at the surface. Hot pressing in vacuum conditions showed that it is possible to obtain higher densities and lower level of impurities. Ductility of the obtained materials is locate in the range of that of wrought materials; unfortunately the oxygen level turned out to be too high for proper M_s temperature control or acceptable characteristics of fatigue resistance. The shock compaction adapts successfully for pre-alloyed powders, gives 98% of theoretical densities, and proceeds with the subsequent sintering, which produces 100% densities.

By emphasizing applications, the received porous structures of Nitinol may have beneficial aspects for medical implants. Bony in-growth abilities combine with shape memory characteristics, good corrosion resistance, and biocompatibility of the material, which are crucial for this particular application.

8.6 Biocompatibility and Corosion Resistance of Nickel Titanium SMA

Biocompatibility is defined as the ability to remain biologically innocuous during the functional period of material inside the host, connect with more rigorous conditions so as to not cause allergic reaction, and not release ions into the blood stream. We attempt to compare the material properties as shown in Fig. 8.27, with known living tissue characteristics that should be as close as possible for better mimic natural reaction of living organism.

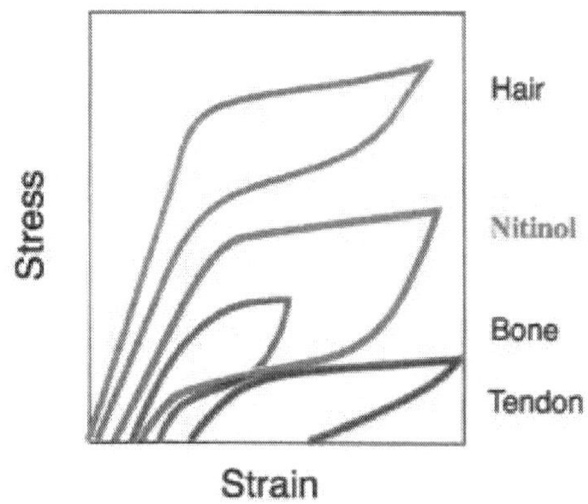

Figure 8.27 Deformation characteristics comparison of living tissues and NiTi placed between them [39]. See also Color Insert.

Generally, allergic reactions are a result of the material surface and inflammatory response of the host, mostly dependent on patient characteristics such as age, immunological state, and health. The surface state profile, porosity or rugosity and also toxic effect of individual elements present in material play an essential role in this situation.

The pH and temperature value are crucial because they affect the rate and progress of the corrosion behavior of the material and can also change the resistivity and solubility of the substance in the operating solution [46].

Experiments show that Nitinol corrosion resistance is significantly affected by the methods of surface preparation. For example, mechanical polishing does not give good results compared with electropolishing as shown in Fig. 8.28. It also decreases nickel release from the material. Results from experiments demonstrate that electropolishing removes excess of nickel from surface and enriches it with titanium [45]. Here, it is crucial to know the functioning surface's behavior, especially the intrinsic properties of titanium inherited by its alloys, such as self-passivating abilities. The formation of a stabile surface oxide layer protects the base material from general corrosion influence. Research by Trapanier and co-workers led first to the conclusion that optimal corrosion and biocompatibility results can be obtained for thin oxide layer. Further, it appears that the uniformity of the oxide layer was more important for material protection, indicating that passivation process of the surface was mandatory for improvement in corrosion resistance.

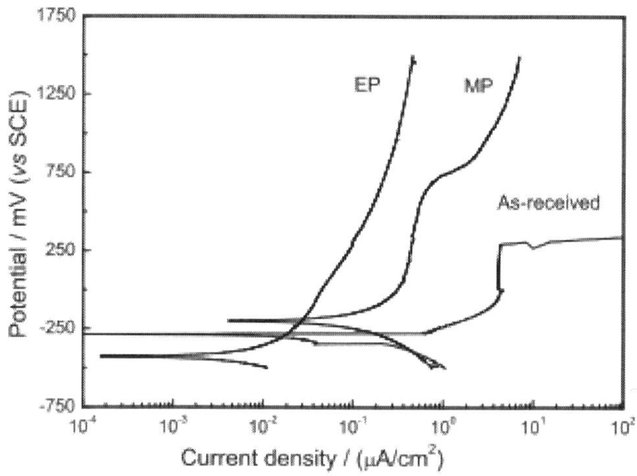

Figure 8.28 Potentiodynamic curves in Hank's solution as received, EP – electropolished, MP – mechanically polished surface of NiTi samples [50].

Various approaches are used for electrochemical anodization [11], which is a well-established surface modification technique for titanium orthopedic and dental implants. Of course, it is the most promising research attempt to combine various kinds of surface modification techniques. For example, heat and alkali treatment results in TiO_2/HA layer formation; the morphology, microstructure, and particle size presented by research as shown in Fig. 8.29, were highly similar to that of the human bone apatite [34].

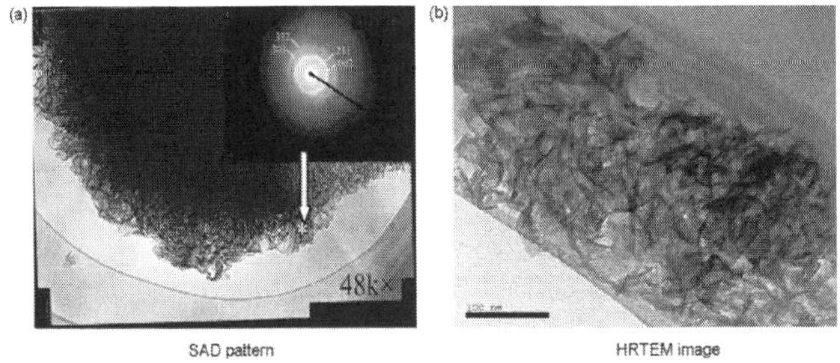

Figure 8.29 TEM photograph and SAED pattern of apatite on the surface of the heat and 3 M NaOH-treated NiTi after soaking in the 1.5 SBF for 7 days (*: center of selected area electron diffraction) [34].

8.7 Shape Memory Materials Application

Most of research based on application in the field of shape memory materials is oriented on biomedical application. In addition, the industry also focuses on valves, damping systems, motion mechanism, fastener rings, connectors, and couplings mentioned earlier. Instantly growing knowledge about this interesting group of the materials seems to be the most effective factor responsible for improvement in properties and the creation of new range of application in other promising areas.

Interesting medical applications include a cardiovascular scenario in which angioplasty and aneurysms treatment procedures are solved by self-expanding implants and devices; orthopedics with spacers; staples for healing acceleration; plates for bone recovery (Fig. 8.30.); and surgical instruments such as non-kinking microsurgical instruments, drills, hingeless instruments, graspers, and needles.

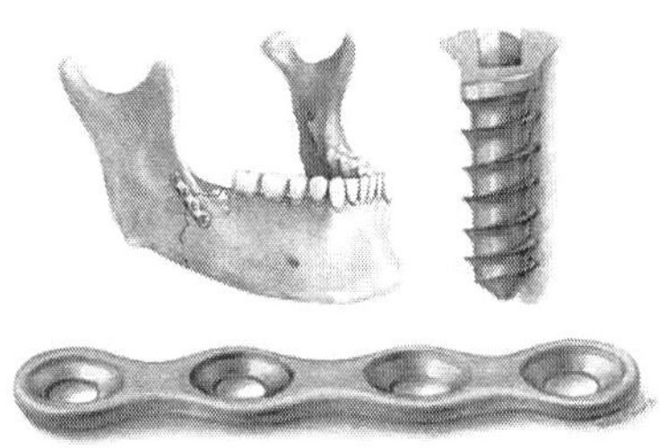

Figure 8.30 Application example of shape memory bone plate fixed on human jaw and detailed view of a four-hole miniplate and miniscrew [10].

For orthodontics, which is relevant to the subject of this book, we discuss the archwire application of NiTi materials.

As it was mentioned in the previous section, the corrosion of NiTi materials may be accelerated by various factors such as electrolyte environment and surface quality. This situation may lead to precocious damage by fatigue mechanism. Corrosion may occur and change the surface state of the material by increasing its roughness. It can result, for example, in the inappropriate force distribution by increasing the friction between the bracket and archwire, which is important in this specific application (see Table 8.2).

Table 8.2 Numerical force value for optimal tooth movement (by permission of DENTSPLY GAC)

Teeth	Short roots	Medium length roots	Long roots
Mandibular incisors	50–55 g	55–65 g	65–70 g
Mandibular canines	85–95 g	95–110 g	110–130 g
Mandibular premolars	70–80 g	80–90 g	90–100 g
Maxillary first molars	280–300 g	300–320 g	320–360 g
Maxillary incisors	65–75 g	75–85 g	85–95 g
Maxillary lateral incisors	60–65 g	65–70 g	70–80 g
Maxillary canines	105–115 g	115–130 g	130–170 g
Premolars, single roots	85–100 g	100–115 g	115–135 g
Premolars, multiroots	100–110 g	120–130 g	130–140 g
Mandibular first molars	230–250 g	250–270 g	270–320 g

Oral cavity should be treated as a changeable environment [43]. Diet, drug use, stress, hygiene, and others factors influence the pH volume, which may varies from 5.2 to 7.8, saliva osmolarity depending on protein concentration and the level of ionized sodium potassium and chloride, temperature from cold ice to hot tea and plaque state.

All these extreme factors play a significant role in non-hostile oral cavity environment. That is why proper material selection for archwire application is so important (Table 8.3).

Table 8.3 Relative levels of important properties for selection of orthodontic wire alloys [4]

Property	Stainless steel	Cobalt-chromium-nickel (Elgiloy blue)	β-Titanium (TMA)	Nickel-titanium
Cost	Low	Low	High	High
Force delivery	High	High	Intermediate	Light
Elastic range (springback)	Low	Low	Intermediate	High
Formability	Excellent	Excellent	Excellent	Poor
Easy of joining	Can be soldered. Welded joints must be reinforced with solder	Can be soldered. Welded joints must be reinforced with solder	Only wire alloy that has true weldability	Cannot be soldered or welded
Archwire-bracket friction	Lower	Lower	Higher	Higher
Concern about bio-compatibility	Some	Some	None	Some

References

1. Ahmad, I. (1988). Smart Structures and Materials, in *Proceedings of U.S. Army Research Office Workshop on Smart Materials, Structures and Mathematical Issues* (ed. Rogers, C.A.), Virginia Polytechnic Institute & State University, Technomic Publishing Co., Inc., pp. 13–16.
2. Beuhler, W.J., Gilfrich, J.V., and Wiley, R.C. (1963). Effect of low temperature changes on the mechanical properties of alloys near composition TiNi , *J. Appl. Phys.*, **34**, pp. 1475–1477.
3. Beuhler, W.J and Wiley, R.C. (1965). US Patent 3 174 851.
4. Brantley, WA., and Eliades, T. (2001). *Orthodontic materials, in Orthodontic Wires: Scientific and Clinical Aspects*, Thieme, New York, pp. 77–103.
5. Chang, L.C., and Read, T.A. (1951). Plastic deformation and diffusionless phase changes in metals — the gold-cadmium beta phase, *Trans. Amer. Inst. Min. Metall. Eng.*, **191**, pp. 47–52.

6. Doonkersloot, H.C., and Vucht, V. (1970). Martensitic transformation in Au–Ti, Pd–Ti and Pt–Ti alloys, *J. Les. Com. Met.*, **20**, pp. 83–91.
7. Duerig, T.W., Melton, K.N., Stockel, D., and Waymanb, C.M. (1990). *Engineering Aspects of Shape Memory Alloys*, Butterworth–Heinemann, London; Kapgan, M., and Melton, K.N., *Shape Memory Alloy Tube and Pipe Couplings*, p. 137; Moberly, W., and Melton, K. (1990). *Ni–Ti–Cu Shape Memory Alloys*, p. 46.
8. Gisser, K.R.C., Geselbracht, M.J., Cappellari, A., Hunsberger, L., Ellis, A.B., Perepezko, J., and Lisensky G.C. (1994), Nickel–titanium memory metal, *J. Chem. Edu.*, **71(4)**, pp. 334–340.
9. Goldstein, D.M., Beuhler, W.J., and Wiley, R.C. (1964), *NOLTR 64-235: Effect of Alloying upon Ceretian Properties of 55.1 Nitinol*, Defense Documentation Center (DDC), Virginia, USA.
10. Haerle, F, Champy, M., and Terry, B. (2009), *Atlas of Craniomaxillo-facial Osteosyntesis, Microplates, Miniplates and Screws, 2nd ed.*, Thieme, New York.
11. Huang, C., Xie, Y., Zhou, L., and Huang, H. (2009). Enhanced surface roughness and corrosion resistance of NiTi alloy by anodization in diluted HF solution. *Smart Mater. Struct.*, **18**, pp. 024003.
12. Jankowska, E., Makowiecka, M., and Jurczyk, M. (2008). Electrochemical performance of sealed Ni–MH batteries using nanocrystalline TiNi-type hydride electrodes, *Renew. Energy*, **33**, pp. 211–215.
13. Janocha, H. (1999). *Adaptronics and Smart Structures: Basics, Materials, Design, and Application*, Springer-Verlag Berlin Heidelberg, pp. 145–146.
14. Jiang, F., Liu, Y., Jang, H., Li, L., and Zheng, Y. (2009). Effect of ageing treatment on the deformation behaviour of Ti–50.9 at.% Ni, *Acta Mater*, **57**, pp. 4773–4781.
15. Eftekhari, A. (2008). *Nanostructured Materials in Electrochemistry*, Wiley, New York.
16. Kiefer, B., and Lagoudas, D.C. (2005). Magnetic field-induced martensitic variant reorientation in magnetic shape memory alloys, *Philo. Mag.*, **85**, pp. 4289–4329.
17. Kurdjumov, G.V., and Khandros, L.G. (1949). First reports of thermoelastic behavior of martensitic phase of Au–Cd alloys, *Doklady Akademii Nauk SSSR*, **66**, pp. 211–213.
18. Kusy, R.P. (1997). A review of contemporary archwires: their properties and characteristics, *Angle Orthod.*, **67**, pp. 197–207.

19. Landlein, A., and Kelch, S. (2002). Shape-Memory Polymers, *Angew. Chem. Int. Ed.*, **41**, pp. 2034–2057.
20. Landlein, A., and Langer, R. (2002). Biodegradable, elastic shape memory polymers for potential biomedical applications, *Science*, **296**, pp. 1673–1676.
21. Leng, J., Lu, H., Liu, Y., Huang, W.M., and Du, S. (2009). Shape memory polymers — A class of novel smart materials, *MRS Biul.*, **34**, pp. 848–855.
22. Logoudas, D.C. (2008). *Shape Memory Alloys Modeling and Engineering Applications*, Springer Science & Business Media LLC., New York.
23. Matsumoto, O., Miyazaki, S., Otsuka, K., and Tamura, H. (1987). Crystallography of martensitic transformation in Ti–Ni single crystals, *Acta Metall.*, **35**, pp. 2137–2144.
24. Melton, K., and Marcier, O. (1978). Deformation behaviour of Ni-Ti based alloys, *Metallic Trans.*, **9A**, pp. 1447–1448.
25. Miller, D.A., and Lagoudas, D.C. (2000). Thermo-mechanical characterisation of NiTiCu and NiTi SMA actuators: Influence of plastic strains, *Smart Mat. Str.*, **9**, pp. 640–652.
26. Miyazaki, S., Mizukoshi, K., Ueki, T., Skuma, T., and Liu, Y. (1999). Fatigue life of Ti–50 at.% Ni and Ti–40Ni–10Cu. (at.%) shape memory alloy wires, *Mat. Sci. Eng. A*, **273–275**, pp. 658–663.
27. Miyazaki, S., Ohmi, Y., Otsuka, K., and Suzuki, Y. (1982). Characteristics of deformation and transformation pseudo–elasticity in Ti–Ni alloys, *J. Phys.*, **43**, pp. 255–260.
28. Miyazaki, S., and Otsuka, K. (1986). Deformation and transition behavior associated with the R-phase in Ti–Ni alloys. *Metall Trans. A*, **17**, pp. 53–63.
29. Mousavai, T., Abbasi, M.H., and Karimzadeh, F. (2009). Thermodynamic analysis of NiTi formation by mechanical alloying. *Mat. Lett.*, **63**, pp. 786–788.
30. Nevit, M.V. (1960). *Trans. Met. Soc. AIME*, **218**, pp. 327–331.
31. Otsuka, K., and Wayman, C.M. (1998). *Shape Memory Materials*, Cambridge University Press, London.
32. Ölander, A. (1932). The crystal structure of Au–Cd, *Z. Krys.*, **83A**, pp. 145–148.
33. Pelton, A.R., Russell, S.M., and DiCello, J. (2003). The physical metallurgy of nitinol for medical applications, *J. Min.*, **55(5)**, pp. 33–37.

34. Qiang, W., Zhen-duo, C., Xian-jin, Y., and Jie, S. (2008). Improving bioactivity of NiTi shape memory alloy by heat and alkali treatment, *App. Sur. Sci.*, **255** pp. 462–465.

35. Rogers, C.A. (1988). Workshop Summary, in *Proceedings of U.S. Army Research Office Workshop on Smart Materials, Structures and Mathematical Issues* (ed. Rogers, C.A.), Virginia Polytechnic Institute & State University, Technomic Publishing Co., Inc., pp. 1–12.

36. Rogers, C.A. (1990). Inteligent materials system and structures, in *Proc. U.S. — Japan Workshop on Smart/Intelligence Materials and Systems* (ed. Ahmad, I., Crowson, A., Rogers, C.A., and Aizawa, M.), Technomic Publishing Co., Inc., Honolulu, Hawaii, pp. 11–33.

37. Rogers, C.A., Barker, D.K., and Jaeger, C.A. (1988). Introduction to smart materials and structures, in *Proceedings of U.S. Army Research Office Workshop on Smart Materials, Structures and Mathematical Issues* (ed. Rogers, C.A.), Technomic Publishing Co., Inc., Virginia Polytechnic Institute & State University, pp. 17–28.

38. Schetky, L. (1979). Shape-memory alloys, *Sci. Amer.*, **241**, pp.74–82.

39. ShabaJovskaya, S. (1996). On the nature of the biocompatibility and medical applications of NiTi shape memory and superelastic alloys, *BioMed. Mater. Eng.*, **6,** pp. 267–289.

40. Siegert, K. (2001). Forming Nitinol — A Challenge, *New Devel. Forg. Tech.*, pp. 119–134.

41. Smith, R.C. (2005). *Smart Material Systems Model Development, Siam Frontiers in Applied Mathematics*, University City Science Center, Philadelphia.

42. Srinivasan, A.V., and McFarland, D.M. (2001). Shape Memory Alloys, in *Smart Structures: Analysis and Design*, Cambridge University Press, UK, pp. 28.

43. Surdacka, A., Strzykała, K., and Rydzewska, A. (2007). Changeability of oral cavity environment. *Eur. J. Dent.*, **1**, pp. 14–17.

44. Tay, B.Y., Goh, C.W., Gu, Y.W., Lim, C.S., Yong, M.S., Ho, M.K., and Myint, M.H. (2008). Porous NiTi fabricated by self-propagating high-temperature synthesis of elemental powders, *J. Mater. Process. Technol.*, **202**, pp. 359–364.

45. Trepanier, C., Venugopolan, R., Messer, R., Zimmerman, I., and Pelton, A.R. (2000). Effect of passivation treatments on nickel release from Nitinol, *Proc. Soc. Biomater.*, pp. 1043.

46. Trepanier, C., and Pelton, A.R. (2006). Effect of temperature and pH on the corrosion resistance of passivated nitinol and stainless stell, *Proc. SMST 2004 Germany*, pp 361–366.

47. Ullako, K., Huang, J.K., Kantner, C., O'Handley, R.C., and Kokorin, V.V. (1996). Large magnetic-field-induced strains in Ni_2MnGa single crystal, *Appl. Phys. Lett.*, **69**, pp. 1966–1968.

48. Wang, F.E., Beuhler, W.J., and Pickart, S.J. (1965). Crystal structure and a unique "martensitic" transition of TiNi, *J. Appl. Phys.*, **36**, pp. 3232–3239.

49. Wu, M.H., and Schetky, L.M. (2000). Industrial application for shape memory alloys, in *Proc. Int. Conf. Shape Memory Sup. Technol.*, Pacific Grove, California pp. 171–182.

50. Wu, W., Liu, X., Han, H., Yang, D., and Lu, S. (2008). Electropolishing of NiTi for improving biocompatibility, *J. Mater. Sci. Techol.*, **24**, pp. 926–930.

Chapter 9

Surface Treatment of the Ti-Based Nanomaterials

Jaroslaw Jakubowicz
Institute of Materials Science and Engineering,
Poznan University of Technology, Sklodowska-Curie 5 Sq., 60-965 Poznan, Poland
jaroslaw.jakubowicz@put.poznan.pl

9.1 Introduction

Among all metallic biomaterials, Ti and its alloys are most widely used because of their great strength, low density, high corrosion resistance, and good biocompatibility [45, 69, 78]. However, the problem for dental as well as orthopedic implant applications is the mismatch of Young's modulus between the bone (10–30 GPa) and metallic part (110 GPa for Ti) [24, 65]. This mismatch may result in the formation of stress and retard the bone healing, which results in increased bone porosity and failure of the implants. One way to reduce Young's modulus of the metallic materials is to introduce the pores [64, 101], minimizing tissue damage, and extended implant life time. Pores introduced into the implant improve its

Bionanomaterials for Dental Applications
Edited by Mieczyslaw Jurczyk

ISBN 978-981-4303-83-5 (Hardcover), 978-981-4303-84-2 (eBook)
www.panstanford.com

fixation with tissue, reducing time between surgical operations, extends implant life time, and improve quality of life of the patients. Rough surface of the implants, with pits and pores, results in tissue growth into these features, which acts as the anchors for the tissue. The strong bonding of the tissue with implant then results in the implant's higher loading-bearing capacity.

There are many possible ways to produce porous implants, for example, powder metallurgical process, metal foaming, and electrochemical etching [24, 28, 93, 101].

Micro- and nanometer diameter pores are useful. As mentioned by Webster and Ejiofor [99], the nanometer structures and molecules in the bone tissue indicate that bone-forming cells are accustomed to interacting with surfaces of nanometer roughness. Conventionally prepared and applied implant surfaces have micro-rough surfaces and are smooth at the nanoscale [39, 40]. For instance, woven bone has an average inorganic mineral grain size of about 10–50 nm [39]. Lamellar bone, has an average inorganic mineral grain size of about 20–50 nm long and is 2–5 nm in diameter [39]. It is commonly accepted that up to this time only rough implants at microscale are utilized and are useful in osseointegration. Now the roughness in nanoscale states a new aspect for implant producers and taking the above assumptions is necessary to find the role of the nanostructures, nanoporosity, and nanoroughness played in the implant materials. For example, nanostructured substrates enhance adhesion of the osteoblasts, decrease adhesion of the fibroblasts, and decrease adhesion of the endothelial cells [95, 99]. Calcium deposition by osteoblasts was four, three, and two times greater on nanophase, compared to conventional microscale alumina, titania, and hydroxyapatite after 28 days of culture, respectively [96, 99]. The latest studies show increased calcium deposition by osteoblasts cultured on alumina nanofibers, carbon nanofibers, poly-lactic-glycolic acid, polyurethane, and composites [15, 42, 73, 82]. Webster [99] found increased osteoblast adhesion on nanophase compared to conventional metals, and osteoblast adhesion occurred preferentially at surface particle boundaries for both nanophase and conventional metals. Since more grain boundaries are present on the surface of nanophase compared to microcrystalline conventional metals, this may be an explanation for the measured increased of osteoblast adhesion [99]. Additional grain boundaries can be prepared in porous materials.

Porous Ti compacts, made by powder metallurgical process, have porosities ranging from 5.0 to 37.1 vol% [65]. In these porous Ti compacts, the rough surface depends on Ti initial powder size, so the porosities and mechanical properties of porous Ti compacts can be controlled by changing powders size and sintering conditions. Oh *et al.* found that Young's modulus and bending strength of porous Ti compacts having the porosity approximately 30 vol% are close to those of human cortical bone [65]. Increasing porosity results in a significant decrease in Young's modulus (Fig. 9.1).

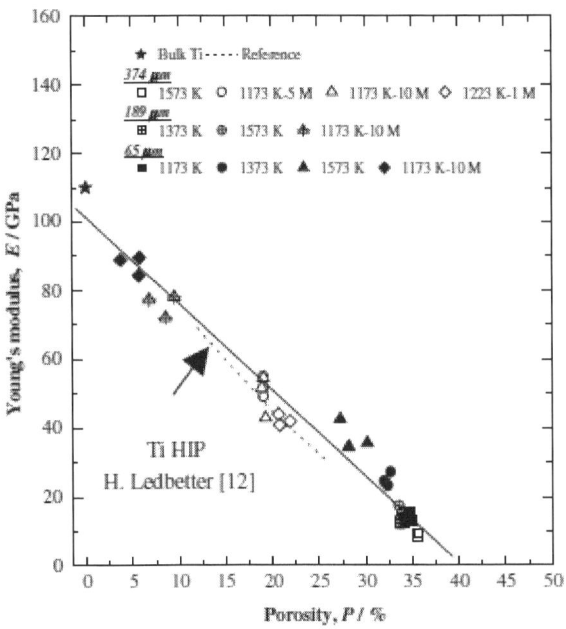

Figure 9.1 Young's modulus of sintered porous Ti compacts as a function of porosity. Porous Ti compacts sintered with applied pressure are abbreviated as temperature (K)–pressure (M) [65].

The Ti metal in all cases during processing (sintering, heat treatment, and etching) in the oxygen-containing atmosphere spontaneously forms a protective TiO_2 layer. When the Ti implant is inserted into the human body, the surrounding tissues directly contact the TiO_2 layer on the implant surface. The surface characteristics of the TiO_2 layer determine the biocompatibility of the Ti implant [112] and it is important to increase the biocompatibility of the

Ti implant for clinical applications, by surface modifications (electrochemical etching, or plasma treatment).

The above-mentioned basic assumptions are relatively easy and simple to achieve and the following section is concentrated on electrochemical ways of the Ti-based surface modifications necessary for improving implant mechanical properties, corrosion resistance, biocompatibility, osseointegration, and other related surface properties. Increase of surface roughness of the hard tissue implant promotes the implant-tissue integration.

9.2 Electrochemical Anodic Oxidation

9.2.1 Electrochemically Grown Porous TiO_2

Anodic oxidation (anodization) is an electrochemical treatment made at potentials positive from the open-circuit-potential (ocp). In the anodic conditions, the metallic implant is the working electrode immersed in the special type of electrolyte. The current flow results in surface oxidation and formation of titania. Depending on the electrochemical conditions, the oxide thickness, its structure (anatase, rutile or amorphous), and adhesion can vary as well as in the surface of formed pores with diameter from nano- to micrometer range.

In the electrochemical etching of titanium, used electrolytes often containing H_3PO_4, CH_3COOH, and H_2SO_4 [50, 59]. In the Ti anodization, the dissolution is enhanced mostly by HF- or NH_4F-containing electrolytes [88]. The current density for these electrolytes is much higher than in electrolyte without HF or NH_4F [88]. Unfortunately, fluoride ions form soluble $[TiF_6]^{2-}$ complexes, results in dissolution of the titanium oxides. In this way the dissolution process limits the thickness of the porous layer [5].

Application of electrolyte containing phosphorus (like H_3PO_4), makes a possible incorporation of phosphorus into the oxide layer. As we know, phosphorus is a basic component of the human bone, so the increase of P-content in the surface layer can improve the osseointegration. The growing anodic oxide layer is mostly porous, which makes it easy to grow a tissue into the implants. For this process, the surface should have sufficient pores diameter and roughness. Sufficiently thick oxide layer restrains corrosion and helps release harmful compounds. Porous implant layer has lower density than respective bulk substrate, and good mechanical strength is provided by bulk substrate. Hence, that material is attractive with respect to bulk titanium alloys or austenitic steels. The porous layer on the Ti substrate is necessary for fast osseointegration with bone [43], which is normally not provided by native oxide. Anodic oxidation usually proceeds at high DC voltages, for example in the range of 90–180 V and spark-discharge occurred at voltages higher than 105 V [109]. On the titanium surface, TiO_2 with anatase structure was observed at 90 V [109], but for higher voltages 155 V and 180 V, mixture of anatase and rutile and single rutile phase was formed, respectively [109].

Generally an unequivocal explanation of pore formation does not exist. The field-assisted dissolution has been considered a predominant mechanism of porous anodic film formation [12, 19, 25, 85, 105]. Transportation of specific ions in oxide is considered as well for explanation of the pore formation [7]. Transformation of amorphous phase into crystalline generates stress, resulting in dissolution of oxides leading to pore formation, too [37]. Most of the models are true, but only for the specific processing conditions is useful for pore formation explanation.

Titania coating can be prepared by a few other methods in addition to anodization, such as heat treatment, sol–gel, chemical treatment, and plasma spraying [22, 27, 28]. Rohanizadeh [76] found that the titanium pretreated in H_2O_2 solution shows the highest adhesion to the titanium substrate.

Shih *et al.* [79] investigated the effect of titanium hydride on the formation of nanoporous TiO_2 on Ti during anodization. They prepared titanium hydride TiH_2 during cathodization (potentials negative to the ocp), followed to oxide layer and nanocrystalline TiO_2 structure after anodization. A multi-nanoporous TiO_2 layer was formed on the titanium. The titanium hydride nanostructure is directly changed to nanoporous TiO_2 by a dissolution reaction during anodization. The nanostructural layer of TiH_2 formed during cathodization plays a crucial role in forming the nanoporous TiO_2 layer. Anodization treatment with cathodic pretreatment transforms the titanium surface into a nanostructured titanium oxide surface.

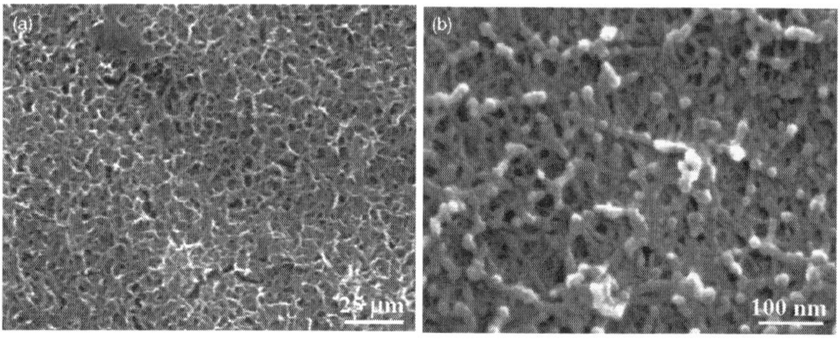

Figure 9.2 Surface morphology of Ti after 15 A/cm^2 anodization (a) and 15 A/cm^2 anodization with 5 A/cm^2 cathodic pretreatment (b) [79].

Figure 9.2a presents the microporous surface morphology of the Ti after anodization at a current density of 15 A/cm^2. The nanoporous structures were obtained on the surface, when the current density of cathodic pretreatment was between 0.1 and 5 A/cm^2 (10 min in 1M H_2SO_4 solution at 25°C). A nanoporous Ti surface was obtained after anodization at 15 A/cm^2 (10 min in 5M NaOH solution) with cathodic pretreatment at 5 A/cm^2 (Fig. 9.2b). The multi-stage treatment results in thicker titanium oxide layer. The higher porous oxide thickness results in the biocompatibility implant improvement. Shih *et al.* [79] found that TiO_2/Ti after anodization with cathodic pretreatment is thicker and more porous than as-machined Ti, cathodic and, anodic Ti sheets. The TiH_2 is important in forming multi-nanoporous TiO_2 layers. Cathodization involved the hydrogen evolution on titanium. Anodization dissolves nano-TiH_2 and forms a thicker nanoporous TiO_2 layer. They suggest that the presence of the TiH_2 phase on titanium is critical in preparing a multi-nanoporous TiO_2 layer [79]. It should be mentioned that TiH_2 (in powders form) also plays a crucial role in the metal foaming process [24, 117], resulting in uniform pore distribution (Fig. 9.3) in whole volume of the implant (foam), which is assured by Ti-hydrides decomposition. The main advantage of that structure is the lowest density, higher porosity (pores in whole volume of the material), and larger pores (in micrometer range) with respect to only porous surface formed during anodization of the bulk materials.

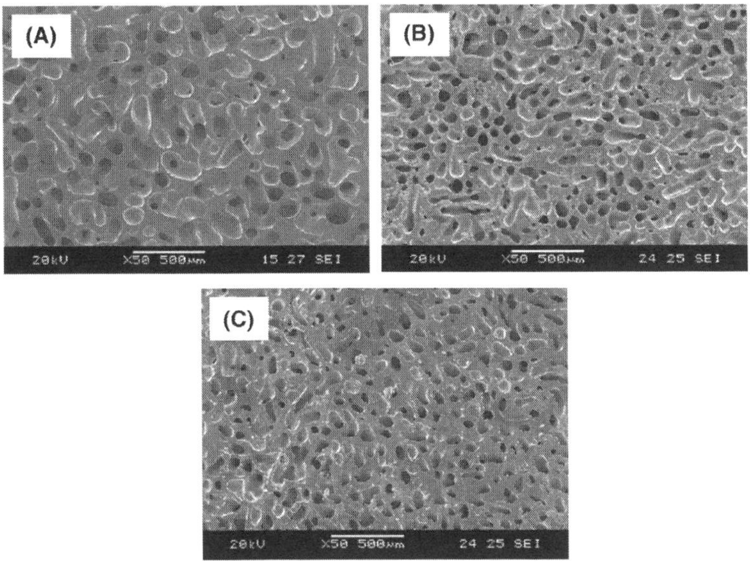

Figure 9.3 SEM micrographs of the porous Ti scaffolds produced with various TiH_2 contents of (A) 15 vol.%, (B) 20 vol.%, (C) 25 vol.% [117].

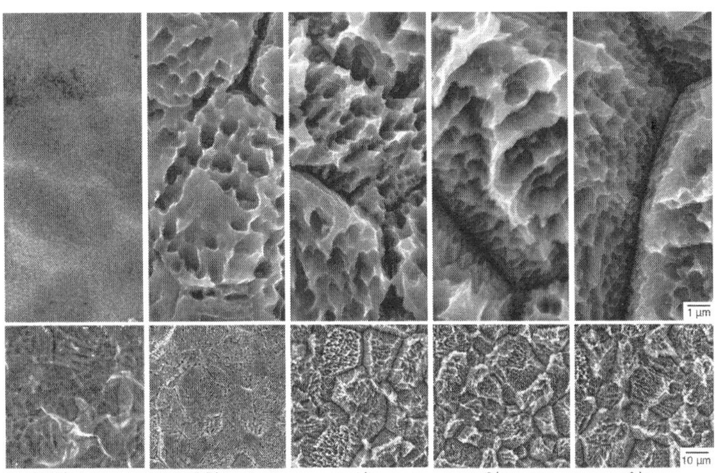

Figure 9.4 SEM images of the Ti after etching in 48% H_2SO_4 at 60°C for 0.25, 0.5, 1, 3 and 8 h. The upper images are 10 times higher magnification of the central region of the lower images [3].

Ban *et al.* [3] investigated the etching behavior of titanium in concentrated sulfuric acid at 60°C for 0.25, 0.5, 1, 3, and 8 h (Fig. 9.4). The etching reveals grain boundaries and the crystal grains are well visible. On the surface are etched pores with diameter less than 1 μm. The depth of the etched grain boundaries and pores increases with etching time [3].

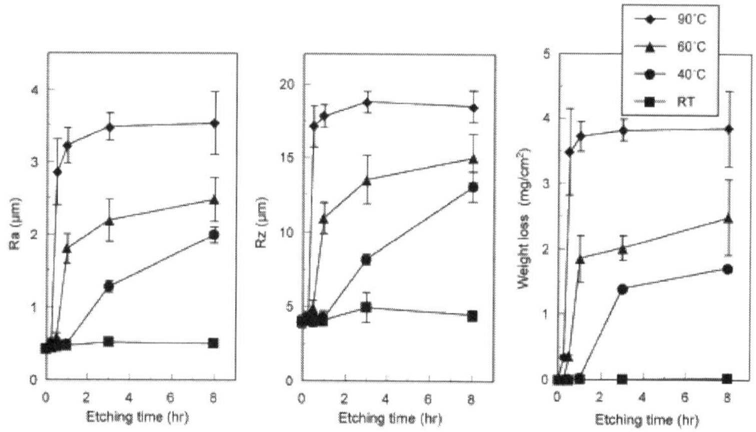

Figure 9.5 Surface roughness R_a and R_z, and weight loss of Ti after etching in 48% H_2SO_4 in the range from RT to 90°C for 0.25–8 h [3].

The surface roughness and weight loss of the Ti increase with the acid temperature and etching time (Fig. 9.5) [3]. A significant difference occurs between the acid temperature and the etching time for roughness and the weight loss. The etching of Ti in concentrated H_2SO_4 involves the following reactions:

$$TiO_2 + 2H_2SO_4 \rightarrow Ti(SO_4)_2 + 2H_2O \tag{9.1}$$

$$Ti + 2H_2SO_4 \rightarrow Ti(SO_4)_2 + 2H_2 \tag{9.2}$$

$$Ti + H_2 \rightarrow TiH_2 \tag{9.3}$$

Surface roughness is strongly correlated with weight loss. At the beginning of the etching, the formation rate exceeds the rate of TiH_2 dissolution, and the relative amount of TiH_2 on the Ti surface increases. The dissolution rate of TiH_2 also increases with time, because the surface area increases with the increase in surface roughness due to the etching.

Xie and Li investigated the Ti etching with respect to photoelectrocatalytic application [108]. They have used two processes done at: low voltages (20–40 V) and long time (6 h) anodization in aqueous 1M H_2SO_4 + 0.3M H_3PO_4 + 0.6M H_2O_2 + 0.03M HF solution and at low voltage (10–50 V) and short time (30 min) anodization in aqueous 0.5M H_3PO_4 + 0.1M HF solution plus post-calcination at 450°C for 2 h. The obtained results could be easily transferred to implant surface modification. The results obtained by Xie and Li confirmed that Ti can be successfully anodized to TiO_2 in both electrolytes at low voltages.

The TiO_2/Ti film prepared at 40 V for 6 h is shown in Fig. 9.6. The film is composed of a multiporous structure with the aggregated particles on its surface (a). The average size of the micropores is up to 1 μm. The thickness of this TiO_2 film was determined to be about 4 μm [108]. The TiO_2/Ti film prepared at 30 V for 30 min with post-calcination at 450°C for 2 h is shown in Fig. 9.7. The nano-structured TiO_2 thin film with a cross-linked multiporous network structure (a) was formed with average size of individual pores to be about 50 nm and interconnection wall thickness was around 30 nm. The thickness of this TiO_2 layer was measured to be 650 nm (b). The anodic oxidation done at higher voltage is beneficial to achieve a higher degree of TiO_2 crystallization. The XRD patterns show that all TiO_2/Ti etched according to first process only exhibited an anatase phase (Fig. 9.8). The nano-structured TiO_2 thin film formed at 30 V for 30 min without calcination has mainly an amorphous structure due to a very short anodizing time, but after calcination at 450°C for 2 h, a well-crystallized TiO_2 film with an anatase-type is formed (Fig. 9.9). The results indicate that for both long time anodization without further thermal treatment and a short time anodization with post-calcination, can successfully form well-crystallized TiO_2 films.

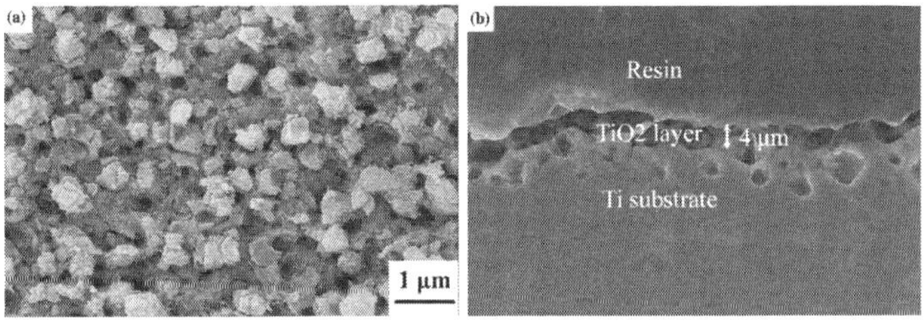

Figure 9.6 Top view (a) and (b) cross-sectional images of TiO_2/Ti film prepared by anodization at 40 V for 6 h [108].

Figure 9.7 Top view (a) and (b) cross-sectional images of TiO_2/Ti film prepared by anodization at 30 V for 30 min and post-calcination [108].

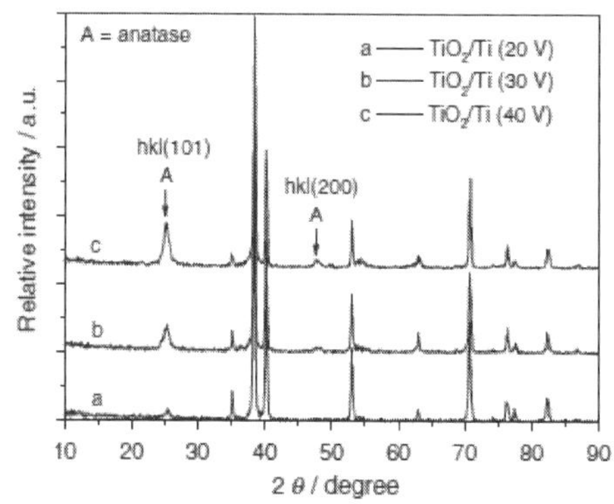

Figure 9.8 XRD patterns of TiO_2/Ti film prepared by anodization at 20, 30, and 40 V, for 6 h [108].

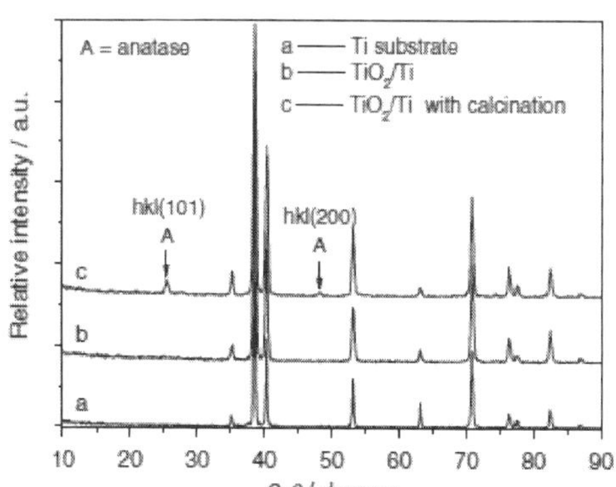

Figure 9.9 XRD patterns of TiO_2/Ti film prepared by anodization at 30 V for 30 min [108].

During the Ti anodization process in aqueous acidic solution, Ti could be oxidized into various types of oxides, such as TiO_2 and Ti_2O_3 with different valence states, as long as the positive potential applied on the titanium anode was higher than its theoretical oxidation potentials (1.630 V for Ti/Ti^{2+}, 1.998 V for Ti/Ti^{3+}, 2.188 V for Ti/Ti^{4+} vs SCE) [108]. The higher applied potential can results in the complete transformation of Ti into TiO_2. During anodic oxidation, a competition between an oxidation reaction (9.4) and dissolution reaction (9.5, 9.6) occurs:

$$Ti + 2H_2O \rightarrow TiO_2 + 4H^+ \tag{9.4}$$

$$TiO_2 + 6HF \rightarrow [TiF_6]^{2-} + 2H_2O + 2H^+ \tag{9.5}$$

$$TiO_2 + H_2O + H^+ \rightarrow [Ti(OH)_3]^+ \tag{9.6}$$

Furthermore, reaction generating oxygen from water occurs (9.7):

$$2H_2O + 4e^- \rightarrow O_2 + 4H^+ \tag{9.7}$$

The voltage applied to titanium anode, the anodizing time, and the composition of the electrolyte are the most important factors to control the surface morphology and crystal structure of the product TiO_2/Ti films [108].

Jakubowicz [28] investigated etching of the Ti-electrode at low voltages in H_3PO_4 electrolyte containing HF addition. Figure 9.10 presents morphology of the commercially pure titanium after anodic oxidation in 1M H_3PO_4 electrolyte containing following amount of HF: 0.5% (a), 2% (b) and 10% (c). The process carried out at relatively low voltage of 10 V vs. ocp for 30 min results in pore formation. When the HF concentration increases, the pore diameter significantly increases, too. For the low HF concentration 0.5% (a), the average pore diameter is in the nanometer range and is approximately 45 nm. For the 2% and 10% HF concentration, the pore diameter significantly increases up to 1.5 μm (b) and 5.2 μm (c), respectively. Depending on the electrochemical conditions, two types of layers can be formed: nanoporous and microporous. Tsuchiya *et al.* [89] found a two-layers structure, with outer nanoporous and underneath with larger diameter tubes layer.

After long etching time, the top nanoporous layer is dissolved. The higher HF electrolyte concentration results in faster dissolution.

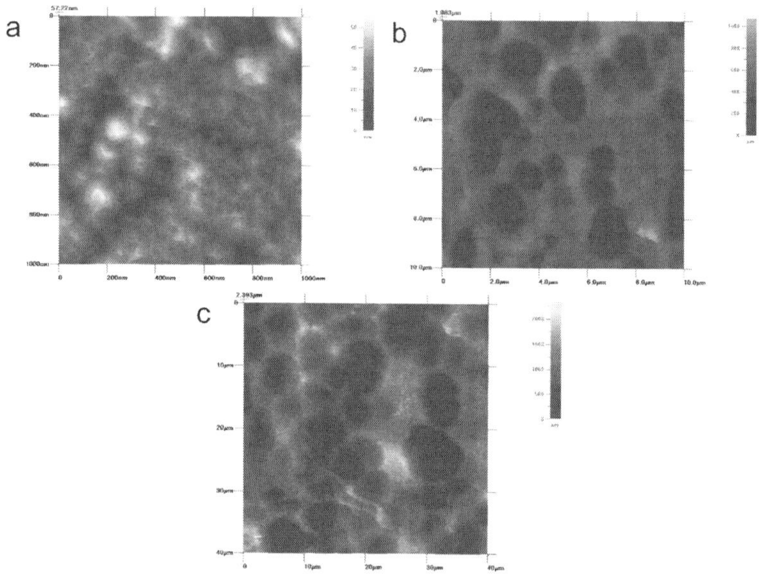

Figure 9.10 Porous TiO_x formed after etching at 10 V for 30 min in 1 M H_3PO_4 electrolyte containing: 0.5% HF (a), 2% HF (b), 10% HF (c); (1 × 1, 10 × 10 and 40 × 40 μm, respective scans size) [28].

On Fig. 9.11 shows the surface of the pure microcrystalline Ti after etching in 1M H_3PO_4 + 10% HF electrolyte at 10 V vs. OCP for 5 min [31]. Etching results in surface roughening and pits formation. The diameter and the depth of the pits are about 4–5 and 1–2 μm, respectively. The pits' dimensions differ for different process conditions, (like H_3PO_4 and HF concentration, etching time and applied voltage or current density) [28] and hence pits' diameter can vary from nano- to micrometer range. Etching at low potential of 1 V vs. OCP for 60 min reveals an initial etching stage (Fig. 9.12a). The surface etched in 1M H_3PO_4 + 0.35% HF has crack-like morphology (Fig. 9.12a) and large remnants of the flat surface are visible [28]. The increase of the HF concentration up to 2% (Fig. 9.12b) results in uniformly etched pores. This different behavior is related with increasing HF concentration, which results in faster etching.

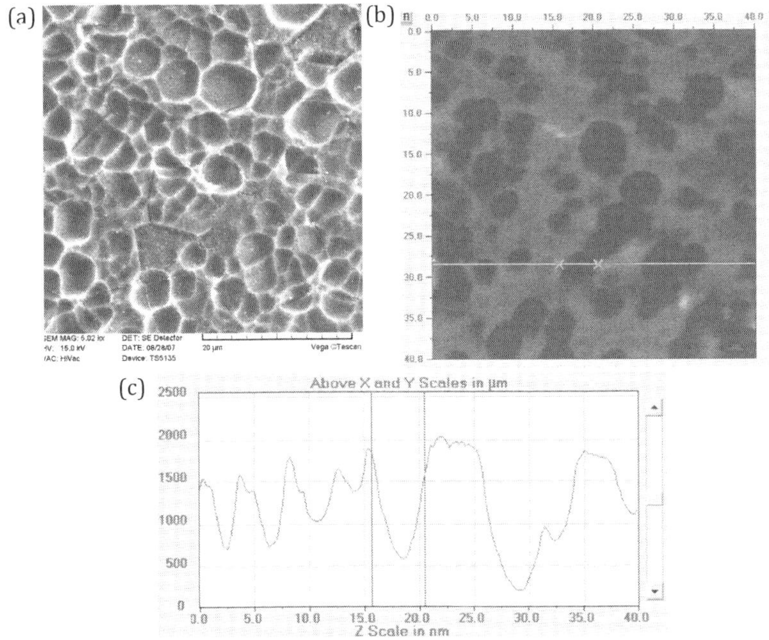

Figure 9.11 SEM (a) and AFM (b) images and cross section (c) of the titanium surface after etching in 1M H_3PO_4 + 10% HF at 10 V for 5 min [31].

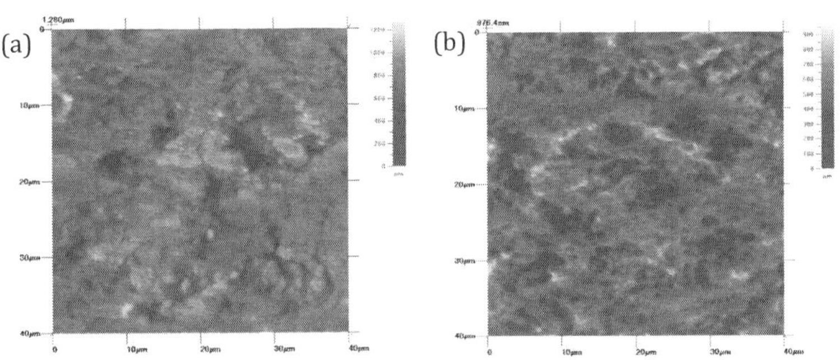

Figure 9.12 Porous TiO_x formed after etching at 1 V for 60 min in 1M H_3PO_4 + 0.35% HF (a) and 1M H_3PO_4 + 2% HF (b) [28].

On Fig. 9.13 is presented variation of the current density for titanium etched for 5, 15, 30, 60, and 120 min. A significant differences in current density between the start (initial time: 5 s) and the end (final time: 5, 15, 30, 60, 120 min) etching, curve (2) and (1), respectively. After etching at constant voltage 10 V and for different time, the initial starting current is almost constant (curve 2), but the end current decreases with the progress of etching time (curve 1). The observed current decay is attributed to a high electrical field oxide formation. For the samples etched in electrolytes containing different HF concentration, the current density increases with increasing HF concentration (Fig. 9.13b). The gap between the current measured at start and end time significantly increases with increasing HF concentration. The HF plays an important role in titanium etching. Tsuchiya *et al.* [88] observed that for higher applied voltages (>80 V), the current significantly increases after initial decreasing stage, which was related with breakdown of the metal oxide. In acidic electrolytes containing HF, the main and minor part of the current is consumed in dissolution reaction and TiO_2 formation, respectively [5, 50]. On the other hand, Macak *et al.* [50] found that using NaF instead of HF leads to the growth of thicker porous layer, because etching of TiO_2 in neutral electrolytes is slower than in acidic electrolytes. Increasing of the HF concentration in the electrolyte, results in high dissolution rate of the oxides. This phenomenon is possible even without anodization conditions (chemical etching).

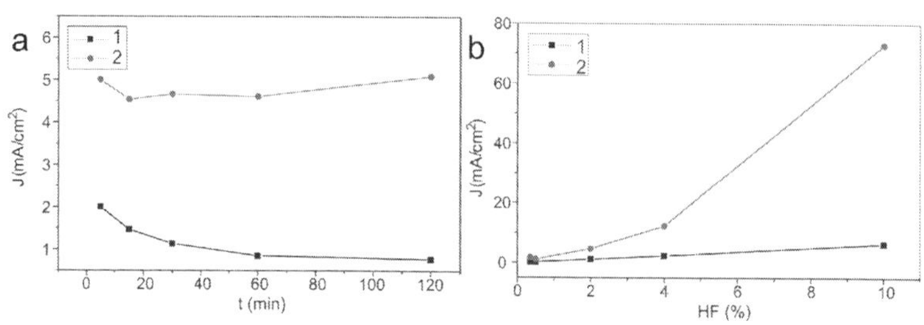

Figure 9.13 Variation of the current density for different etching time (1M H_3PO_4 + 2% HF at 10 V) (a) and for different HF content (1M H_3PO_4, 10 V for 30 min) (b); lines number: 1 and 2 present current density after end and start etching time, respectively [28].

The polarization curves clearly indicate the etching behavior. The polarization curves for Ti immersed in 1M H_3PO_4 with different HF content is shown in Fig. 9.14 [4]. The current densities of the polarization curves, recorded by Bauer *et al.* [4], increase two decades with HF concentrations and increasing from 0.0 to 1.0 wt%.

In the absence of the fluorides, the surface is clearly spontaneously passive. Addition of fluorides leads to the shift of the open circuit potential to the more negative values and the occurrence of an active passive transition. For the concentrations of 0.2 and 0.3 wt%, a second current increase occurs at about 1200 mV, which is related with anodic pore formation [4].

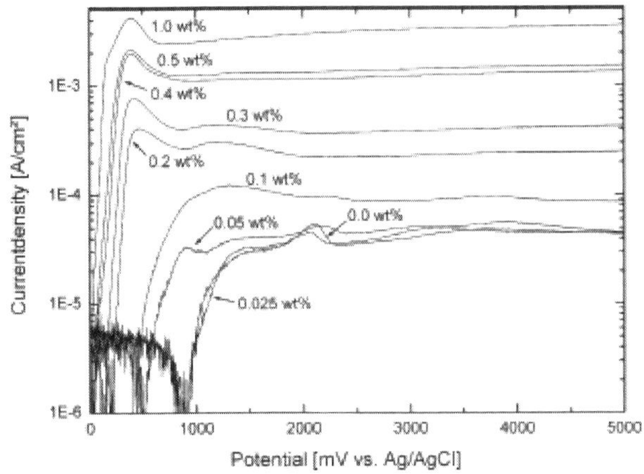

Figure 9.14 Polarization curves recorded from 0 V to 5 V at a sweep rate of 5 mV/s for different concentrations of HF [4].

During anodization, it is possible to find some characteristic stages of anodic oxidation on the recorded current curve (Fig. 9.15) [118]. Generally, the curve can be divided into four stages:

(i) In the first stage, a compact oxide barrier layer is formed, which leads the current to decrease significantly due to the low conductivity of metal oxide.

(ii) In the second stage, some cracks and narrow slits appear on the surface due to field-enhanced dissolution of the oxide layer, and the current starts to increase.

(iii) In the third stage, the current reaches a stable state, which corresponds to the random formation of porous structure in the slits and cracks. In this process, pore formation and dissolution of the oxide layer is possible as well (equilibrium of the pore formation with the pore dissolution).

(iv) In the fourth stage, when the dissolution rate is larger than the pore formation rate, the porous structure is consumed and current density decreases.

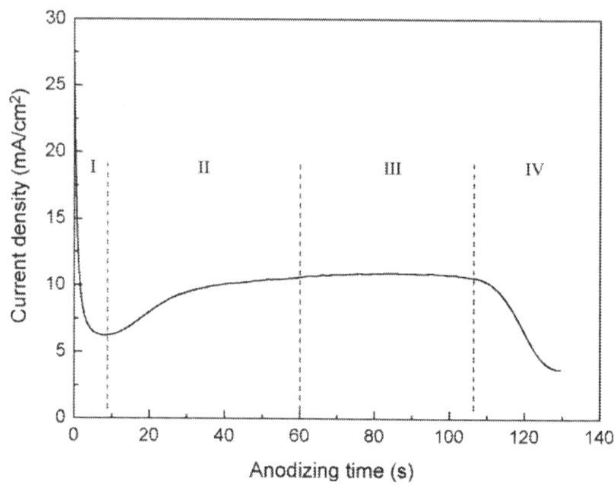

Figure 9.15 Current transient recorded during anodization [118].

Jakubowicz *et al.* [30, 32] investigated the biocompatibility of the porous surface obtained after etching of Ti in 1M H_3PO_4 + 10% HF electrolyte (results of human osteoblasts culture see chapter 12).

Mixture of H_3PO_4 + HF is a most often used electrolyte for Ti surface modification. Kim *et al.* etched Ti for 1 h in different concentration of H_3PO_4 + HF and voltage conditions [44]. In the osteoblast cells cultured for 1, 3, and 7 days (Fig. 9.16), they found that samples a, b, and g show a good initial attachment of the cells compared to the others, whereas proliferation levels for 7 days are not good compared to the others. For example, samples a and g show that the cell density for 1 day is almost the same as that for 7 days [44]. Samples d–f show distinct proliferation levels for the period of 7 days and in the case of sample e, a linear increase in the cell density is observed through 7 days, and the highest cell density was observed after culturing for 7 days [44].

The samples prepared in HF containing electrolytes allow incorporation of F^- ions in the oxide. Kim *et al.* suggest that the initial attachment of cells is much stimulated on a titanium oxide layer containing F^- ions [44]. The enhanced proliferation levels of cells in titanium oxides containing PO_4^{3-} (especially with higher concentration of H_3PO_4 electrolytes) can be attributed to the intrinsic properties of phosphate anionic groups that facilitate the pre-adsorption of proteins to promote cell attachment/growth on biomaterial surfaces [17, 20, 44, 119]. The morphological changes during etching in H_3PO_4 electrolytes (Fig. 9.17) are attributed to the fast oxide dissolution by F^- ions delayed via a competition reaction between F^- and PO_4^{3-} ions. This leads to a local dissolution of the oxide, allowing for the formation of nanotubes [44].

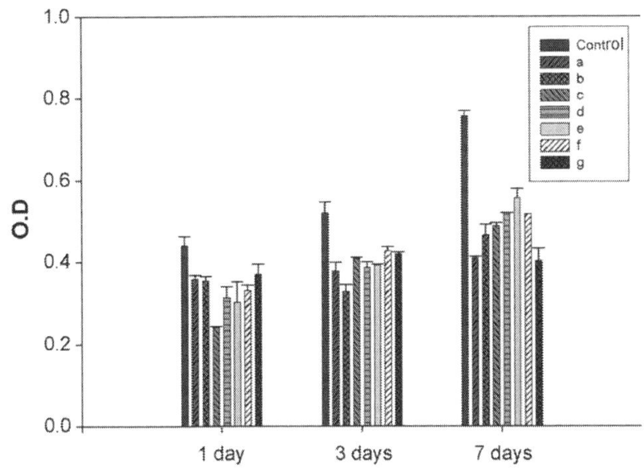

a: aqueous 1 wt% HF at 60 V.
b: aqueous 1M H_3PO_4 + 1 wt% HF at 60 V.
c: aqueous 5M H_3PO_4 + 1 wt% HF at 60 V.
d: aqueous 10M H_3PO_4 + 1 wt% HF at 60 V.
e: aqueous 1M H_3PO_4 at 60 V.
f: aqueous 1M H_3PO_4 at 200 V.
g: aqueous 1M H_3PO_4 + 1 wt% HF at 20 V

Figure 9.16 MTT results of osteoblasts cultured for 1, 3, and 7 days, respectively [44].

Figure 9.17 developed by Kim *et al.* is a map showing a different surface morphologies formed at different anodic conditions [44]. In 1 wt% HF electrolyte at 60 V, dot-like structures are produced due to a fast dissolution of the formed oxide (Fig. 9.17a). In 1M H_3PO_4 nanopowder consisting of granules is formed on the dot-like structures (Fig. 9.17b). Increase of H_3PO_4 concentration results in formation of single nanopowders (Fig. 9.17c) and coexistence of nanopowders with nanotubes (Fig. 9.17d). The microporous structures can be formed in a single H_3PO_4 (without HF) electrolyte above the breakdown potential (Fig. 9.17f). Barrier oxide layers with bursts and cracks are synthesized below the breakdown potential (Fig. 9.17e) and nanotubular structures can be formed in a mixture of aqueous 1M H_3PO_4 and 1 wt% HF at a moderate potential (Fig. 9.17g). As shown by Kim *et al.* [44], the

F^- ions show very good initial attachment of cells (group I), and PO_4^{3-} ions show a good cells growth for 7 days (group II) and these results are attributed to a low surface potential for favoring the cell attachment caused by F^- ions and osteoblast growth by phosphate ions, respectively [44].

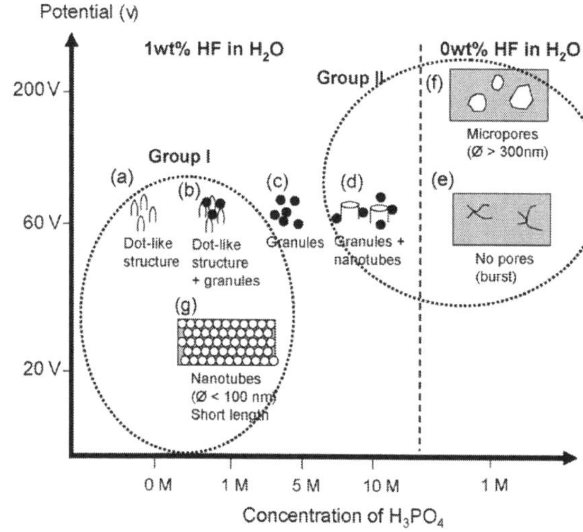

Figure 9.17 Morphology maps of anodic titanium oxides prepared at different anodization conditions; (a–f) are the results of samples denoted in Fig. 9.16. Group I shows a good attachment of cells at an initial stage and group II exhibits the good proliferation of cells for the period of 7 days [44].

Yang *et al.* [112] investigated electrochemical anodization of Ti to produce a nano/submicron-scale network oxide layer for biomedical implant application. Biological species (blood and cells) have various dimensions ranging from nm to μm. Yang *et al.* suggests that a mixed nano- and submicron Ti implant surface may have better initial responses to blood and cells [112]. SEM images of the Ti specimens with and without anodization obtained by Yang *et al.* are shown on Fig. 9.18. A multilayered nano/submicron-scale network is clearly visible. The lateral pore size for I_1 and I_2 specimens (both below 0.2 A, but $I_1 < I_2$) were approximately 20–110 and 30–160 nm, respectively. The thickness of the multilayered network was about 180 nm for I_1 and 320 nm for I_2. Increase of anodic current led to an increase in network pore size and thickness. This surface network layer structure is build mainly from anatase TiO_2-type structure.

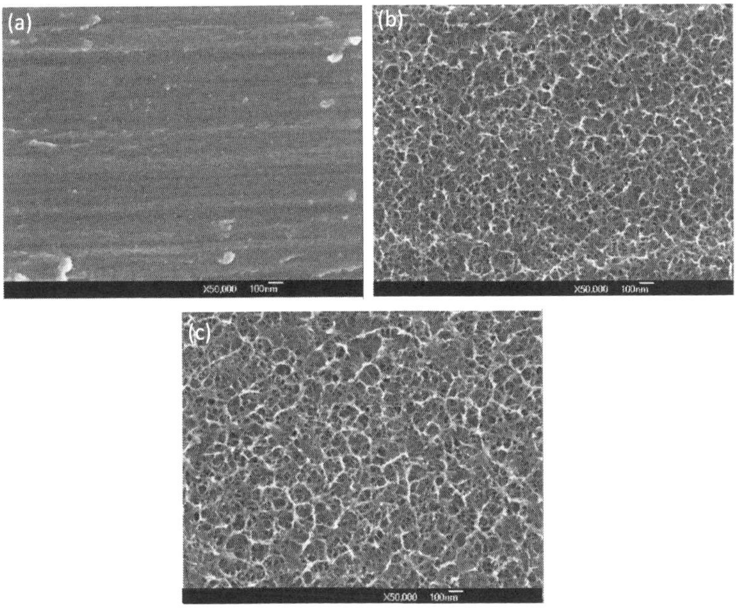

Figure 9.18 SEM images of Ti with and without electrochemical anodization treatment: untreated I_0 (a); anodized in 5M NaOH (b, c); current density I_1 for b is lower than I_2 for c [112].

Biological tissues interact mainly with the outermost atomic layers of the biomaterials when biomaterials are implanted into the body environment [112]. The optical photographs (Fig. 9.19) of the Ti specimens after 10 min of blood clot formation revealed that the blood clotted more significantly on the anodized I_1 and I_2 specimens than on the untreated I_0 specimen. The nano/submicron-scale of the TiO_2 network layer enhances the blood coagulation of the Ti surface [112].

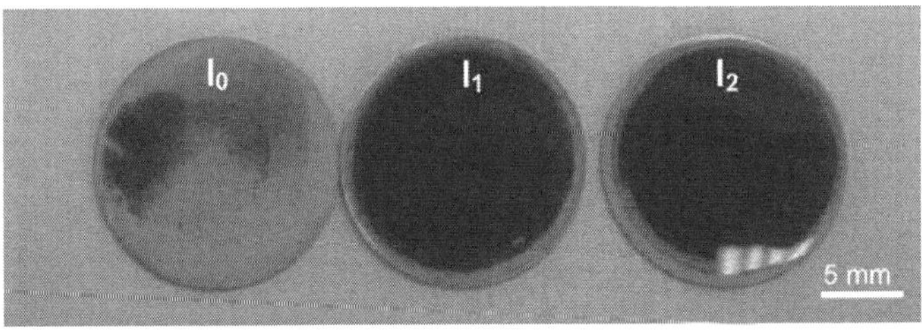

Figure 9.19 Optical photographs of the Ti specimens (I_0, I_1, and I_2) after 10 min blood clot formation, indicating the blood clotted more significantly on the anodized I_1 and I_2 specimens [112].

Figure 9.20 shows the SEM images of the attached human bone marrow stem cells (hBMSCs) on the Ti specimens after 24 h of cell culture [112].

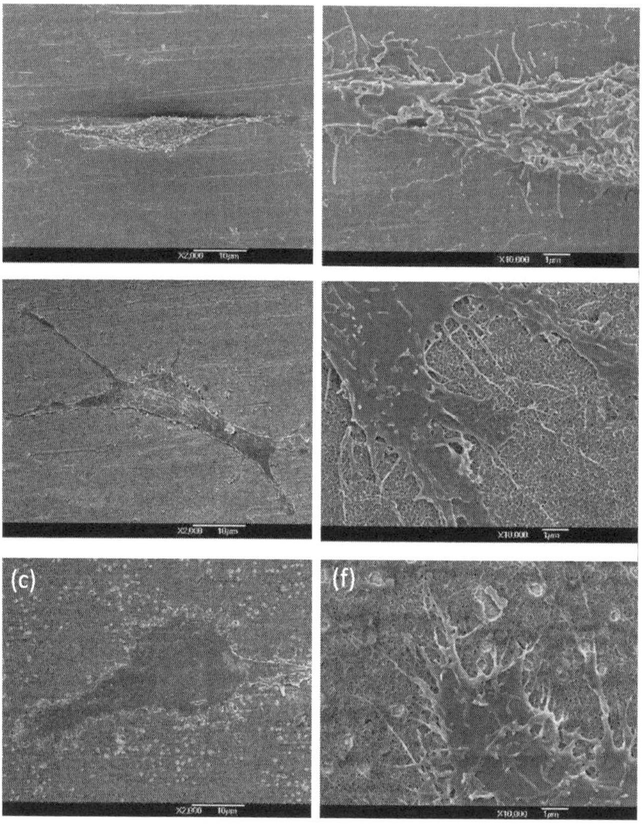

Figure 9.20 SEM images of the attached human bone marrow stem cells on the Ti specimens after 24-h cell culture (a, b, and c: I_0, I_1, and I_2 with magnification of 2000×; d, e, and f: I_0, I_1, and I_2 with magnification of 10000×) [112].

The attached hBMSCs exhibit a better spreading appearance on the anodized I_1 and I_2 specimens compared to the untreated I_0 specimen, which states a positive sign for the biocompatibility of the anodized nano/submicron-scale TiO_2 network surfaces [112]. The cell proliferation of hBMSCs on the anodized I_1 and I_2 specimens was significantly faster than the untreated I_0 specimen [112]. The XPS investigations of the Yang *et al.* show that higher proteins content existed on the anodized Ti specimen with respect to the untreated Ti specimen [112]. The results are consistent with those reported by Woo *et al.* [104]. They found that scaffolds with nano-fibrous pore walls adsorb more proteins than scaffolds with solid pore walls [104]. The nanoscale topography surface enhances the cell adhesion of fibroblast cells.

Oh *et al.* [63] showed that anodic oxidation of Ti at a very high voltages result in a surface attractive for biomedical applications. They kept anodization at a constant voltage of 180 V up to 60 min in H_2SO_4 + H_3PO_4 electrolytes. After anodizing at high voltages, a porous titania surface is formed (Fig. 9.21) with a mixture of the anatase and rutile phases on the Ti substrate (Fig. 9.22) [63].

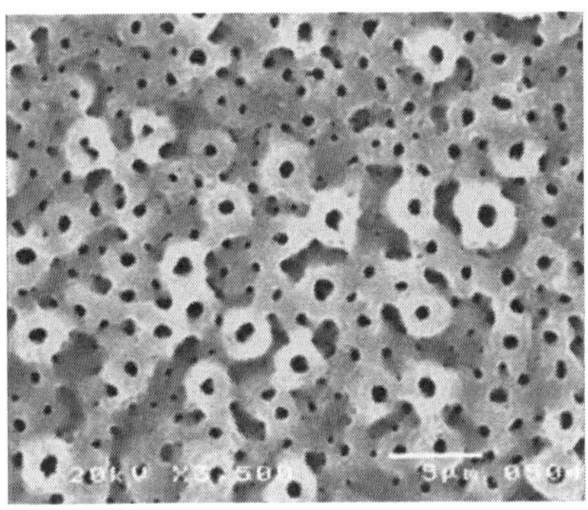

Figure 9.21 SEM image of the surface morphologies of the anodic TiO_2 film prepared at 180 V in 0.9M H_2SO_4 + 0.1M H_3PO_4 for 30 min [63].

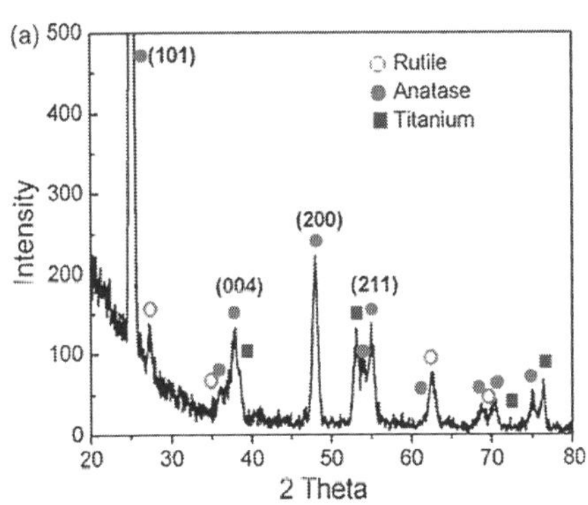

Figure 9.22 X-ray diffraction pattern of the anodic titania film formed at 180 V in 0.9M H_2SO_4 + 0.1M H_3PO_4 for 30 min [63].

The cross section of the anodic titanium oxide film obtained by Oh *et al.* [62] formed by electrochemical method at 180 V for 30 min in 1.5M H_2SO_4 + 0.3M H_3PO_4 + 0.3M H_2O_2 electrolyte is shown in Fig. 9.23. The variation of pore layer thickness and diameter is shown, too. The growth of the pore diameter of cell structure with anodic time increases rapidly in the beginning stage of anodization. The pore diameter and layer thickness of TiO_2 increases with anodization time. The anodic film thickness is dependent on anodic time with a rate of 3.15×10^{-2} μm/min at 180 V [62].

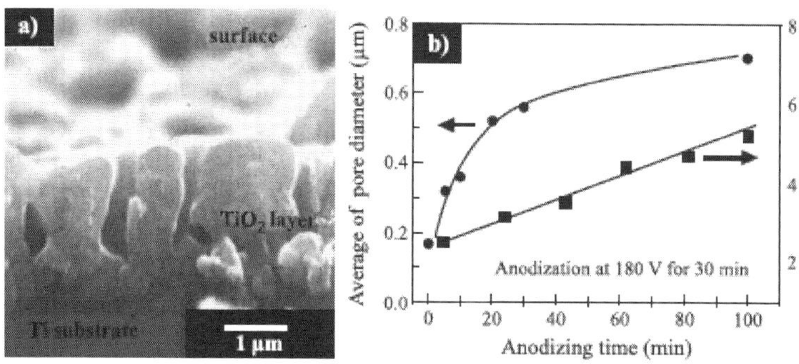

Figure 9.23 Cross-sectional image of TiO_2 film (a) and the relationship between average pore diameter and anodic film thickness with anodizing time at 180 V in 1.5M H_2SO_4 + 0.3M H_3PO_4 + 0.3M H_2O_2 electrolyte (b) [62].

The bioactivity of the as-prepared surface is improved with the immersion in SBF for 3 days, resulting in the formation of surface Ca-P compounds. The nucleation of Ca-P compounds is made convenient by ions introduced to the surface layer during anodization [47, 63]. The depth profiles of phosphorus concentration in anodic titania (Fig. 9.24) show that the degree of the residual phosphate concentration in anodic oxide layer increases with concentration of H_3PO_4 in electrolyte [47]. The species containing phosphate ions infiltrate into the oxide film during anodization [55, 62] and these ions in electrolyte penetrate more easily into the oxide/electrolyte interface with the phosphoric acid concentration [47]. In the surface layer, phosphorus is in the forms of mainly HPO_4^-, PO_4^-, and PO_3^-. Therefore, these negatively charged species on the anodic titania surface can act as preferential nucleation sites of calcium phosphate by attractive interaction with Ca^{2+} ions in SBF [47].

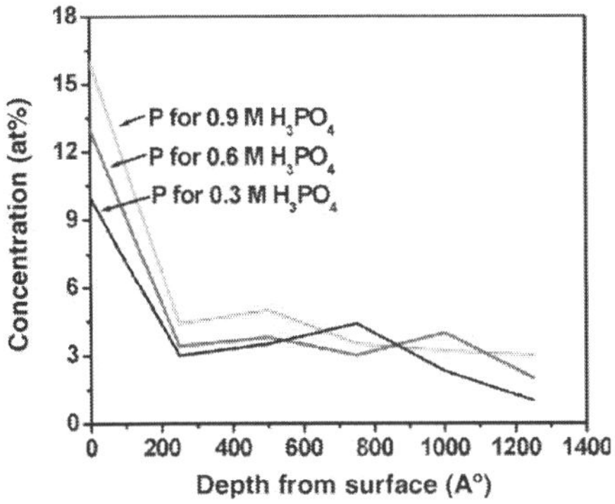

Figure 9.24 Depth profile of anodic TiO_2 layer formed in electrolytes [47]: 1.5M H_2SO_4 + 0.3M H_3PO_4, 1.5M H_2SO_4 + 0.6M H_3PO_4, 1.5M H_2SO_4 + 0.9M H_3PO_4. See also Color Insert.

Yu *et al.* [118] investigated formation of nanostructured TiO_2 thin films deposited on silicon, and silicon carbide substrates by DC magnetic sputtering, with subsequent anodization in acidic 0.1–2.0 wt% HF, and neutral electrolytes containing 1M $(NH_4)_2SO_4$ and 0.5 wt% NH_4F. They described the effect of the electrolytes and substrates on the microstructure of the anodized thin films. Nanoporous films (Fig. 9.25) formed on silicon substrates with an average pore diameter of 25 nm and interpore distance of 40 nm was obtained [118]. Yu *et al.* showed the possibility of the formation of anodic film on silicon carbide substrate, too [118]. They did the anodization in 0.3 wt% HF electrolyte at 2.5 V. The obtained film was composed of the individual clusters that exhibited a laminated structure (Fig. 9.26). The unique microstructure is related with different orientation of the titanium film deposited on silicon carbide substrate.

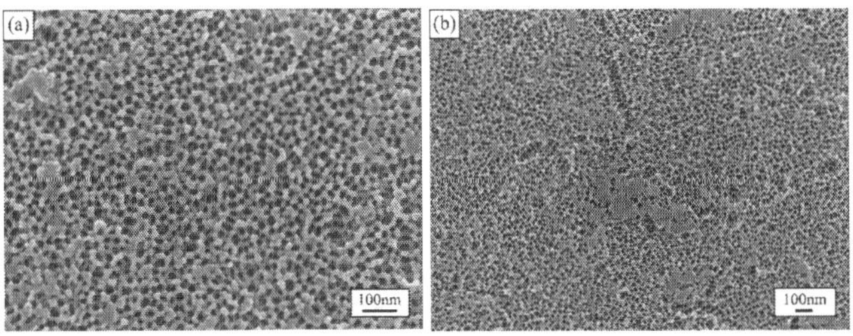

Figure 9.25 SEM images of nanoporous TiO_2 thin film on silicon substrate formed in 0.5% HF electrolyte at 3 V; high (a) and low (b) magnifications [118].

Figure 9.26 SEM images of the anodic film with a layered structure on silicon carbide substrate [118].

Anodization of Ti is not limited to dental or orthopedic implant surfaces only. The process can be applied to dental arch wire, too [110], resulting in nanosurface topography. The arch wire provides torsion and bending forces, and through the brackets, the forces are exerted on the teeth. Stainless alloys, NiTi-based alloys, and β-Ti-based alloys are most widely used materials for the dental arch wires [103, 110]. NiTi-based alloys and β-Ti-based alloys are the shape memory alloys and have unique super elastic properties [70, 72]. By carefully choosing the anodization conditions, the arch wires can be produced with red, yellow, green, and other kinds of colors. The colors of the wires are caused by light interferences through the oxide layers, indicating oxide thickness. Yang *et al.* [110] used different electrolytes for respective surface preparation. They applied 1M H_2SO_4, 1M H_3PO_4, 3 wt% $Na_2SO_4 \cdot 10H_2O$, and 5 wt% $Na_3PO_4 \cdot 12H_2O$. The anodization was conducted at 10–60 V, and at a constant temperature of 25°C. The anodization time was 1 and 10 min for the β-Ti wires, and 120 min for the NiTi wires. The colors of the anodized wires vary with the applied voltages, and the wires of a wide spectrum of colors were obtained (Fig. 9.27, Fig. 9.28). The TiO_2 layers are found on both kinds of anodized wires, and no nickel element was detected on the surface of the anodized NiTi wire [110]. The thickness of the oxide layers increases, with the increasing voltages and time of anodization (Table 9.1). The oxide layer is uniformly formed on the anodized β-Ti wires, but the surface roughness of the NiTi wires increases after anodization (Fig. 9.29). The NiTi wires are not as bright as the β-Ti wires, which is related to oxide layer uniformity (more uniform for β-Ti wires). Since anodization only modifies the surfaces, the tensile properties of both the two anodized wires remain unchanged [110].

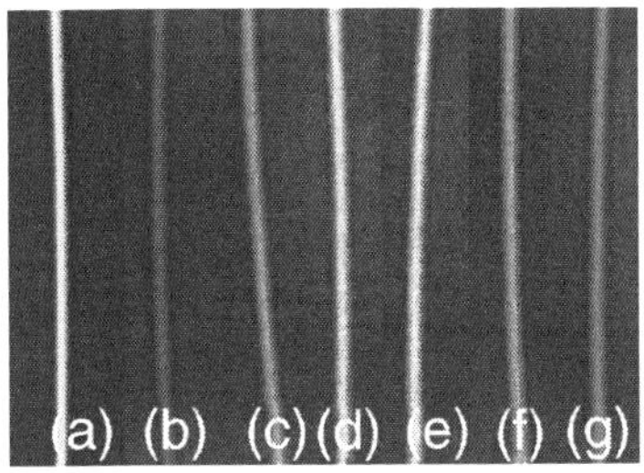

Figure 9.27 The optical micrographs of the β-Ti wires anodized in 1M H_2SO_4 electrolyte for 10 min: (a) original (non-anodized); (b) at 10 V, (c) at 20 V, (d) at 30 V, (e) at 40 V, (f) at 50 V, and (g) at 60 V [110]. See also Color Insert.

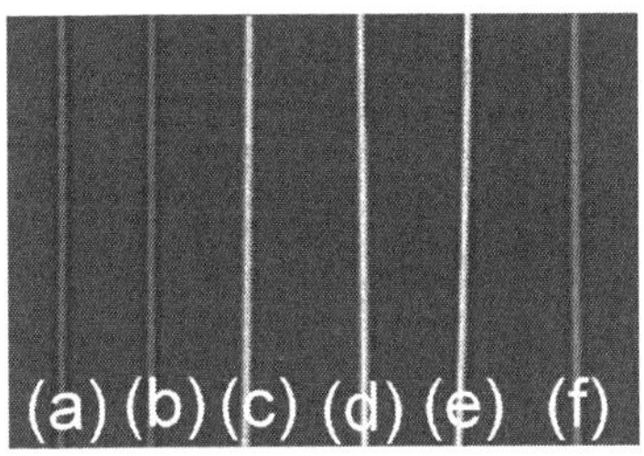

Figure 9.28 The optical micrographs of the β-Ti wires anodized in 1M H_3PO_4 electrolyte for 10 min: (a) at 10 V, (b) at 20 V, (c) at 30 V, (d) at 40 V, (e) at 50 V, and (f) at 60 V [110]. See also Color Insert.

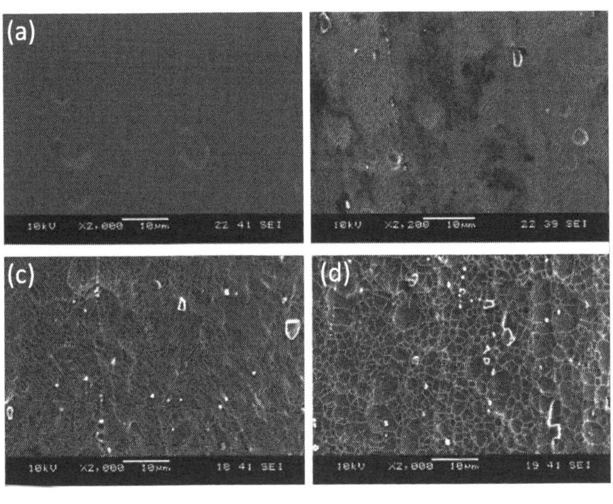

Figure 9.29 The SEM images of the NiTi wires anodized in a 1M sodium sulfate electrolyte for 120 min: (a) original (non-anodized), (b) at 10 V, (c) at 20 V, (d) at 30 V, and (e) at 40 V [110].

Table 9.1 Thickness of the oxide layers of the β-Ti wires anodized for 10 min [110]

Electrolyte	**Voltage (V)**	**Oxide color**	***d*(AES) $m\lambda$ = 2ndcosθ (Å)**	**$n(TiO_2)$ = 2.4**	***d* (interference) (Å)**
H_2SO_4	20	Blue	654	$m = 1, \lambda_{blue}$ = 4625 (Å)	964
H_2SO_4	40	Yellow	1221	$m = 1, \lambda_{yellow}$ = 5675 (Å)	1182
H_2SO_4	60	Purple	2380	$m = 2, \lambda_{purple}$ = 4100 (Å)	1708
H_3PO_4	50	Gold yellow	1631	m=1, λ_{blue} = 4625 (Å)	1182
				$m = 1, \lambda_{red}$ = 4625 (Å)	1422
H_3PO_4	60	Pink	2117	$m = 2, \lambda_{orange}$ = 4625 (Å)	2844
				$m = 2, \lambda_{orange}$ = 4625 (Å)	2531

Sul [81] investigated implants that contain sulphur (S), phosphorus (P) and, calcium (Ca) anions electrochemically incorporated into the TiO_2 matrix from sulphuric acid, phosphoric acid, calcium containing mixed electrolyte system, respectively (Table 9.2). The implants showed two types of surface morphologies (Fig. 9.30): a nonporous structure in control implants and a porous structure in test S, P and, Ca implants. The pore diameter of test groups was 0.2 μm in S-containing implants, 1.5 μm in P-containing implants, and 1.3 μm in Ca-containing implants.

The quantitative *in vivo* results of the implant loosening torque were obtained by Sul [81] through the removal torque test, reflecting the interfacial shear strength. All oxidized implant groups showed higher mean peak values of removal torque than control groups. Ca implants revealed the highest torque values of all implant groups (Fig. 9.31).

Bone to metal contact (BMC) measurements in all implants demonstrated significant differences in comparison with the paired control groups. Figure 9.32 shows comparisons of the bone contact in all threads between test groups and the paired control groups. The increase of mean BMC values was 186% in S implants, 232% in P implants, and 272% in Ca implants compared to the paired control groups. The mean BMC values were 39% in S implants vs. 21% in the paired controls, 44% in P implants vs. 19% in the paired controls, and 49% in Ca implants vs. 18% in the paired controls [81].

Table 9.2 Surface oxide characteristics of the control and tests of the screw-shaped cp titanium implants [81]

Oxide characteristics	Turned implant	S implant	P implant	Ca implant
Chemical composition	Primarily TiO_2 Contaminant C, Traces Ca, Na, Si < 1%	Primarily TiO_2 and S ≤ 1 at.% Contaminant C, Traces Ca, Na, Si<1%	Primarily TiO_2 and P ≤ 8 at.% Contaminant C, Traces Ca, Na, Si < 1%	Primarily TiO_2 and Ca ≤ 11 at.% Contaminant C, Traces Ca, Na, Si < 1%
Oxide thickness	17 ± 6 nm	1080 ± 324 nm	1224 ± 144 nm	1296 ± 225 nm
Morphology	Nonporous structure being with turned grooves ≤10 μm	Porous structure being with a number of craters	Porous structure being with a number of craters ≤1.5 μm by length	Porous structure being with a number of craters ≤1.3 μm by length
Pore size distribution		≤0.2 μm by length	≤1.5 μm by length	≤1.3 μm by length
Crystallinity	Amorphous	Strong anatase	Amorphous	Anatase
Roughness Sa	0.83 ± 0.32 μm	1.04 ± 0.42 μm	0.83 ± 0.29 μm	0.85 ± 0.32 μm
Roughness Scx	9.78 ± 1.4 μm	12.05 ± 3.74 μm	11.19 ± 2.33 μm	9.83 ± 1.07 μm

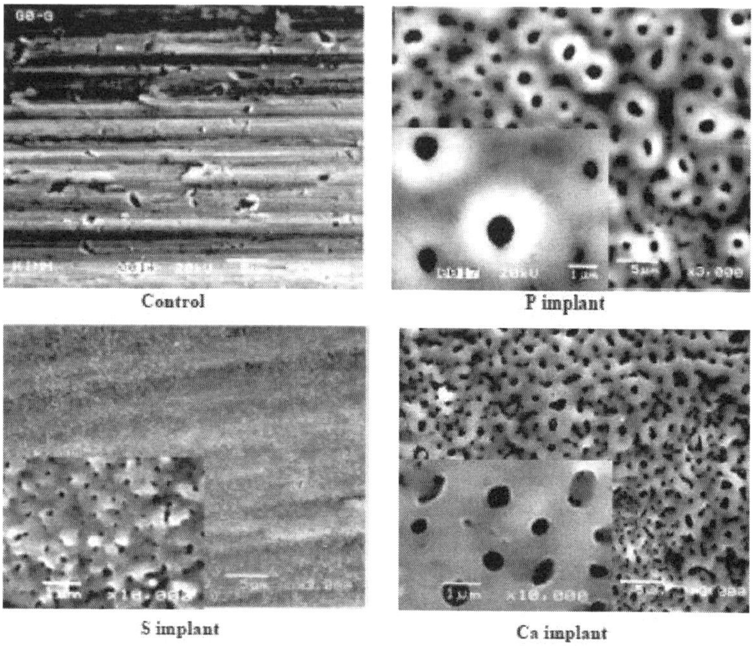

Figure 9.30 SEM images at ×3000 (×10000 in box) show non-porous structure of the turned screw implants and porous structure of test S-, P-, and Ca-containing implants [81].

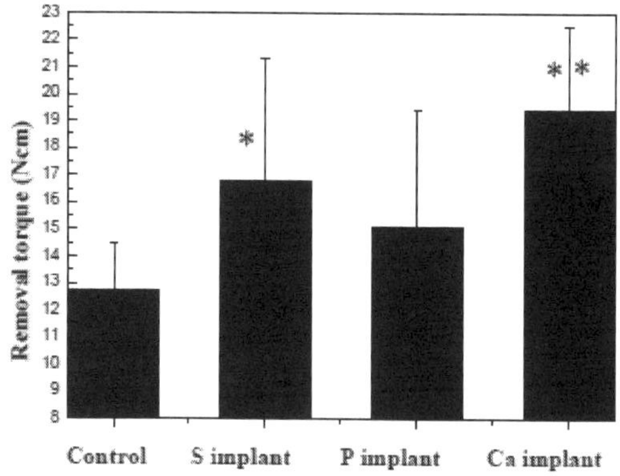

Figure 9.31 Mean removal torque values (N · cm) after 6 weeks of healing time, demonstrating statistically significant differences between S implants and controls, and Ca implants and controls, respectively (*P < 0.005; **P < 0.001) [81].

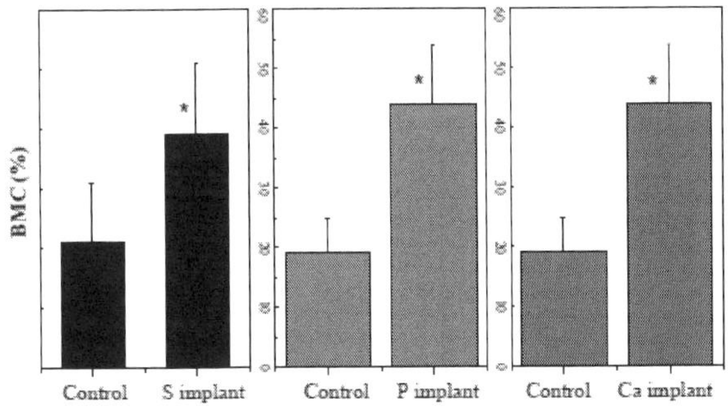

Figure 9.32 Comparisons of the bone to metal contact (BMC) in all threads of control and test S, P and, Ca implants after 6 weeks of healing time (*P < 0.005) [81].

Sul [81] shows that the electrochemically oxidized nanoporous Ca-containing implant demonstrated the strongest bone response regarding removal torque test and histomorphometrical quantifications between the four groups. S and P implants also showed significantly stronger bone tissue reactions than the controls. The improved bone reactions of S-contained implants are most likely due to the topographical properties (porous structure) while the strong bone reactions of P and Ca implants may be attributed to the chemical properties [81]. Electrochemically oxidized implants of 200 nm or less oxide thickness showed no significant differences in bone response in comparison to controls but implants with 600 nm or more oxide thickness showed a significantly enhanced bone response in a rabbit model [81]. For obtaining thick micrometer anodic oxide layer, phosphoric/sulphuric acid electrolytes are definitely useful.

Das *et al.* [14] investigated *in vitro* cell–materials interactions using human osteoblast cells on the anodized titanium in H_2SO_4, HF and H_3PO_4 electrolytes at anodizing voltage of 20 V. Figures 9.33 and 9.34 show the SEM images of the cell morphologies on Ti-control (without oxidation) and anodized surfaces after 3 and 11 days of cell culture, respectively. Cells on Ti-control (Fig. 9.33a) show an elongated, flattened morphology. Cells on the H_2SO_4 anodized surface show a rounded morphology (Fig. 9.33b), whereas on the H_3PO_4 oxidized surface, they show good cell attachment (Fig. 9.33c). The HF anodized surface is uniformly covered with cells and filopodia extensions were observed (Fig. 9.33d). After 11 days of culture, the cells on the Ti-control surface have flattened elongated morphology with many filopodia extensions from the cell to the substrate (Fig. 9.34a). In H_2SO_4 oxidized surface (Fig. 9.34b), the surface area of cells attachment is minimized. The H_3PO_4 oxidized surface shows excellent cell spreading and proliferation (Fig. 9.34c), forming a three-dimensional fibril network. In the HF treated surface, the colonization of the cells was noticed, with cells developing a double net-like layer structure (Fig. 9.34d) [14].

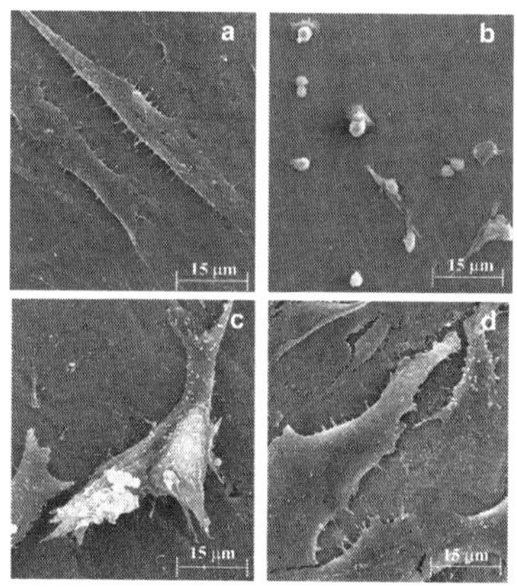

Figure 9.33 SEM images illustrating OPC1 cell adhesion after 3 days of culture on (a) control-Ti and oxide surfaces anodized in (b) H_2SO_4, (c) H_3PO_4 and (d) HF electrolytes [14].

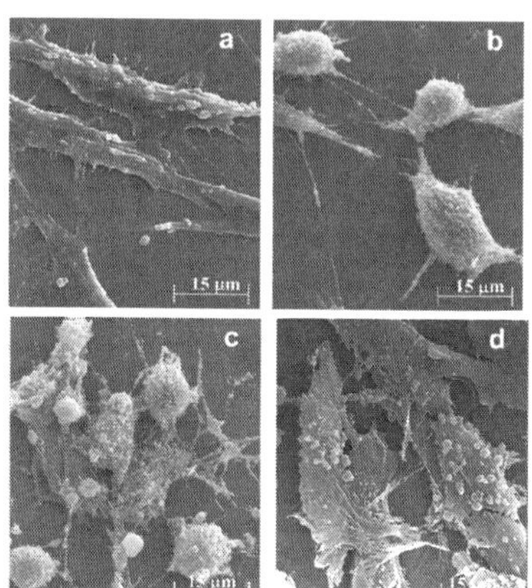

Figure 9.34 SEM images showing the cellular adhesion and proliferation on (a) control Ti, oxide surfaces anodized in (b) H_2SO_4, (c) H_3PO_4 and (d) HF electrolytes after 11 days of incubation [14].

A comparison of cell densities on the different anodized surfaces for 5, 11, and 16 days was shown, as well (Fig. 9.35) [14]. For all days of culture, surfaces etched in H_3PO_4 electrolytes show the highest cell density. Cell density for the anodic oxide increases in the order of H_3PO_4 < HF < H_2SO_4 < Ti-control. The HF and H_3PO_4 oxidized surfaces are rougher than H_2SO_4 and polished Ti-control surfaces. Therefore, the surface morphology and properties, such as high roughness, low values of contact angles, high wettability, and high surface energy plays a crucial role in cell attachment. The rough H_3PO_4 surface with improved wettability enhances the cell attachment process [14].

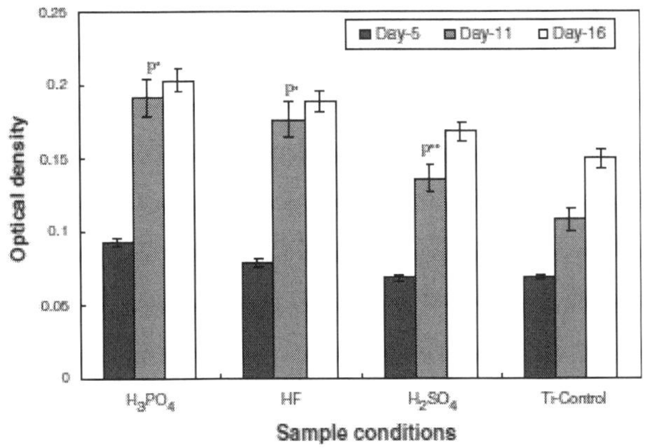

Figure 9.35 Optical density measured after culture for 5, 11, and 16 days [14].

9.2.2 TiO_2 Nanotubes and Nanofibers

Nanotubes or nanofibers have shown good biological behavior, for example, improved cell adhesion and differentiation, and therefore exhibit potential in osseointegration for dental and bone implants applications.

Anodization of titanium can result in pores [28] or nanotubes formation [66, 122]. Beside anodization [19, 51], the titanium oxide nanotubes can be prepared by various techniques such as sol–gel method [46], electrophoretic deposition [56], hydrothermal method [87], seed growth [86], and template-assistant deposition [77]. Uniform TiO_2 nanotube arrays dimensions can be carefully controlled by adjusting electrolyte composition, pH, and voltage

[10, 18, 84, 120]. It is known that an increase in nanotube length enhances the effective surface area and the resistance of the oxide film, which is promising in implant applications.

Yang *et al.* [113] have used a two-step anodization procedure at potential 20 V for the nanotube formation (Fig. 9.36). The two-step anodization consists of the following:

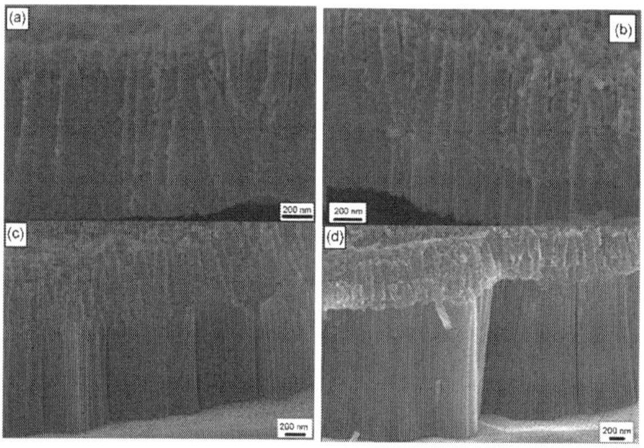

Figure 9.36 SEM cross-sectional images of TiO_2 nanotube arrays fabricated by two-step anodization for different step-2 anodization duration: (a) 0.5 h, (b) 1 h, (c) 2 h and (d) 4 h [113].

(i) Anodization of titanium foil in an aqueous electrolyte (1M H_3PO_4 + 0.5 wt% HF) for 2 h, followed by rinsing with DI water and drying in ambient air

(ii) Anodization of the specimen in a nonaqueous electrolyte (glycerin containing 0.5 wt% NH_4F)

They found that TiO_2 nanotubes formed in aqueous electrolyte (H_3PO_4 + HF) have a much wider diameter compared to nanotubes formed in nonaqueous electrolyte (glycerin + NH_4F). The morphologies of the TiO_2 nanotubes formed by anodization are strongly dependent upon the electrolyte composition. Figure 9.36 shows the SEM cross-section images of the TiO_2 nanotubes arrays fabricated by two-step anodization. After anodization for 0.5 h, another layer of TiO_2 nanotubes with 150 nm in length grows underneath the already formed nanotubes. The extension of the anodization time to 4 h results in the 2 μm TiO_2 nanotubes growth.

The two different layers of nanotubes connect closely, exhibiting any exfoliation. Yang *et al.* [113] suggest that the new layer of the TiO_2 nanotubes grows directly from the bottom of the already formed nanotubes. The growth mechanism of graded TiO_2 nanotube arrays is schematically presented in Fig. 9.37. After the step-1 anodization (in H_3PO_4 + HF), the highly ordered TiO_2 nanotubes grow upon a barrier layer of the titanium oxide and hydroxide (Fig. 9.37a). This layer of TiO_2 nanotubes has rough side-wall and wide diameter. The formation of rough side-wall may be due to the voltage oscillations during the anodization process in aqueous electrolytes. After immersing the already formed TiO_2 nanotubes in glycerin + NH_4F electrolyte, the electrochemical environment alters due to the changes of the electrolyte composition. Dissolution and breakdown of the barrier layer at the bottom of the already formed TiO_2 nanotubes occur in the initial stage of the step-2 anodization (Fig. 9.37b). The formation of the breakdown sites is due to the high electric field intensity at the bottom of the already formed TiO_2 nanotubes. The breakdown sites act as seeds to the growth of a new layer of TiO_2 nanotubes (Fig. 9.37c) and the new nanotubes grow directly from the breakdown sites, which are the bottom of the already existing TiO_2 nanotubes. By extending the time of the step-2 anodization, graded TiO_2 nanotube arrays can be formed (Fig. 9.37d). The anodization in step-1 can produce higher electric field intensity and hence faster chemical dissolution rate. In the step-2 anodization, at lower electric field intensity, a slower chemical dissolution rate dominates.

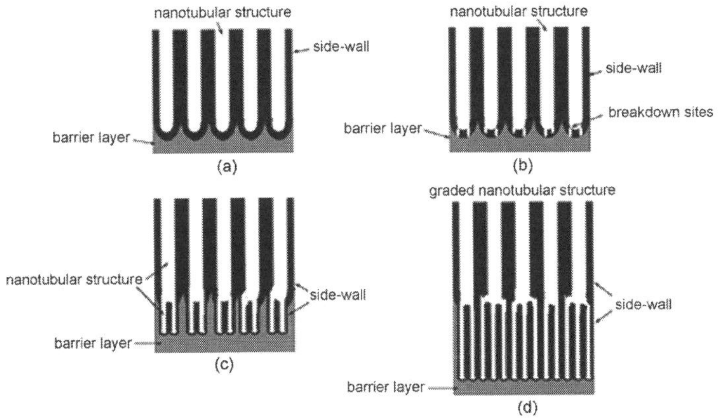

Figure 9.37 Growth mechanism of graded TiO_2 nanotube arrays: (a) already formed TiO_2 nanotubes by the step-1 anodization, (b) breakdown of the barrier layer in the bottom of the already formed TiO_2 nanotubes, (c) growth of a new layer of TiO_2 nanotubes, (d) graded nanotubular structure [113].

A possible growth mechanism of the nanotubes compared with the nanoporous structure was proposed by Yu *et al.* [118] and is presented in Fig. 9.38 (based on the assumptions of Fig. 9.26). For samples anodized in aqueous HF solutions, a compact oxide layer is formed at the initial stage of anodization, followed by the random generation of small pits on it (Fig. 9.38a). Then at steady-state stage of dissolution (Fig. 9.26), pores grow on the basis of such pits (Fig. 9.38b) and a porous structure forms. For samples anodized in F^--containing neutral electrolytes (with NH_4F), the interpore regions are also attacked by the F^- ions (Fig. 9.38c). Slits are generated at these parts and cause the formation of the nanotubes (Fig. 9.38d) [118].

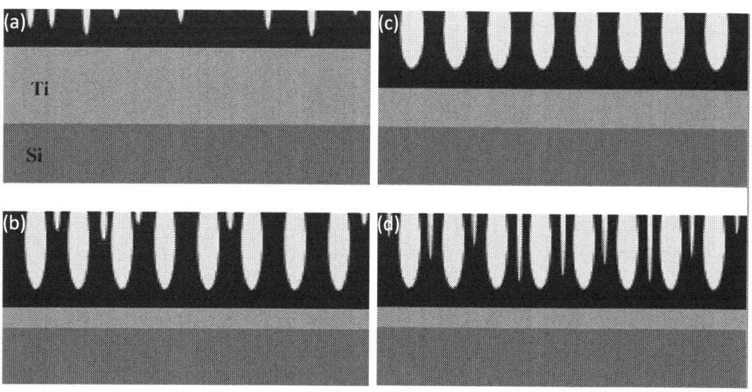

Figure 9.38 Schematic diagram of (a) pits formed on the compact oxide, (b) growth of nanoporous structures, (c) dissolution at the inter-pore region and (d) formation of discrete nanotubes [118].

Crawford and Chawla [13] also proposed a comparable, more evaluated mechanism of surface changes during anodization of Ti (Fig. 9.39). They fabricated the TiO_2 porous and tubular coatings by anodic oxidation in 1M H_2SO_4 + 0.1M NaF solution. They found that oxidized film can consist of large (1 to 20 μm) pores and small nanotubes (50 nm diameter). For the bioactive applications, the nanoporous structure enhances bioactivity and osteoblast function [41, 67, 68, 90, 114] and microporous structure improve a bone ingrowth and mechanical fixation [8].

Crawford and Chawla described the evolution of the TiO_2 microstructure in four stages (Fig. 9.39). In the initial stage 1 of anodization, a large decrease in the current density with time is observed due to the formation of a dense oxide layer by the dissolution of Ti:

$$Ti + 2H_2O \rightarrow TiO_2 + 4H^+ + 4e^- \quad (9.8)$$

The final thickness of this oxide layer is related to the applied potential. In stage 2, in the presence of F^-, chemical dissolution of TiO_2 is possible:

$$TiO_2 + 4H^+ + 6F^- \rightarrow TiF_6^{2-} + 2H_2O \quad (9.9)$$

Stage 2 is characterized by the formation of nanopores across the surface of the TiO_2 layer. The formation of nanopores results in an increase in current density, local dissolution of TiO_2, and increasing the electric field intensity at the bottom of the pore. The increased current density drives the formation of new oxide at the metal/oxide interface while also increasing the rate of chemical dissolution at the oxide electrolyte interface [13].

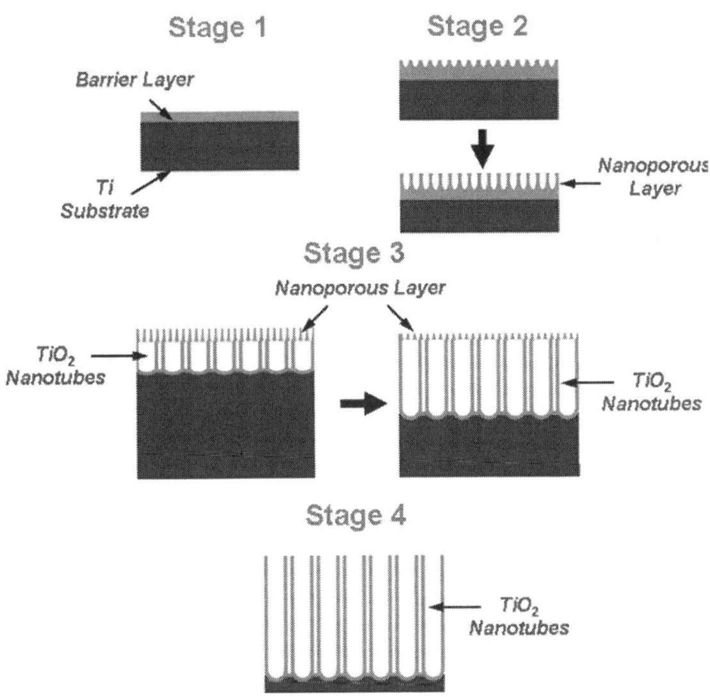

Figure 9.39 Schematic showing the mechanism for TiO_2 nanotube formation [13].

In stage 3, the nanopores continue to grow into the oxide layer and there is an abrupt transition from nanopores to nanotubes, so a bi-layered structure is visible [13]. The top and bottom layers consist of the nanopores and the nanotubes, respectively. In this stage, nanotubes grow deeper into the substrate due to the competition of the oxide growth and dissolution at the bottom of the nanotube. As nanotubes grow, the nanoporous layer is chemically dissolved, resulting in thinning or disappearance. In Crawford and Chawla's experiments, the nanoporous layer disappears after 90 min of anodization [13]. In stage 4, once the nanoporous layer has undergone complete dissolution, the coating consists of an ordered array of TiO_2 nanotubes. With anodization time, the nanotubes continue to grow in length (Fig. 9.40), and the current density continues to drop, until the both current density and coating thickness stabilize. Evolution of four distinct layers within TiO_2, (barrier layer, nanoporous layer, TiO_2 nanotube layer, and precipitate layer) measured by Crawford and Chawla [13] is shown in Fig. 9.40. They have observed a thick, relatively dense precipitate layer on the surface of the nanotube coating (Fig. 9.41). The pre-cipitate layer grows in thickness rapidly between 2 and 4 h (Fig. 9.40). The presence of this precipitate layer is likely associated with the rapid dissolution of Ti. For samples anodized at more acidic pH <5.0 they did not observe a precipitate layer [13]. For longer etching times (20 h), the layer disappears.

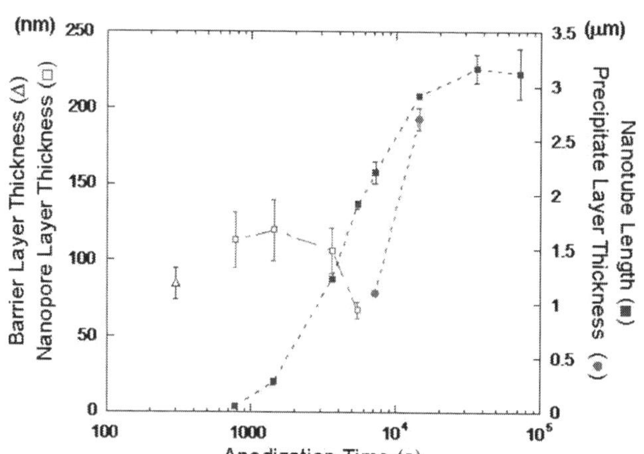

Figure 9.40 Layer thickness versus anodization time for the barrier layer, nanoporous layer, precipitate layer, and nanotube layer [13].

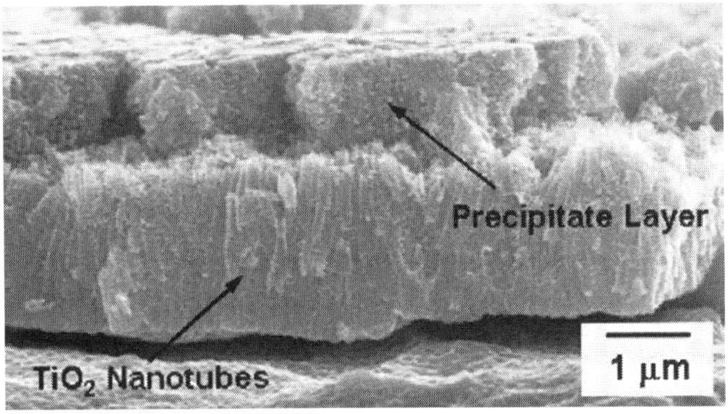

Figure 9.41 SEM cross-section of the TiO_2 nanotube coating and precipitate layer formed by anodization of Ti for 2 h in 1M H_2SO_4 + 0.1M NaF solution at a constant potential of 20 V and pH 5.0 [13].

The pH and temperature plays an important role in the nanotube length (Fig. 9.42, Fig. 9.43). The pH has no noticeable effect on the diameter of the TiO_2 nanotubes, but the nanotube length increases drastically with increasing electrolyte pH. This effect is related to the enhanced chemical dissolution rate of TiO_2 in acidic solutions. With increasing electrolyte temperature, the rate of chemical dissolution increases, resulting in a thinner film of the nanotubes.

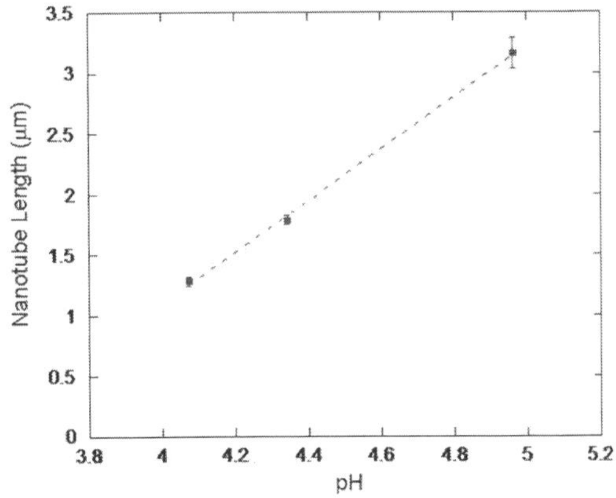

Figure 9.42 Nanotube length (coating thickness) vs. pH for the anodization of Ti for 10 h in 1M H_2SO_4 + 0.1M NaF solution at a constant potential of 20 V, at 24°C [13].

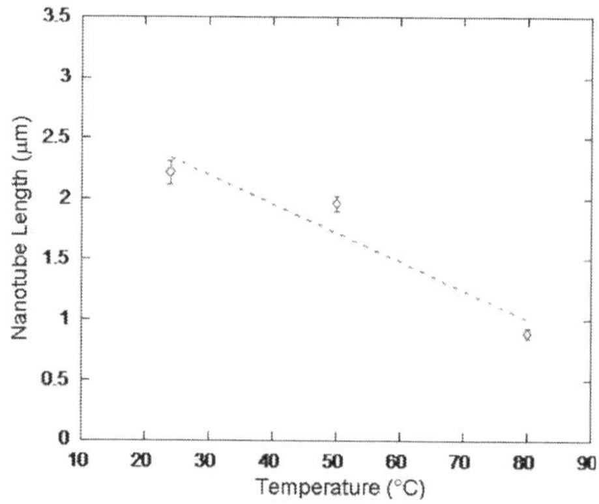

Figure 9.43 Nanotube length (coating thickness) vs. electrolyte temperature for the anodization of Ti for 2 h in 1M H_2SO_4 + 0.1M NaF solution at a constant potential of 20 V, and pH 5.0 [13].

Chemical dissolution and field-assisted dissolution were considered to play key roles in the formation of nanotubes [74]. Mor *et al.* [58] suggest that unanodized materials part could exist between the pores and field-assisted oxidation/dissolution of these interpore regions, resulting in the formation of voids. The growth of voids in equilibrium with pores has been considered to render nanotubular structure. Slight altering electrochemical condition could result in pores or tubes formation. Both structures can be produced in comparable electrolytes. The fluoride ions are necessary in the formation of nanoporous structure [123], but when anodization is conducted at higher potentials (>200 V), the porous film is possible to do in non-fluoride containing solutions [121]. Because the dielectric breakdown occurs, the pores are not uniform with cracks in the oxide film. Formation of nanoporous layer in fluoride solutions required a much lower voltage (<40 V) while the pores are well separated and self-ordered.

The anodic oxidization is a relatively simple technique for preparing highly oriented uniform nanotube arrays. The anodization can be conducted at either high or low potentials of typically [5, 10, 58, 80, 92]. At low potential, chemical dissolution of the titanium oxide play a key role in the formation of nanotube arrays. The formation is determined by the balance between the electrochemical oxidization and the chemical dissolution [10]. When the electrochemical oxidization proceeds faster than the chemical dissolution, the barrier layer grows, which in turn reduces the electrochemical oxidization process to the rate determined by the chemical dissolution [11]. The chemical dissolution rate determines the nanotube length [10]. When the chemical dissolution does not occur, the further anodization is obstructed by the oxide layer and higher potential up to hundreds volts are needed to break the barrier layer [11].

Raja *et al.* [74] investigated a nanotube formation during anodization of Ti foil at constant 20 V in the following electrolytes: 0.5M H_3PO_4 + 0.138M HF, 0.5M H_3PO_4 + 0.138M NaF, 0.5M H_3PO_4 + 0.138M NaCl, 0.5M H_3PO_4 + 0.138M NaBr. They show evolution of the nanoporous structure during etching in 0.5M H_3PO_4 + 0.14M HF at constant potential of 20 V (Fig. 9.44). After 10 s etching are observed, a microcracks acts as a nucleation points for nanopores during anodization for longer period.

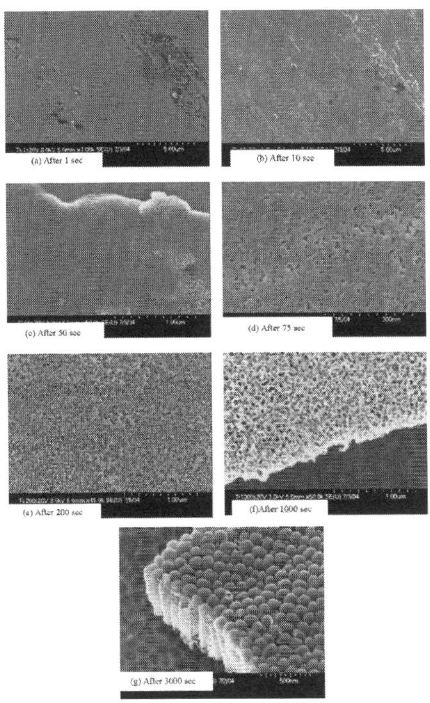

Figure 9.44 Stages of nanotubular film evolution during anodization at 20 V in 0.5M H_3PO_4 + 0.14M HF solution: from micro-cracks through pores to a thick oxide with self-organized nanotubes [74].

Complete porous oxide morphology is observed after 200 s anodization. Extending time of anodization to 1000 s results in separation of individual tubes from the nanopores and the outer oxide layer is estimated to be about 150 nm thick. A clear self-ordered nanotubular structure appears after 3000 s of anodization. The thickness of the oxide layer was about 500 nm with about 100–120 nm outer diameter tubes.

In the acidic electrolyte solutions containing HF, the strong dissolubility of hydrofluoric acid limits the nanotube length to a maximum of about 500 nm [58].

Raja *et al.* [74] tried to explain mechanism of self-ordered nanotubular oxide formation. The sequence of nanotubular oxide layer formation during anodization is described as follows:

(i) Formation of a passive inner barrier-type film during the first few seconds of anodization
(ii) Thickening of barrier layer and subsequent microfissuring (formation of easy paths)
(iii) Secondary oxide nucleation through these easy paths and pore nucleation
(iv) Coverage of the secondary oxide on the entire surface and growth of pores
(v) Pore separation to form individual and self-ordered nanotubes.

Pore separation leading to formation of individual nanotubes occurs under specified environmental condition that results in higher anodic current density. In acidic solutions, the current density is higher and steady, leading to nanotubular structure (Fig. 9.45a,b). When Ti^{4+} ions dissolve into solution, cation vacancies are generated [74]. These cation vacancies migrate along the electric field, reaching the metal/oxide interface. If the vacancies are not annihilated, they condense to form voids. When dissolution in acidic solution proceeds fast, the density of cation vacancies is very high. Dissolution occurs mainly near the inner wall of the pores containing adsorbed anions, but in this case, the field strength may not be as high as that in the barrier layer. Therefore, Raja [74] considered vacancy transportation in the radial direction, which results in rows of vacancies reaching centers of the interconnected porewalls from the two neighbor pores (Fig. 9.45c). If the charges have the same sign, they repel and keep an equilibrium distance. To maintain electrical neutrality, oxygen vacancies are generated. If the dissolution rate is much higher than the generation of oxygen vacancies at the metal/barrier oxide interface, the repelling forces of the cation vacancies results in separation of the neighboring pores leading to individual nanotubes [74]. Addition of bromide or chloride ions to H_3PO_4 electrolyte initiates pitting, but did not result in nanopores [74]. The electrolyte must be acidic (with HF or NaF addition) to form ordered nanopores layer.

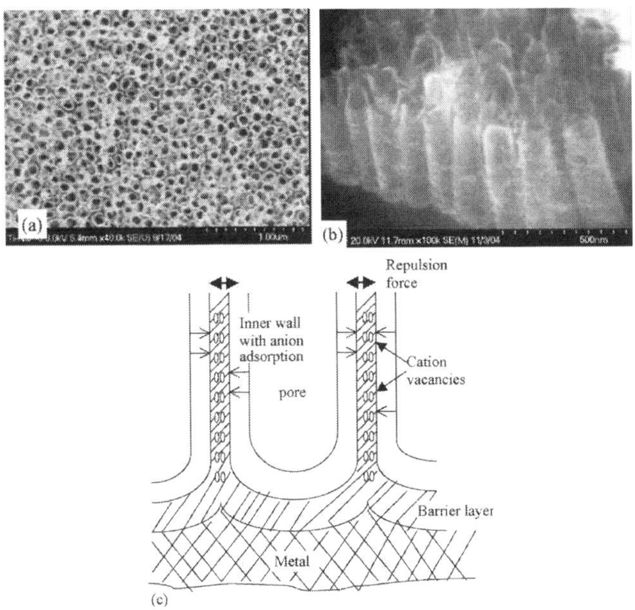

Figure 9.45 Separation of nanopores to form individual nanotubular oxide layer during anodization of Ti: (a) formation of individual nanotubes during anodization, (b) side-view of the nanotubes after anodization, (c) schematic illustration of pore separation mechanism. Cation vacancies generated by dissolution of Ti cations are transported radially from the two sides of common wall of the neighbor pores. Charges of similar polarity repel and when electrical neutrality is not maintained this repulsion causes separation of pores into individual nanotubes [74].

Oh [85] produced vertically aligned TiO_2 nanotubes on Ti sheet (0.25 mm thick, 99.5% purity) using anodization technique at 20 V for 30 min in electrolyte consisting 0.5% HF in water. Titanium oxide nanotubes were then chemically treated with NaOH solution to enhancing theirs bioactivity. In chemical treatment, the nanotubes were immersed in a 5 mol NaOH solution at 60°C for up to 60 min. The nanotube samples were then heat-treated at 500°C/2 h to crystallize the amorphous TiO_2 nanotubes into the anatase structure (Fig. 9.46). The anatase structure of TiO_2 is much more efficient in nucleation and growth of hydroxyapatite than the rutile phase of the TiO_2, because of the better lattice match with hydroxyapatite phase [91].

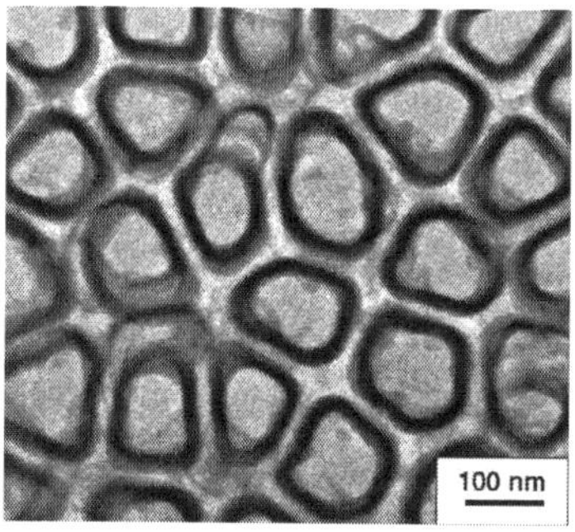

Figure 9.46 Top-view TEM micrograph of the anatase TiO_2 nanotubes [66].

When exposing the TiO_2 nanotubes to NaOH solution, an additional, extremely fine, and predominantly nanofiber-like structure is formed on the top of the TiO_2 nanotubes. These nanofibers are more sodium titanate compounds, such as $Na_2Ti_5O_{11}$ or $Na_2Ti_6O_{13}$ [66]. The preferential occurrence of nanofibers at the top of nanotubes is explained by surface-tension mechanism, resulting in difficulties of NaOH solution getting into nanopores within and in-between TiO_2 nanotubes (Fig. 9.47).

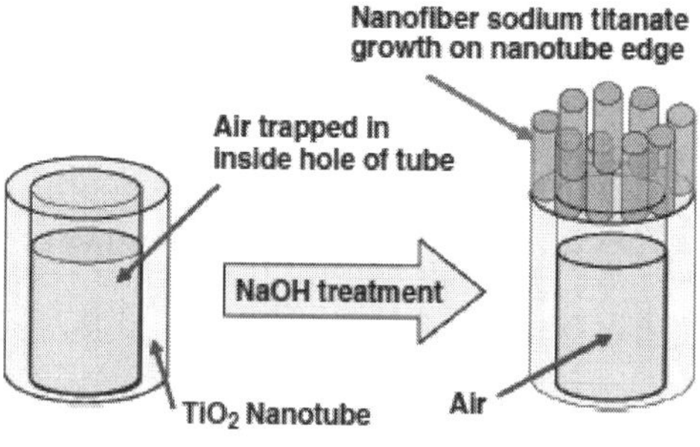

Figure 9.47 Effect of trapped air bubble on NaOH-treated TiO_2 nanotube array [66].

Lim *et al.* [48] investigated TiO_2 nanofibers fabricated by the electrospinning method using a mixture of Ti(IV)isopropoxide and poly(vinyl pyrrolidone) (PVP) in acidic alcohol solution. As-prepared nanofibers (Fig. 9.48) were immobilized on Ti plate treated by NaOH/HCl by inducing the alcohol condensation reaction of Ti(IV) isopropoxide with Ti–OH group on the titanium surface and subsequent calcination (500–1000°C). They found that the diameter of the TiO_2 nanofibers can be controlled within the range of 20–350 nm by changing amounts of Ti(IV)isopropoxide and PVP.

Figure 9.48 SEM images of remaining nanofibers on NaOH/HCl-treated Ti plate with alcohol condensation reaction [48].

Macak *et al.* [53] as well as Yang *et al.* [111] shows formation of nanotubular layer in the sputtered Ti. Macak *et al.* [53] used anodization of Ti deposited on Si, in 1M H_2SO_4 + 0.15 wt% HF electrolyte at potentials between 1 V and 25 V, which results in tubes with diameter ranging from 30 to 100 nm. They investigated the formation of nanotubes at different temperatures (–2 to 20°C) [53]. At room temperature, an incubation time for self-organization of up to 1 h is required. The higher is the electrolyte temperature, the higher is the overall current density and thin film dissolution. The lower the etching electrolyte temperature, the higher is the current efficiency. At a lower temperature, the chemical dissolution rate of TiO_2 is drastically reduced. They found that a temperature of 2°C states an optimum to suppress the chemical dissolution rate sufficiently while not causing drastic precipitation of the oxy-hydroxide (covering Ti surface in the early stage oxidation)

[53]. The cooling of the electrolytes results in the dissolution rate of the TiO_2 that can be significantly lowered. They found that 1 h of anodization results in formation of nanotubular self-organized TiO_2 structure with high degree of homogeneity, flatness, and precipitates free. Different anodization voltages (Fig. 9.49) clearly shows that at 1 V, there was only some pore formation observed, but considerable dissolution of Ti takes place due to the fact that at this potential, the Ti is in active dissolution region [53]. At potentials equal or higher than 5 V, a self-organized nanotubular layer is formed. The layer thickness and the tube diameter increases with increasing applied potential [53].

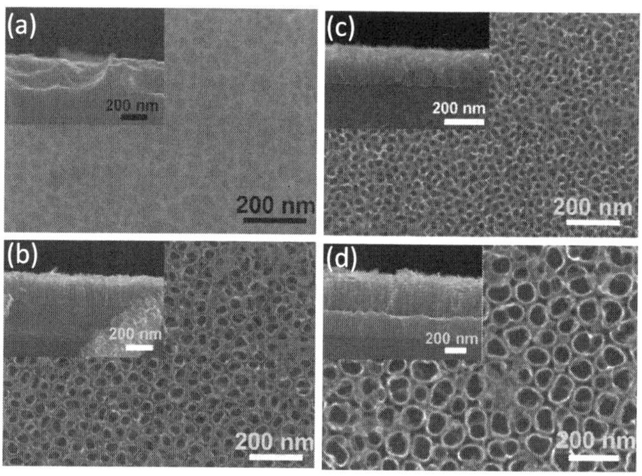

Figure 9.49 SEM images of sample anodized at 2°C in 1M H_2SO_4 + 0.15 wt% HF electrolyte at potentials: 1 V (a), 5 V (b), 10 V (c), and 20 V (d). The insets show the side views of respective structures [53].

The influence of applied potential on the resulting diameter and length of the nanotubes after 1 h of anodization is shown on Fig. 9.50. At 5 V, the tubes have diameter of 30 nm and length of 180 nm, which increases at 20 V to 100 nm and 400 nm, respectively. At higher potentials (25 V), no regular tubes were formed in the Macak *et al.* setup [53]. Comparable results were achieved for 1M H_3PO_4 + 0.3 wt% HF electrolytes [4]. The tube length and diameter can be controlled in a wide range of applied potentials. They found that for potentials between 1 and 25 V, tubes could be grown with any desired diameter ranging from 15 to 120 nm combined with tube length from 20 nm to 1 μm [4]. The diameter and the length depend linearly on the voltage, too. So, by careful choice of anodic oxidation parameters, it is possible to control the growing nanotubes.

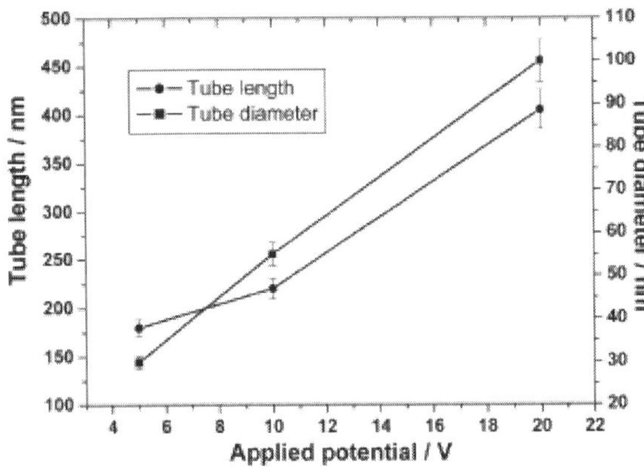

Figure 9.50 Dependence of the tube diameter and length on the applied potentials [53].

Yang *et al.* [111] showed the formation of nanotubular Ti on sputtered material (Fig. 9.51). They deposited titanium layers of 863 nm thickness on silicon substrates by DC magnetron sputtering. The anodization procedure consisted etching in aqueous solution (1M Na_2SO_4 + 0.2M $C_6H_8O_7 \cdot H_2O$ with the addition of 0.4 wt% NaF) and glycerol electrolyte (0.5 wt% NH_4F in glycerol). The potential was ramped by 0.1 V/s from the ocp to 5–30 V and then kept at constant value for various times. After anodization, the standard rinsing with deionized water and drying with a nitrogen stream was applied. Much higher current densities are observed in the aqueous electrolyte than in the glycerol electrolyte, because higher diffusivity and concentration of ions in the aqueous solution. The nanotubes are open on the top and closed at the bottom [111]. The morphologies of nanotube anodized in glycerol electrolyte were different from those in aqueous electrolyte (Fig. 9.51). The outer diameter of the nanotubes increases with the increasing anodization voltage for both types of electrolytes. The nanotube diameter was remarkably affected by the electrolyte and at the same applied voltage; the nanotubes with 100 nm and 40 nm diameters were obtained in the aqueous and glycerol electrolyte, respectively.

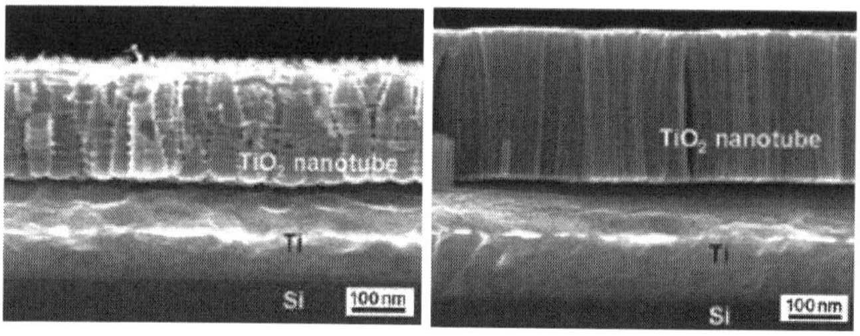

Figure 9.51 SEM images of nanotube/Ti/Si anodized in aqueous solutions (1 M Na_2SO_4 + 0.2 M $C_6H_8O_7 \cdot H_2O$ with the addition of 0.4 wt.% NaF) at 20 V for 600 s (in left) and in glycerol electrolyte (0.5 wt.% NH_4F in glycerol) at 10 V for 2 h (in right) [111].

Figure 9.52A presents the thickness of the nanotube as a function of the anodization time for the aqueous and glycerol electrolyte. Higher anodizing voltage leads to longer nanotubes for both electrolytes. Yang *et al.* [111] calculated the conversion ratio L/T, where L is the length of the nanotube and T is the thickness of the consumed titanium (Fig. 9.52B). They found that conversion ratios of the samples anodized in glycerol electrolyte are more than 1 and those in aqueous electrolyte are nearby 1. The volume expansion occurs when the titanium metal converts into titanium oxide. They concluded that a violent etching of nanotube simultaneously occurred at the interface of electrolyte and nanotube surface during anodization. The etching made the final titanium nanotube short. The anodization in aqueous electrolyte has a faster growing rate and a larger loss in length of nanotube than in the glycerol electrolyte due to higher diffusivity of ions.

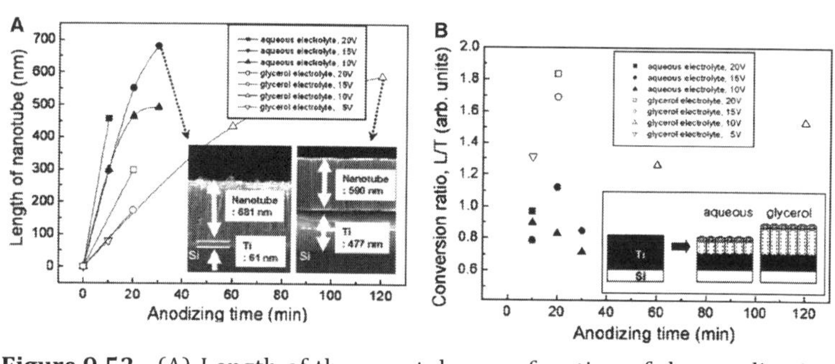

Figure 9.52 (A) Length of the nanotube as a function of the anodization time for the aqueous and glycerol electrolyte. (B) Conversion ratio, L/T (where L is the length of the nanotube grown by anodization and T is the thickness of the consumed titanium) [111].

Cai *et al.* [11] investigated formation mechanism of self organized TiO_2 nanotube arrays prepared by anodic oxidization of the pure titanium in electrolyte containing potassium fluoride and sulfate. They investigated anode mass, current density, and surface topography changes during the anodization. They found that with the protection of the oxide layer, long nanotubes could be formed in electrolyte solutions with relatively high pH [11]. Figure 9.53 shows an example of the anodic current density, the mass lost in anode, and the surface morphology during the anodization in moderately acidic electrolytes (pH 2.8). The time-dependent anodic current density, especially during the first minutes characterized the property of the electrolyte. The anodization results in a fast drop, followed by a small increase, and then slow decrease in current density [11]. The anodization can be divided into three stages:

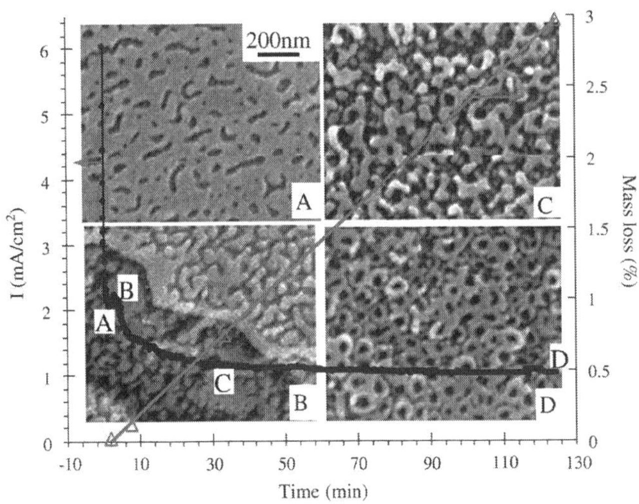

Figure 9.53 Anodization of titanium in pH 2.8 electrolyte. The plots show the anodic current density and the loss in anode mass during the anodization, respectively. The SEM images show the surface morphology of the samples anodized at different times as marked on the *I*–*t* plot [11].

(i) The electrochemical oxidization of titanium surface that corresponds to the first current drop

(ii) The nanotube formation that corresponds to the current increase

(iii) The growth of nanotubes that corresponds to the slow current decrease.

Cai *et al.* [11] observed that with a decreasing chemical dissolubility (increasing pH) of the electrolytes, the current peak disappears, and more time was therefore required for the nanotube formation.

The electrochemical oxidation forms a dense oxide film (barrier layer), which is typically confirmed by the oxide evolution. The increase in the thickness of the barrier layer gave rise to an ohmic resistance proportional to the thickness of the barrier layer. The ohmic resistance is also related to the surface morphology. Cai *et al.* [11] observed that the oxide layer formed in pH 2.8 electrolyte (Fig. 9.53) is less dense than those formed in stronger acidic electrolyte (pH $<$ 1) and even in pH 4.2 electrolyte. The less density of the oxide film formed in pH 2.8 electrolyte is related with the relatively weaker hydrolysis ability compared with pH 4.2 electrolyte, or the relatively weaker chemical dissolubility compared with the strong acidic solution [11]. The decrease in anode mass during the etching indicates that the chemical dissolution is significant in the presence of F^- (Eq. 9.9). The nanotube formation strongly depends on the chemical dissolubility of the electrolytes. Increasing pH reduces the dissolubility of the electrolytes. More time was desired for the formation of nanotubes in higher pH electrolytes [11]. In pH 2.8 electrolyte, nanotubes were not completely formed after anodization for 125 min. In absence of chemical dissolution, anodization would finally be terminated with the barrier layer growing, while no nanotubes could be formed at low anodic potential [11]. Long nanotubes can be fabricated with the protection of the oxide layer because the chemical dissolution proceeded mainly at the outer surface of the oxide layer while the nanotubes were formed under the barrier layer. By increasing the electrolyte pH, an oxide layer can be maintained along the anodization [11].

Bestetti *et al.* [6] investigated the electrochemical formation of nanotubular titanium oxide films in 1M H_2SO_4 + 0.05–0.4 wt% HF electrolytes. In Fig. 9.54, the corresponding pore diameter and spacing as a function of voltage are presented [6]. Increasing the anodization voltage from 10 to 15, 20, and 25 V results in pore diameter increasing to 28, 60, 92, and 103 nm, respectively. For voltages of 30 V or higher, nanotubular structure disappears [6].

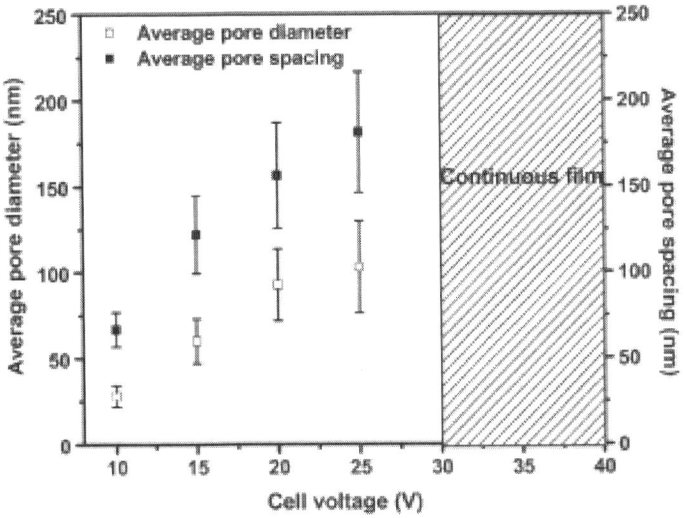

Figure 9.54 Pore diameter and interpore distance of nanotubular titanium oxides films after anodization in 1M H_2SO_4 + 0.15 wt% HF at RT as a function of voltage [6].

Hydrofluoric acid concentration plays a major role in pore formation and dissolution. Steady-state nanotubular formation is possible, when the concentration of HF is in the range of 0.15–0.4 wt% [6]. After 24 h of anodization, the thickness of nanotubular titanium oxide films reaches 400–500 nm. The cell voltage plays a key role in titania structure. Using an electrolyte 1M H_2SO_4 + 0.15 wt.% HF, it is possible to obtain a rutile structure at low voltages and a mixture of anatase and rutile at higher voltages [6]. Annealing the titanium oxide films result in the formation of anatase or rutile crystal structure (Fig. 9.55, 9.56) as well. XRD patterns of the titanium oxide, obtained after anodizing in 1M H_2SO_4 + 0.15 wt% HF at 15 V, and annealing at temperature increasing from 200 to 600°C, are shown on Fig. 9.55 [6]. Both rutile and anatase crystal forms have tetragonal structure. The anatase phase prevails by annealing at 300 and 400°C, while at 600°C, the rutile structure appears. Annealing time also affects the crystallization process (Fig. 9.56). The anatase phase formation proceeds after 180 min of annealing, while for longer time annealing, the rutile phase appears [6].

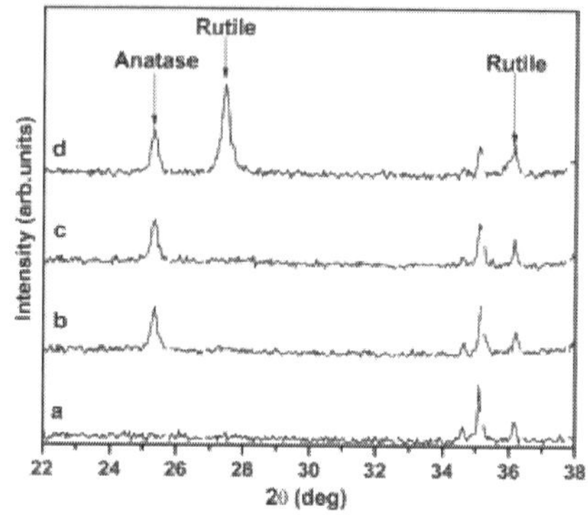

Figure 9.55 XRD patterns of porous titanium oxide film obtained by anodization in 1M H_2SO_4 and 0.15 wt% HF at 15 V and annealed at 200°C for 2 min (a), 300°C for 3 h (b), 400°C for 3 h (c), and 600°C for 3 h (d) [6].

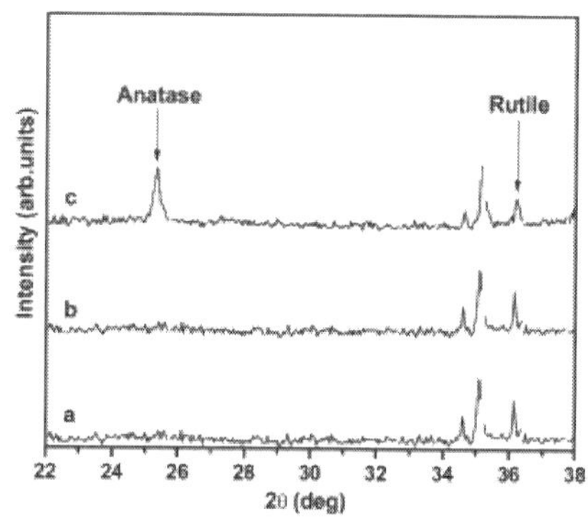

Figure 9.56 XRD patterns of anodic porous titanium oxide film obtained by anodization in 1M H_2SO_4 and 0.15 wt% HF at 15 V and annealed at 300°C for 2 min (a), 30 min (b), and 3 h (c) [6].

Electrochemical formation of nanotubes is also possible in Ti-alloys such as Ti–Al–V [49, 52], Ti–Al–Nb [52], Ti–Zr [115], Ti–Mn [57], and Ti–Zr–Nb [16]. For example, Luo *et al.* [49] investigated oxide nanotube arrays fabricated by anodization of Ti–6Al–4V alloy in 0.3M H_3PO_4 + 0.14M NH_4F electrolyte. They prepared nanotubes with 90–180 nm in diameter and 10–20 nm in wall thicknesses (Fig. 9.57). The obtained nanotubes on Ti-alloy background have morphology comparable to the etched pure Ti (see previous pages). The nanotubes were prepared by etching for 2 h with applying 20–40 V, but for 10 V and 50 V, the nanoporous structure was obtained (Fig. 9.57). Nanotube samples prepared using 20 V, 30 V, and 40 V have average inner diameters of 90 nm, 110 nm, and 180 nm, respectively and average tube wall thicknesses of 10 nm, 15 nm, and 20 nm, respectively. Length of the nanotubes is approximately 400 nm. The Ti–6Al–4V alloy shows a dual phase of $\alpha + \beta$ microstructure. The β-phase region is enriched with vanadium, which is etched preferentially by the electrolyte (high solubility of V-oxides), resulting in some pits on the surface (Fig. 9.57f). X-ray photoelectron spectroscopy revealed that the etched samples are predominately titanium and oxygen, with traces of aluminum and phosphorus. The sample etched at 20 V without annealing exhibits an amorphous structure [49]. Annealing of the nanotubes at 400°C, results in anatase phase, whereas at 500°C it results in a mixture of anatase and rutile. After 600°C calcination temperature, the rutile phase dominates [49].

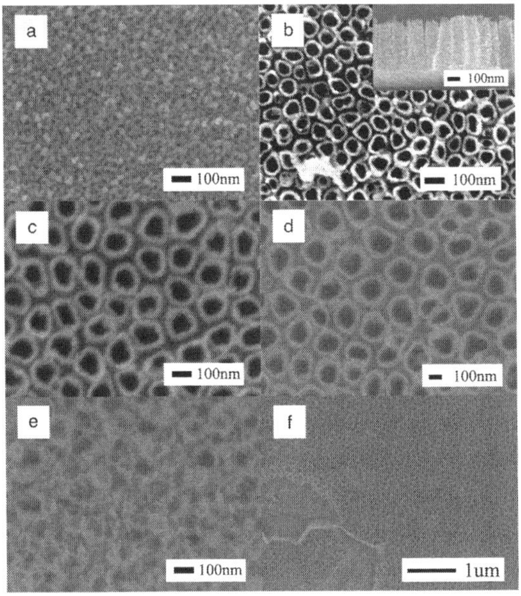

Figure 9.57 SEM images of top-view and cross section of the Ti–6Al–4V arrays obtained at different voltages: (a) 10 V, (b) 20 V, (c) 30 V, (d) 40 V, (e) 50 V after annealing at 400°C; (f) the bi-phase region of the film obtained at 20 V [49].

Jeong *et al.* suggest application of Ti–Hf alloys for dental application after the nanotubular surface formation [36]. The nanotubes were formed on Ti–Hf alloys by etching in 1M H_3PO_4 + 0.5 wt% NaF electrolyte followed by crystallization in the Ar atmosphere (1 h at 550°C). Uniform nanotubes were formed for Hf contents up to 20 wt%, and for higher Hf content the irregular nanotubes were found [36]. Increasing the Hf content in the Ti alloy led to the formation of nanotubes with more narrow size. The nanotubes diameter ranging from 80–120 nm and a length of approximately 1.7 μm was obtained. The nanotube morphology of the Ti–Hf alloys can be controlled by varying the amount of Hf. The anatase is observed predominantly with increasing Hf content [36].

The biomedical application possibilities of the nanotubes were presented by Schmuki's group [54, 71] (Fig. 9.58). The cell adhesion, proliferation, and migration are significantly affected by the nanotube size [71]. Geometries with a spacing of 15 nm were most stimulating for cell growth and differentiation, whereas diameters of approx. 100 nm led to a drastically increased cell apoptosis (Fig. 9.58).

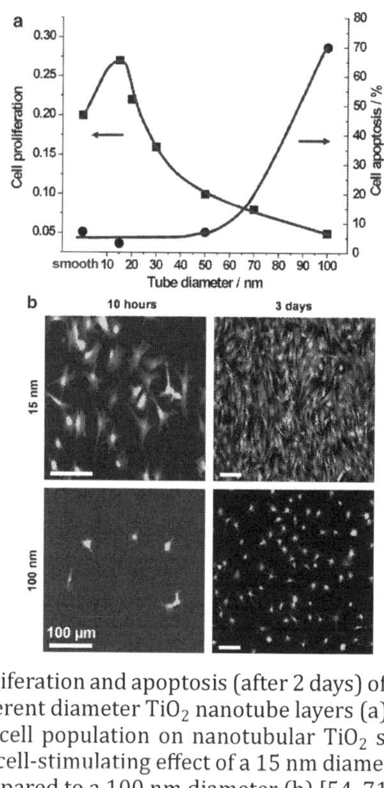

Figure 9.58 Proliferation and apoptosis (after 2 days) of cells dispersed on different diameter TiO_2 nanotube layers (a); optical images of the cell population on nanotubular TiO_2 surfaces indicating the cell-stimulating effect of a 15 nm diameter nanotube layer compared to a 100 nm diameter (b) [54, 71].

9.2.3 Ti-Based Nanocomposites

Ti-ceramic nanocomposites are a new class of biomaterials with interesting properties for future medical applications [38, 61]. The bionanocomposites are based on Ti-ceramic phase mixture with ultra fine grains. The nanograins essentially improve mechanical properties, whereas the ceramic phase (hydroxyapatite or bioglass, for example) improves osseointegration. For that reason, the bionanocomposites are promising in implant applications. The nanocomposites can be electrochemically etched, like conventional micro-Ti-based biomaterials. During the surface modification by the electrochemical etching, a large volume of the grain boundaries improve penetration of the nanograins by the electrolyte, and consequently enhances the pores formation. Jakubowicz *et al.* applied the electrochemical anodic oxidation process for the surface modification of the mechanically alloyed bionanocomposites [31].

Ti–hydroxyapatite (HA) and Ti–glass (45S5 Bioglass) bio-nanocomposites were prepared using mechanical alloying (MA) [31]. After milling the nanocomposite, powder mixture was uniaxially pressed at pressure of 500 MPa and sintered at 1150°C for 2 h. After the process, the grain size reaches a value of about 20–40 nm. The prepared Ti–HA and Ti–glass nanocomposites were electrochemically etched at 10 V vs. ocp in 1M H_3PO_4 + 2–10% HF electrolyte [31]. Preparation of Ti–HA and Ti–glass nanocomposites by MA + powder metallurgical process, results in porous compacts (Fig. 9.59a) with density of about 90% of the theoretical value. For the implant application it is a very attractive property, but not sufficient for the surface tissue bonding. On the other hand, the mechanical strength is probably deteriorated by relatively poor density. To improve the surface porosity necessary for the human tissue growth and strong fixation, the compacts were electrochemically etched in 1M H_3PO_4 + 2% HF electrolyte at 10 V vs. ocp for 15, 30, and 60 min [31]. The etching of Ti–HA for 15 min results in fast Ti atoms removal (Fig. 9.59b). In the nanocomposites, the large surface of the grain boundaries facilitates the atoms removal and hence the HF concentration can be lower for etching nanocomposites, than bulk microcrystalline Ti [28]. The surface after etching is rough with sponge pores with diameter of 1–6 μm and significant depth, and deeper than in the case of pure microcrystalline Ti (Fig. 9.11). On the Ti–glass nanocomposites, the surface was etched for 15, 30, and 60 min (Fig. 9.60). The increasing etching time results in fast roughening of the surface and sponge pore formation. The pore diameter increases up to 15 μm after 60 min etching. For these different etching times, the surface is significantly different.

(a)

(b)

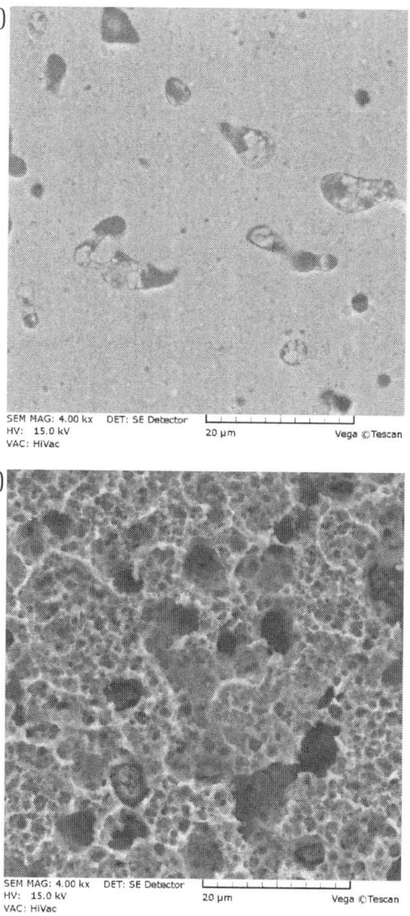

Figure 9.59 SEM images of Ti–HA nanocomposites before (a) and after (b) etching in 1M H_3PO_4 + 2% HF at 10 V for 15 min [31].

The Ti–glass nanocomposites presented in Fig. 9.60 were investigated by EDS analysis and the results are shown on Fig. 9.61 [31]. For all cases, the surface composition is comparable. In all samples, phosphorus is observed (incorporated to the oxide layer from the H_3PO_4 electrolyte). Phosphorus, together with calcium (from the glass) is attractive for osseointegration. It is significant that longer etching time results in decreasing oxygen content from 4.36 to 1.83 wt% for 15 and 60 min etching time, respectively. This means that longer etching results not only in porous oxide formation but also in oxide dissolution. So, after initial fast oxidation, the TiO_x etching rate, exceed TiO_x formation, or in the electrolytes containing HF the oxides are preferentially etched.

The HF is used to enhance etching and the current density is significantly greater, than for electrolyte without HF. The HF content in the electrolyte results in trace of fluorine, detected on sample etched for 30 min.

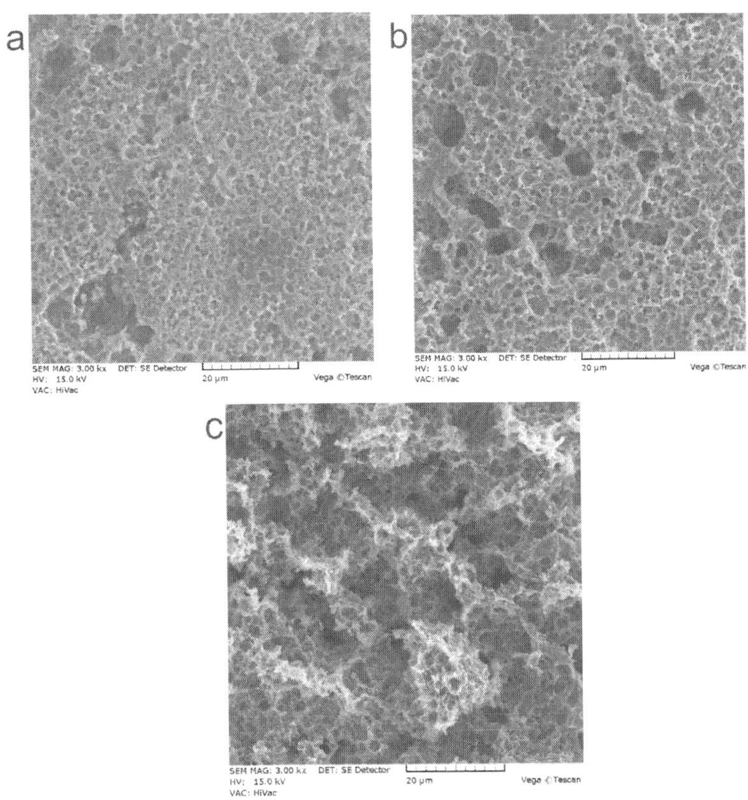

Figure 9.60 SEM images of the Ti–glass nanocomposites after etching in 1M H_3PO_4 + 2% HF at 10 V for 15 (a), 30 (b) and 60 (c) min [31].

a

Element	Wt%	At%
Ti	92,75	84,06
P	0,71	1,00
Si	0,84	1,30
Ca	0,90	0,97
Na	0,45	0,85
O	4,36	11,82
Total	100,00	100,00

b

Element	Wt%	At%
Ti	94,41	87,18
P	0,23	0,33
Si	0,56	0,88
Ca	0,72	0,80
Na	0,25	0,48
F	0,58	1,35
O	3,25	9,00
Total	100,00	100,00

c

Element	Wt%	At%
Si	0,70	1,13
Ti	95,23	90,77
Ca	1,01	1,15
P	0,37	0,55
Na	0,21	0,42
O	1,83	5,23
Fe	0,40	0,32
Al	0,25	0,42
Total	100,00	100,00

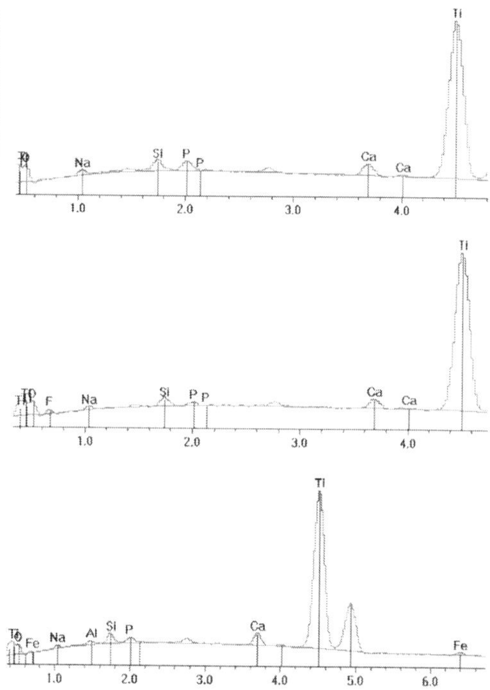

Figure 9.61 EDAX analysis of the Ti–glass nanocomposites after etching in 1M H_3PO_4 + 2% HF at 10 V for 15 (a), 30 (b), and (60 min) [31].

9.2.4 Ti-Based Nanoalloys

The Ti–6Al–4V was one from the firstly used Ti-based biomedical alloys. This alloy shows good mechanical and chemical properties, respective for implant applications [1, 9, 26, 83].

Ti-based alloys generally have excellent properties, but for the hard tissue replacement application, an additional surface treatment improving surface roughness is required. Bulk Ti–6Al–4V microcrystalline alloy was used for the etching as a reference material. As a new approach Jakubowicz and Adamek [29] applied mechanical alloying for the preparation of the nanocrystalline alloy, followed by pressing, sintering, and electrochemical etching, respective for the surface roughening. In the mechanical alloying, pure elements Ti, Al, and V with particle size of about 300 mesh were mechanically milled in protective high-purity argon atmosphere for 48 h [29]. After the process, the powder was uniaxially pressed (400 MPa) and sintered at 1000°C for 60 min. In the following stage, the electrochemical etching was done using 1M H_3PO_4 + 2% HF electrolyte at 10 V vs. ocp for 30 min [29]. Mechanical alloying is a simple and cheap process for nanomaterials synthesis. In the process, the mixture of microcrystalline pure elemental powders (Fig. 9.62a) in the stoichiometric ratio was milled for 48 h to achieve Ti–6Al–4V alloy composition. The XRD data in Fig. 9.62b clearly shows that after milling, the Ti–Al–V mixture is ultra fine grained (18 nm). After sintering (Fig. 9.62c), the structure is comparable to the microcrystalline bulk alloy (Fig. 9.62d), but with definitely smaller grains (35 nm). Due to the structure, this material exhibits quite different properties.

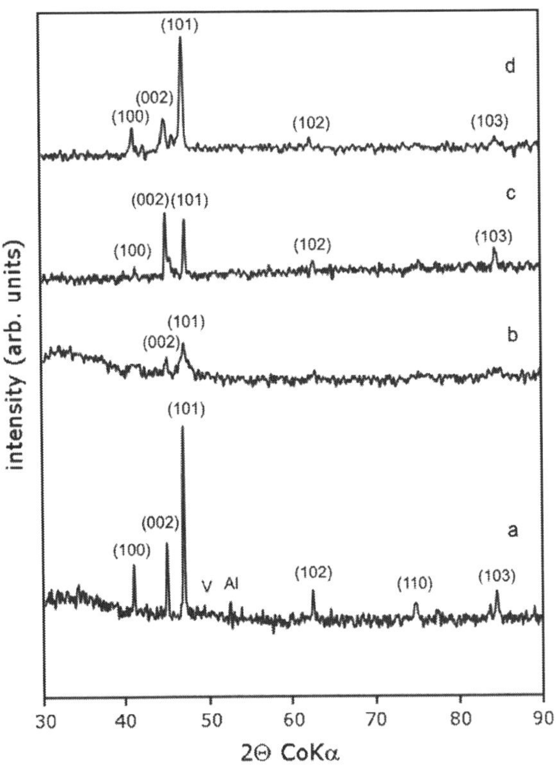

Figure 9.62 X-ray data for mixture of the microcrystalline Ti, Al, and V powder before MA (a) after MA (b), after sintering (c), and for comparison bulk microcrystalline Ti–6Al–4V alloy (d) [29].

Generally mechanical alloying consists of repeated fracture, mixing, and cold welding of a fine blend of elemental particles, resulting in size reduction and chemical reactions. The sintering results in compacts, used for the next treatment, respective for the pores formation. The compacts with relatively low density (90% of the theoretical density) are attractive material for the electrochemical etching, due to the large volume of the grain boundaries, which state the easy paths for the electrolyte penetration and diffusion. For that reason the nanocrystalline compacts are etched fast and easy.

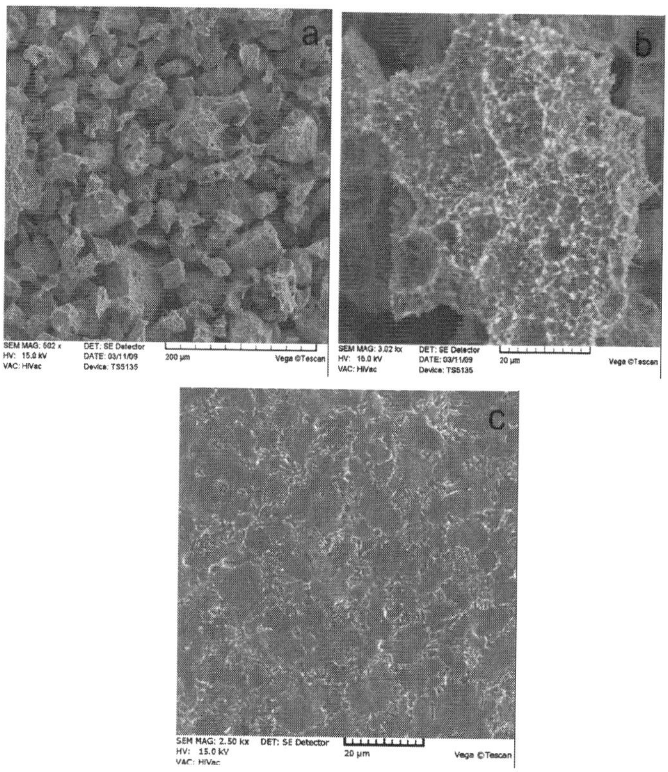

Figure 9.63 Mechanically alloyed Ti–6Al–4V after sintering and electrochemical etching (a, b — different magnifications); for comparison is shown electrochemically etched bulk microcrystalline sample (c) [29].

The electrolyte applied by Jakubowicz and Adamek [29] contained H_3PO_4 with small HF addition, enhancing dissolution [28] and resulting in porous Ti–6Al–4V compacts (Fig. 9.63a,b). After pressing, sintering, and electrochemical etching, the nanocrystalline compacts achieved density of about 80% of thc bulk ingots, which is related to pores in the compacts. The electrolyte composition as well as etching conditions also works very well during the etching of pure microcrystalline Ti as well as Ti/ceramic nanocomposites [28, 31], but does not work for microcrystalline Ti–6Al–4V alloy ingots (Fig. 9.63c). The surface is slightly rough, revealing two-phase microstructure.

The etched nanocrystalline alloy is very rough with pores, with diameter up to 60 μm (Fig. 9.63a), which is very useful for the tissue growth and fixing. Larger magnification and closer inspection of the porous nanocrystalline compacts reveals that the remaining surface material (particles) is also etched, with pores inside with diameter of about 0.1–1 μm (Fig. 9.63b).

The EDS spectrum for the porous surface after etching is shown on Fig. 9.64. The measured amount of Al is slightly larger from the initial loaded stoichiometric 6% of Al powder. Probably it is related with losses of some powders (mainly Ti) during the milling, when a part of the powder strongly adheres to the milling container and balls. The phosphorus is presented in the etched compacts, as a remnant from the electrolyte. This effect was observed for previously etched microcrystalline Ti as well as Ti/ceramic nanocomposites [28, 31], too. The phosphorus content in the alloy surface is useful for the osseointegration, because phosphorus is a main component of the bone.

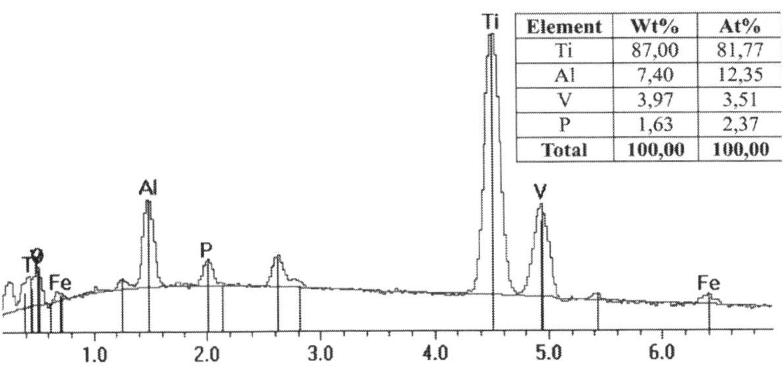

Element	Wt%	At%
Ti	87,00	81,77
Al	7,40	12,35
V	3,97	3,51
P	1,63	2,37
Total	**100,00**	**100,00**

Figure 9.64 EDS spectra for the nanocrystalline Ti–6Al–4V after electrochemical etching [29].

9.3 Formation of Electrochemically Grown Ca-P Layer

Titanium is a metallic bioinert material, and from that point of view, it is desirable to find possible methods to improve the bioactivity as well as corrosion resistance of the porous titanium and its alloys. To improve the bioactivity of the porous titanium alloy implants, a bioactive layer of hydroxyapatite (HA) can be formed on the implant surface. Hydroxyapatite can easily bond to living tissues. The HA layer improve the biocompatibility, while the metallic background provide good mechanical properties. Hydroxyapatite can be deposited using processes of mineralization in SBF, sputtering, sol–gel, laser deposition, thermal spraying, electrochemical deposition, or biomimetic [14, 21, 23, 102].

Vertically aligned titanium oxide nanotubes grown on the surface of titanium substrate by anodization in HF electrolyte show promising properties for implant applications [66]. The array of TiO_2 nanotubes adherent to Ti implant surface can be useful for accelerated bone growth in orthopedic/dental applications [66]. The additional chemical treatment with NaOH makes them more bioactive, which results in significantly accelerated kinetics of the hydroxyapatite growth by a factor as much as 7. Figure 9.65. shows SEM image (a) of the nanotubes after soaking the SBF for 1 day. The EDX spectrum (b) shows presence of Ca, P, and O in the formed layer, indicating the hydroxyapatite formation. The nanostructured HA has feature dimension of about 25 nm (Fig. 9.65a) and taking into account results of Webster *et al.* [96, 98], the nanostructured ceramics significantly improves osteoblast adhesion. So the Oh and Jin [66] obtained comparable results to Webster *et al.* [96, 98]. The adhesion/growth of osteoblast cells is also significantly accelerated by the topography of the TiO_2 nanotubes, with the filopodia of the growing cells going into the nanotube pores, producing a locked-in cell structure [66]. The number of the adhered cells to the TiO_2 nanotubes increases by 400% in comparison with microcrystalline Ti.

In vitro cell culture using MC3T3-E1 osteoblast cells on pure Ti (with native oxides) and vertically aligned anatase TiO_2 is shown on Fig. 9.66. The growth of cells and the propagation of filopodia are much faster in the TiO_2 nanotubes (b) (2 h incubation) compared with the Ti sample (a) (12 h incubation) [66].

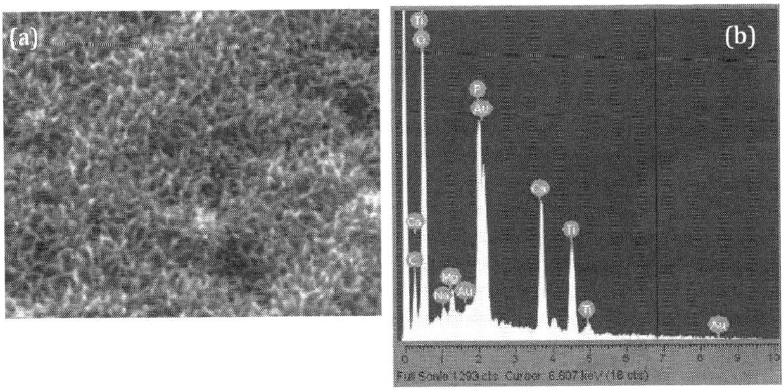

Figure 9.65 SEM image (a) and EDX spectrum (b) showing the nanoscale hydroxyapatite grown on TiO_2 nanotubes [66].

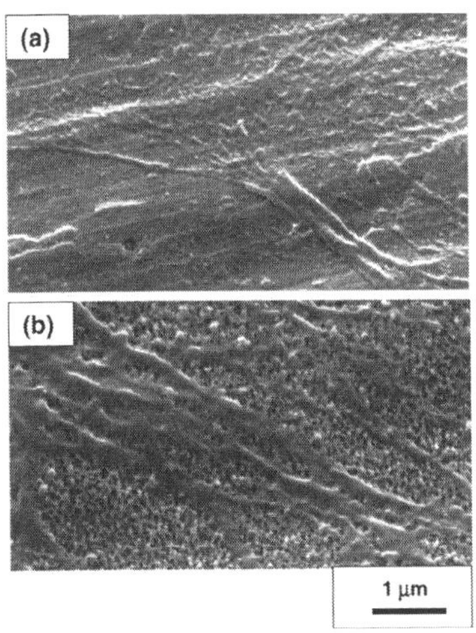

Figure 9.66 SEM images of osteoblast cells growth on (a) Ti surface (after 12 h) and (b) TiO_2 nanotube surface (after 2 h) [66].

Micrometer-size hydroxyapatite layer coated Ti surface exhibits interfacial failures. The nanostructured bioactive hydroxyapatite or the aligned TiO_2 nanostructures form strongly bonded and stable nanoporous layer, which can reduce interfacial fracture [66]. Natural bone consists of nanophase hydroxyapatite in the collagen matrix, too, and hence nanostructured surface layer has similar structure to the natural bone. Oh and Jin show that the MC3T3-E1 cell adhesion and theirs growth are significantly accelerated on TiO_2 nanotubes compared with the pure Ti (Fig. 9.67) [66].

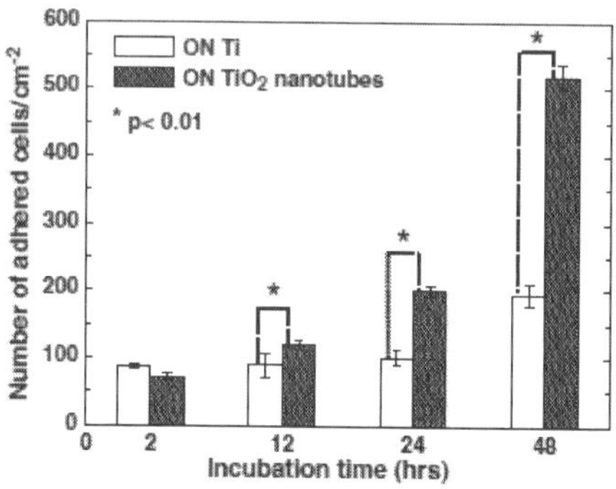

Figure 9.67 The counted number of adhered cells as a function of incubation period on the surface of Ti and anatase TiO_2 nanotubes [66].

Narayanan *et al.* [60] deposited nano-grained calcium phosphate coatings on titanium alloy substrates using aqueous electrolyte consisting 0.042M $Ca(NO_3)_2$ + 0.025M $(NH_4)_2HPO_4$ with pH of 4.1. This solution has a molar Ca/P ratio of 1.67. Calcium phosphate coatings with grain size in the range of 50–100 nm were produced on the surface of Ti–6Al–4V alloy (Fig. 9.68).

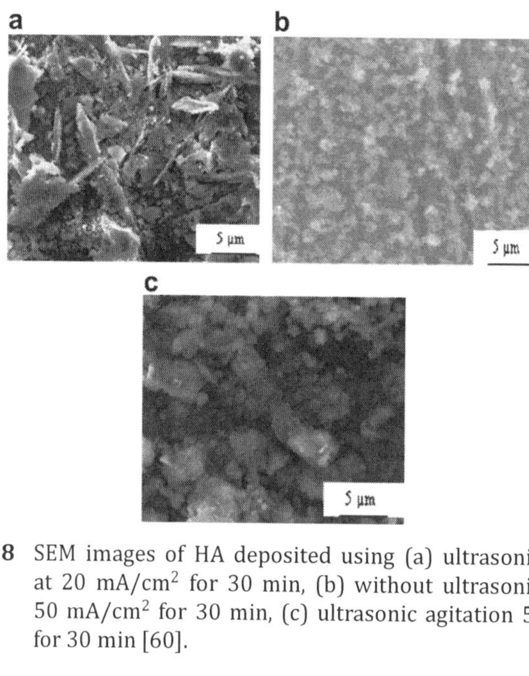

Figure 9.68 SEM images of HA deposited using (a) ultrasonic agitation at 20 mA/cm^2 for 30 min, (b) without ultrasonic agitation 50 mA/cm^2 for 30 min, (c) ultrasonic agitation 50 mA/cm^2 for 30 min [60].

Bone composed of nano-fiber HA in a collagen matrix and cells are preferentially interacting with nanostructured surfaces, and osteoblast functions increase on nano-structured materials [2, 60, 94]. Hence, Ti surface containing nano-grained HA provides better cell bioactivity than the surface containing coarse-grained HA. By employing current densities higher than 20 mA/cm^2 and using ultrasonated bath, it is possible to produce coatings containing nano-grained HA. Ultrasonics helps in the formation of nano-grained calcium phosphate and also increases the evolution rate of hydrogen at the cathode surface. This evolved hydrogen dispels the calcium ions from the vicinity of the cathode and resulting in coatings with a Ca/P ratio close to unity [60]. Increase of the current density from 20 to 50 mA/cm^2 leads to the increase of the hydrogen evolution rate, too, and also transports more calcium ions to the implant surface.

Xiao *et al.* [106] improve the bonding strength of the HA coating, by modifying hydrothermal–electrochemical (HTEC) method. In this process, the Ti particles are introduced to the

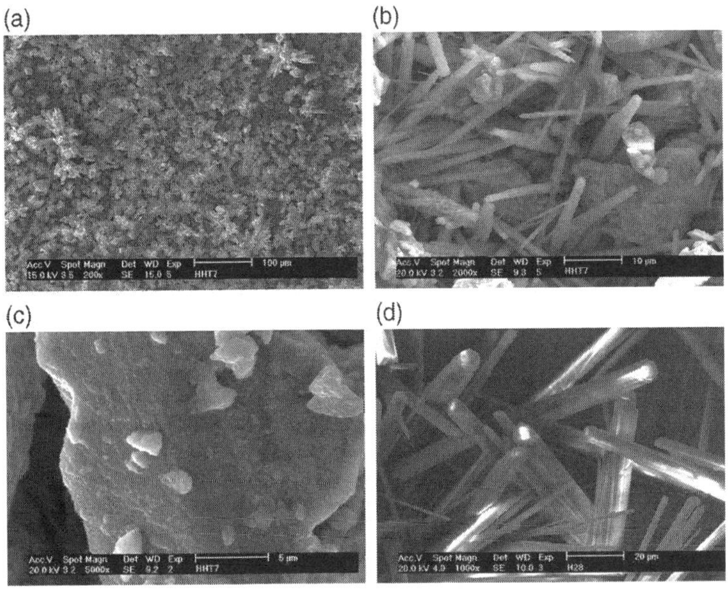

Figure 9.69 Surface morphologies of HA/Ti composite coatings at different magnification: (a) ×200, (b) ×2000, (c) ×5000; HA coating (d) ×1000 [106].

electrolyte so that the HA and Ti particles could be codeposited on a titanium substrate [106]. The electrolyte used for deposition of coatings contained 0.0105M $Ca(NO_3)_2$, 0.0063M $NH_4H_2PO_4$, 0.1M $NaNO_3$, and 0–100 g/L Ti particles. The pH of the electrolyte was adjusted to 4.60. The HTEC deposition was carried out at –0.4 mA/cm^2 and 200°C for 120 min. The pressure was about 1.60 MPa. The SEM images of the HA/Ti coatings (Fig. 9.69) reveals the Ti particles and a needle-like HA. HA and Ti particles are uniformly distributed in the coating with a rough surface. Some small HA particles are deposited on Ti particles surface. The pure HA coating (Fig. 9.69d) has much larger particle size than that in the HA/Ti composite coating (Fig. 9.69a–c), indicating the function of Ti particles as the HA crystal growth inhibitor during HTEC deposition [106].

Because the hydrogen evolution on the surface occurs during the process, the HA coatings prepared by electrodeposition are porous, and the addition of Ti leads to a decrease in porosity and an increase in density of the composite coatings. Smaller porosity and higher density of coating lead to higher cohesive strength of the coatings [106].

In the process, the Ti particles are deposited on the cathode surface, and around them are build HA crystals incorporated of Ti particles as reinforcement within the coating, which improves the bonding strength between coating and metal substrate. The addition of Ti reduces the HA crystallite size. The bonding strength of HA/Ti composite coatings increases with addition of Ti [106]. The results of bonding strength of HA/Ti composite coatings after heating at 500°C for 2 h are shown in Fig. 9.70. The pure HA coating shows the lowest bonding strength and the bond strength of the composite coating increases with the addition of Ti content. The strengthening mechanism of HTEC codeposited HA/Ti coatings is related to the dispersion strengthening by homogeneous distribution of Ti particles in the HA [106].

The surface morphology of the HA/Ti composite coatings heated at 500°C for 2 h after additional soaking in SBF for 7 days is shown in Fig. 9.71. The surface of the coating is covered by a newly formed granular layer with many cracks (often observed in many works). The addition of Ti reduces the HA content in the coating, but it does not distinctly affect the formation of bone-like apatite layer in SBF [106].

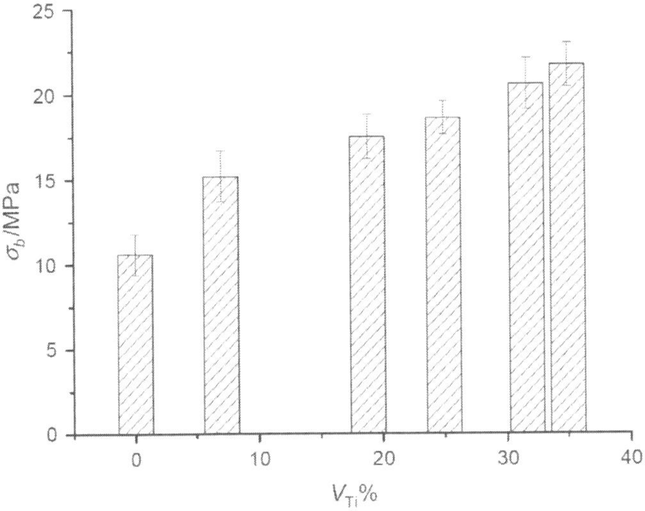

Figure 9.70 Relationship between the bonding strength of coatings and the content of Ti in the coatings [106].

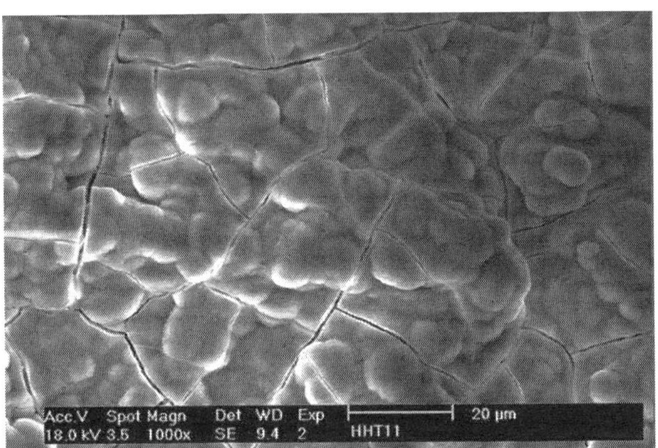

Figure 9.71 Surface morphology of HA/Ti composite coatings after soaking in SBF for 7 days [106].

Jakubowicz *et al.* investigated formation of HA layer on micro-Ti [33], Ti-45S5 nanocomposites [35], and micro-/nanocrystalline Ti–6Al–4V alloys [34]. They applied two different types of electrolytes for hydroxyapatite (HA) deposition. The source of Ca and P was: 0.1M HCl + 0.005M HA and 0.042M $Ca(NO_3)_2$ + 0.025M $(NH_4)_2HPO_4$ + 0.1M HCl. In the first type of electrolyte, the commercially available powder of HA was dissolved in 0.1M HCl solution. The second electrolyte consist 0.042M $Ca(NO_3)_2$ + 0.025M $(NH_4)_2HPO_4$ + 0.1M HCl, which is often reported in the literature [60, 116] (but they did not reports using of HCl). The molar ratio of Ca/P is 1.68 which corresponds to HA. In the second electrolyte, the addition of HCl dissolve the reaction product (calcium-phosphates particles) of the $Ca(NO_3)_2$ + $(NH_4)_2HPO_4$ and make the electrolyte homogenous for cathodic deposition at −1.5 V for 60 min, for example, in both electrolytes. In these depositing conditions, the formed layers have good mechanical strength and adhere strictly to the porous surface.

Anodic etching of titanium results in surface roughening [33] attractive in osseointegration. During 30 min of anodization in 1M H_3PO_4 + 10% NH_4F at relatively low 10 V potential, the Ti atoms are effectively removed from the surface, resulting in pits formation with diameter in the range of 7–12 μm [33]. Additionally, etching in H_3PO_4 electrolyte results in some phosphorous deposition on the etched surface, useful for the osseointegration [31].

The surface after calcium phosphate deposition, with using the 0.1M HCl + 0.005M HA electrolyte, is shown on Fig. 9.72. For the flat and porous surface, the Ca-P layer has comparable spongy morphology. In the case of flat surface (a), a lot of cracks or grain boundaries in the layer are visible in spite of porous sample, where the layer strictly covers the titanium background (b). The positions of pores are still visible as the darker circular spots. Deposition of calcium-phosphate with using the 0.042M $Ca(NO_3)_2$ + 0.025M $(NH_4)_2HPO_4$ + 0.1M HCl electrolyte, results in significantly different morphology (Fig. 9.73) than presented on Fig. 9.72. The deposited HA layer is rough with large lamellas features. When comparing the small spongy structure from Fig. 9.72 to large lamellas on Fig. 9.73, the former one could be an initial stage of grow of the thick HA layer and the spongy structure (Fig. 9.72) is also composed from the lamellas, but in different (smaller) scale than on Fig. 9.73. The initial porous surface is not visible, and both porous and flat samples after HA deposition looks comparable, but with expected stronger adhesion of the HA layer to the porous surface. The HA morphology presented on Fig. 9.73 is in good agreement with structures obtained by Yen and Lin [116].

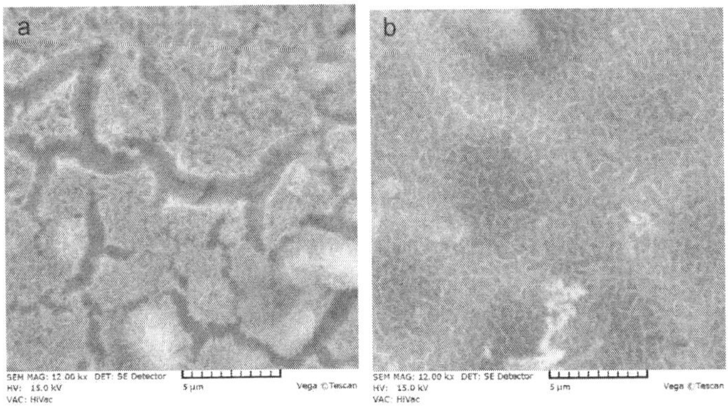

Figure 9.72 Calcium-phosphates deposited from HA + HCl electrolyte on flat (a) and porous Ti (b) [33].

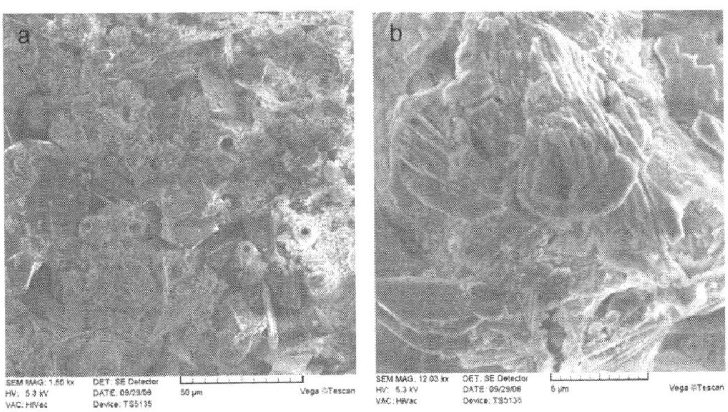

Figure 9.73 Calcium phosphates deposited from $Ca(NO_3)_2$ + $(NH_4)_2HPO_4$ + HCl electrolyte; different magnification on (a) and (b) [33].

Fig. 9.74 shows EDS analysis of the calcium-phosphate deposited on flat (a) and porous (b) surface with using the 0.1M HCl + 0.005M HA electrolyte. For the porous sample, the deposited layer has larger Ca/P ratio corresponding to HA (1.68). Porous sample shows higher oxygen, calcium and phosphorus content, with respect to flat surface. The surface composition should results in good corrosion resistance as well as osseointegration. The trace of silicon is a contamination introduced from the silicon rubber used for the seal of the electrochemical cell. Comparable results (presented on Fig. 9.72) of HA morphology (Fig. 9.75, 9.76) was obtained by Xie *et al.* [107] and Wei *et al.* [100], respectively.

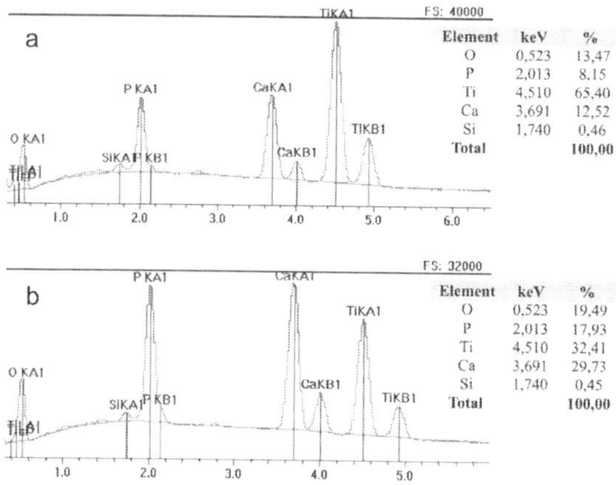

Figure 9.74 EDS analysis of the calcium-phosphate layer on flat (a) and porous (b) Ti [33].

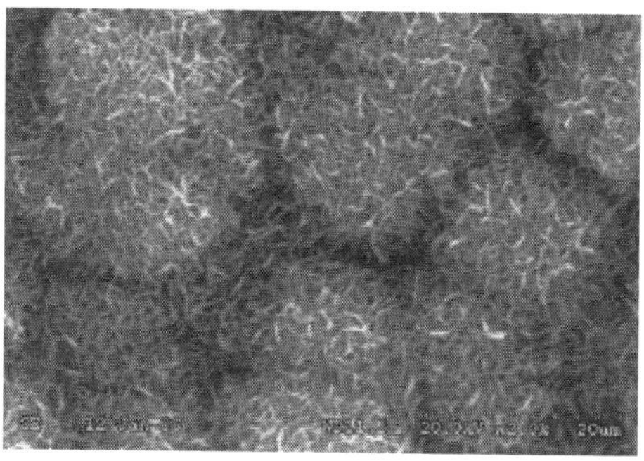

Figure 9.75 SEM surface morphologies of a hydroxyapatite coating formed on the activated porous Ti substrate [107].

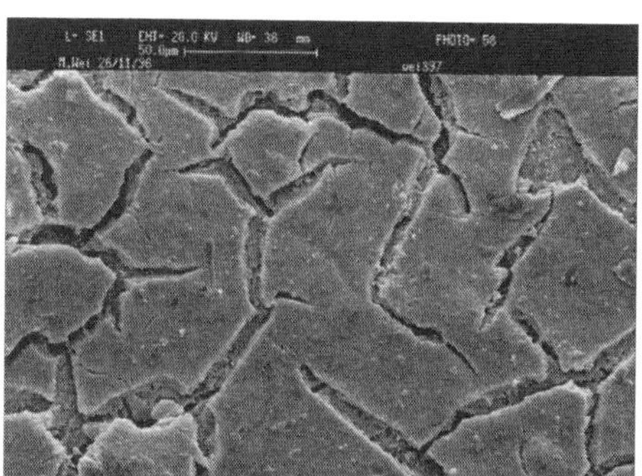

Figure 9.76 Surface morphologies of HA coating on Ti substrate [100].

The anodic oxidation of the new class of the biomaterials: Ti-45S5 nanocomposites, results in porous background attractive for the Ca-P deposition [35]. The hydroxyapatite was cathodically deposited by Jakubowicz *et al.* using two types of the electrolytes, too [35]. The cathodic deposition results in surface coverage by Ca-P compounds. Ca-P morphology after deposition from HA-HCl electrolyte is shown on Fig. 9.77. Deposited Ca-P, using second type of electrolyte, gave comparable results. The Ca-P after deposition on non etched (flat) surface (Fig. 9.77a) results in particles, which loosely adhere to the surface. The rough surface results in Ca-P deposits that grown inside the pores and surface after deposition is still rough (Fig. 9.77b, c). The deposited Ca-P is build mainly from the small lamellas (Fig. 9.77c). Such prepared rough surface with deposited HA should be very attractive for the osseointegration process.

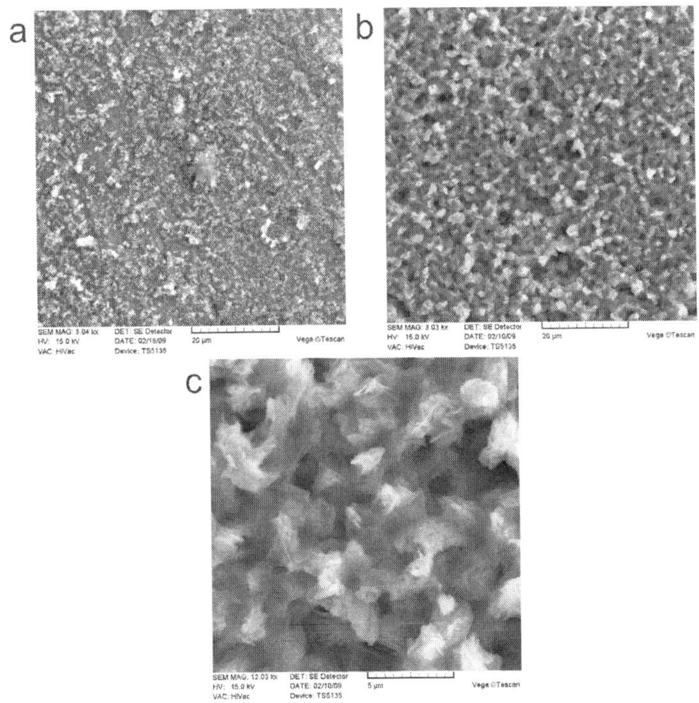

Figure 9.77 HA layer deposited on Ti–glass nanocomposite on flat (a) and porous etched (b,c) sample (a, b: 3000 × magnification; c: 12000 × magnification) [35].

The EDS spectra (Fig. 9.78) shows distinctive difference in surface samples composition after etching (a) and additional Ca-P deposition (b). The ratio of 1.67 of the Ca/P suggests that the grown layer is the hydroxyapatite.

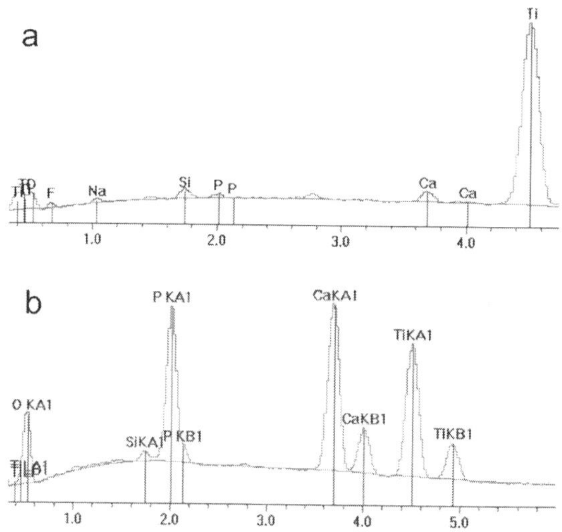

Figure 9.78 EDS analysis of the Ti–glass nanocomposite after etching (a) and after additional Ca-P deposition (b) [35].

Deposition of HA on micro- and nanocrystalline Ti–6Al–4V using two different electrolytes investigated by Jakubowicz *et al.* [34] results in distinctly different morphology (Figs. 9.79 and 9.80). In the microcrystalline alloy, porous structure is hardly formed (at them experimental conditions) and the surface is relatively flat [29]. Hydroxyapatite deposited ($Ca(NO_3)_2$ + $(NH_4)_2HPO_4$ + HCl electrolyte) on that background formed large lamellas features with size up to 200 μm (Fig. 9.79a). Completely different behavior was observed for the nanocrystalline alloy prepared by mechanical alloying, pressing, and sintering. Porous structure is easily formed in that alloy, due to a large volume of the grain boundaries which are penetrated by electrolyte [29] and lots of the HA nucleation sites, which are a Ti–Al–V nanograins. On that surface with pores up to 60 μm, HA uniformly covered the alloy surface and the HA fuzzy particle size is not larger than 0.5 μm. In the HA layer are cracks, often observed for different Ti-backgrounds as well as processing conditions [100, 106].

The HA deposited from HA + HCl electrolyte has significantly different morphology on micro-Ti–6Al–4V with respect to second electrolyte. The HA layer has many cracks with small spherical particles on top (Fig. 9.80a). For nano alloy, the HA layer covers the alloy surface, but pores are still visible (Fig. 9.80b) [34].

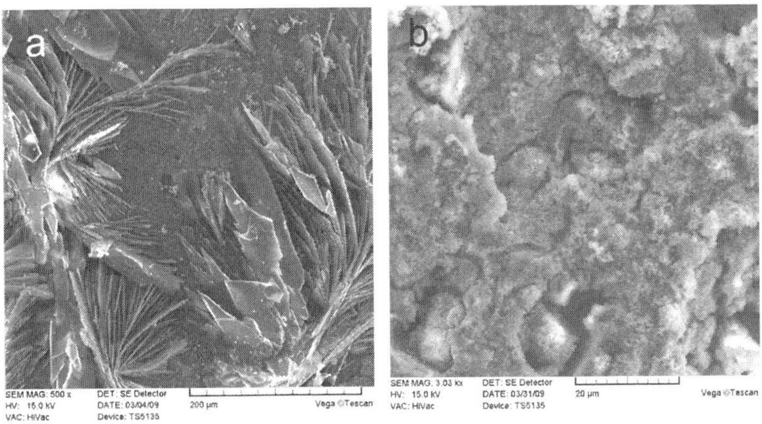

Figure 9.79 HA layer deposited on etched micro- (a) and nanocrystalline (b) Ti–6Al–4V; ($Ca(NO_3)_2$ + $(NH_4)_2HPO_4$ + HCl electrolyte) [34].

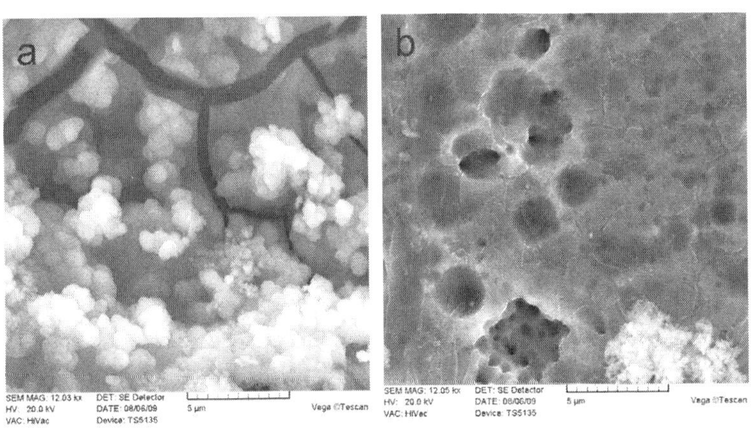

Figure 9.80 HA layer deposited on etched micro- (a) and nanocrystalline (b) Ti–6Al–4V; (HA + HCl electrolyte) [34].

Raja *et al.* deposited calcium-phosphate compounds on the nanotubular Ti surface (diameter of the nanotubes was in the range of 60–100 nm and the length was in the range of 400–500 nm) and investigated theirs bond strength [75]. They electrodeposited calcium phosphate at 65°C from the 0.04M $Ca(NO_3)_2$ + 0.027M $NH_4H_2PO_4$ + 0.5M NaCl electrolyte. The deposition was carried out by applying a constant potential of –1200 mV and by a two-step process, of first pulsing the potential between –1200 mV and –200 mV for 60 cycles, and then applying a constant potential of –1200 mV for a different time. During the potential pulsing between –1200 mV and –200 mV, the potential of –1200 mV was applied for 0.2 s in a cycle time of 10.2 s. At pH ~4, flaky type HA coating was observed (Fig. 9.81), whereas a hexagonal columnar growth (crystals with diameter of 100–200 nm and 2 μm long) was observed at pH ~6 (Fig. 9.82) [75]. Bond strength of the coatings varied from 16 MPa to 19 MPa.

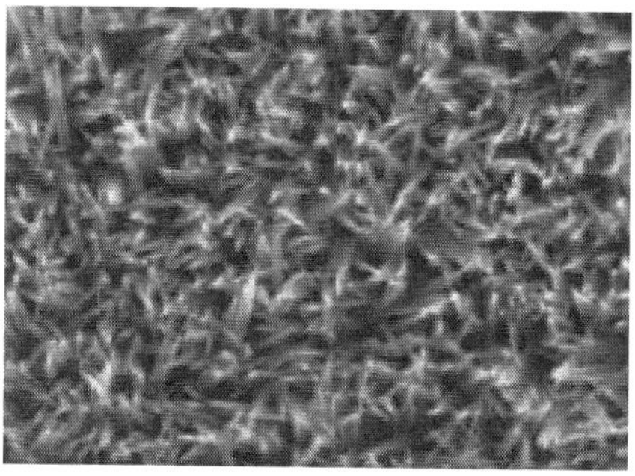

Figure 9.81 Flaky morphology of the calcium phosphate coating deposited at –1200 mV. The pH of the solution was 4 [75].

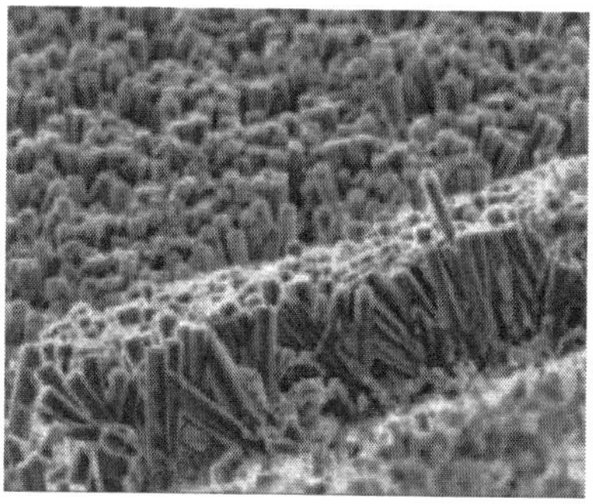

Figure 9.82 Vertically oriented Ca-P crystals grown on the nanoporous surface of the anodized Ti during a two-step electrodeposition process. The pH of the solution was 6 [75].

The EDS analysis of the surface after Ca-P removing shows Ca and P components (Fig. 9.83), and Raja *et al.* suggest that calcium phosphate grows from the bottom of the nanotubes rather than just sticking on the top surface of the titanium oxide [75]. Calcium phosphate coating on a polished flat titanium surface is extremely weak, not uniform and is easily removed after washing [75].

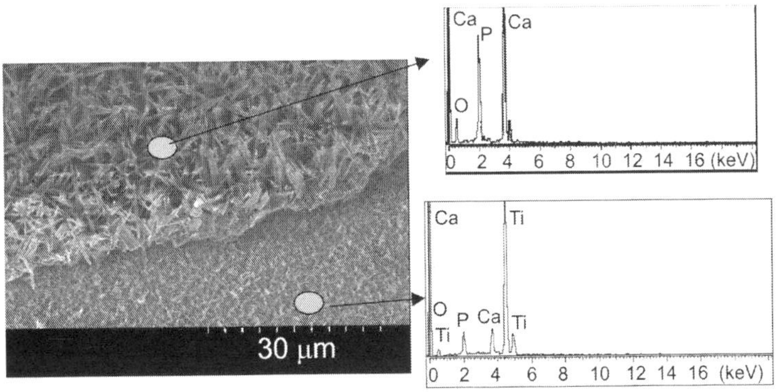

Figure 9.83 EDS analyses of the tensile tested surface where the coating was partially removed. The upper spectrum is for the Ca-P coating. The lower spectrum is for the surface, where the coating was removed (the peaks of Ca and P are still observed, indicating that the calcium phosphate nucleated at the bottom of the nanotubes) [75].

References

1. Anselme, K., Linez, P., Bigerelle, M., Le Maguer, D., Le Maguer, A., Hardouin, P., Hildebrand, H.F., Iost, A., and Leroy, J.M. (2000). The relative influence of the topography and chemistry of Ti–Al6–V4 surfaces on osteoblastic cell behaviour, *Biomaterials*, **21**, pp. 1567–1577.

2. Balasundaram, G., Sato, M., and Webster, T.J. (2006). Using hydroxyapatite nanoparticles and decreased crystallinity to promote osteoblast adhesion similar to functionalizing with RGD, *Biomaterials*, **27**, pp. 2798–2805.

3. Ban, S., Iwaya, Y., Kono, H., and Sato, H. (2006). Surface modification of titanium by etching in concentrated sulfuric acid, *Dental Materials*, **22**, pp. 1115–1120.

4. Bauer, S., Kleber, S., and Schmuki, P. (2006). TiO_2 nanotube: tailoring the geometry in H_3PO_4/HF electrolytes, *Electrochem. Commun.*, **8**, pp. 1321–1325.

5. Beranek, R., Hildebrand, H., and Schmuki, P. (2003). Self-organized porous titanium oxide prepared in H_2SO_4/HF electrolytes, *Electrochem. Solid-State Lett.*, **6**, pp. B12–B14.
6. Bestetti, M., Franz, S., Cuzzolin, M., Arosio, P., and Cavallotti, P.L. (2007). Structure of nanotubular titanium oxide templates prepared by electrochemical anodization in H_2SO_4/HF solutions, *Thin Sol. Films*, **515**, pp. 5253–5258.
7. Bradhurst, D.H., and Leach, J.S.L. (1966). The Mechanical Properties of Thin Anodic Films on Aluminum, *J. Electrochem. Soc.*, **113**, pp. 1245–1249.
8. Branemark, P.I. (1983). Osseointegration and its experimental background, *J. Prosthet. Dent.*, **59**, pp. 399–410.
9. Brunette, D.M., Tengvall, P., Textor, M., and Thomsen, P. (2001). *Titanium in Medicine*, Springer-Verlag, Berlin, Heidelberg.
10. Cai, Q., Paulose, M., Varghese, O.K., and Grimes, C.A. (2005). The effect of electrolyte composition on the fabrication of self-organized titanium oxide nanotube arrays by anodic oxidation, *J. Mater. Res.*, **20**, pp. 230–236.
11. Cai, Q., Yang, L., and Yu, Y. (2006). Investigations on the self-organized growth of TiO_2 nanotube arrays by anodic oxidization, *Thin Solid Films*, **515**, pp. 1802–1806.
12. Choi, J., Wehrspohn, R.B., Lee, J., and Gosele, U. (2004). Anodization of nanoimprinted titanium: a comparison with formation of porous alumina, *Electrochim. Acta*, **49**, pp. 2645–2652.
13. Crawford, G.A., and Chawla, N. (2009). Porous hierarchical TiO_2 nanostructures: Processing and microstructure relationships, *Acta Mater.*, **57**, pp. 854–867.
14. Das, K., Bose, S., and Bandyopadhyay, A. (2007). Surface modifications and cell–materials interactions with anodized Ti, *Acta Biomater.*, **3**, pp. 573–585.
15. Elias, K.E., Price, R.L., and Webster, T.J. (2000). Enhanced functions of osteoblasts on nanometer diameter carbon fibers, *Biomaterials*, **23**, pp. 3279–3287.
16. Feng, X.J., Macak, J.M., Albu, S.P., and Schmuki, P. (2008). Electrochemical formation of self-organized anodic nanotube coating on Ti–28Zr–8Nb biomedical alloy surface, *Acta Biomater.*, **4**, pp. 318–323.
17. Fujibayashi, S., Neo, M., Kim, H.M., Kokubo, T., and Nakamura, T. (2004). Osteoinduction of porous bioactive titanium metal, *Biomaterials*, **25**, pp. 443–450.

18. Ghicov, A., Tsuchiya, H., Macak, J.M., and Schmuki, P. (2005). Titanium oxide nanotubes prepared in phosphate electrolytes, *Electrochem. Commun.*, **7**, pp. 505–509.

19. Gong, D., Grimes, C.A., Varghese, O.K., Hu, W., Singh, R.S., Chen, Z., and Dickey, E.C. (2001). Titanium oxide nanotube arrays prepared by anodic oxidation, *J. Mater. Res.*, **16**, pp. 3331–3334.

20. Granja, P.L., Pouysegu, L., Deffieux, D., Daude, G., De Jeso, B., Labrugere, C., Baquey, C., and Barbosa, M.A. (2001). Cellulose phosphates as biomaterials. Surface chemical modification of regenerated cellulose hydrogels, *J. Appl. Polym. Sci.*, **82**, pp. 3354–3365.

21. Gu, Y.W., Khor, K.A., and Cheang, P. (2004). Bone-like apatite layer formation on hydroxyapatite prepared by spark plasma sintering (SPS)., *Biomaterials*, **25**, pp. 4127–4134.

22. Gu, Y.W., Tay, B.Y., Lim, C.S., and Yong, M.S. (2005). Characterization of bioactive surface oxidation layer on NiTi alloy, *Appl. Surf. Sci.*, **252**, pp. 2038–2049.

23. Gu, Y.W., Tay, B.Y., Lim, C.S., and Yong, M.S. (2005). Biomimetic deposition of apatite coating on surface-modified NiTi alloy, *Biomaterials*, **34**, pp. 6916–6923.

24. Gu, Y.W., Yong, M.S., Tay, B.Y., and Lim, C.S., (2009). Synthesis and bioactivity of porous Ti alloy prepared by foaming with TiH_2, *Mater. Sci. Eng. C*, **29**, pp. 1515–1520.

25. Hoar, T.P., and Mott, N.F. (1959). A mechanism for the formation of porous anodic oxide films on aluminium, *J. Phys. Chem. Solids*, **9**, pp. 97–99.

26. Hsu, R.W.-W., Yang, Ch.-Ch., Huang, Ch.-A., and Chen, Y.-S. (2004). Electrochemical corrosion properties of Ti–6Al–4V implant alloy in the biological environment, *Mater. Sci. Eng. A*, **380**, pp. 100–109.

27. Huang, P., Xu, K., and Han, Y. (2007). Formation mechanism of biomedical apatite coatings on porous titania layer, *J. Mater. Sci., Mater. Med.*, **18**, pp. 457–463.

28. Jakubowicz, J. (2008). Formation of porous TiO_x biomaterials in H_3PO_4 electrolytes, *Electrochem. Commun.*, **10**, pp. 735–739.

29. Jakubowicz, J., and Adamek G. (2009). Preparation and properties of mechanically alloyed and electrochemically etched porous Ti–6Al–4V, *Electrochem. Commun.*, **11**, pp. 1772–1775.

30. Jakubowicz, J., Jurczyk, K., and Jurczyk, M. (2009). Formation and properties of electrochemically grown porous TiO_x biomaterials, *Proc. V Int. Mater. Symp. MATERIAIS*, Portugal, CD-Rom, **101**, pp. 1–6.

31. Jakubowicz, J., Jurczyk, K., Niespodziana, K., and Jurczyk, M. (2009). Mechanoelectrochemical synthesis of porous Ti-based nanocomposite biomaterials *Electrochem. Commun.*, **11**, pp. 461–465.
32. Jakubowicz, J., Jurczyk, K., and Jurczyk, M. (2010). Electrochemical formation and corrosion properties of porous TiO_x biomaterials, *Mater. Sci. Forum*, **636–637**, pp. 15–21.
33. Jakubowicz, J., and Adamek, G. (2010). Hydroxyapatite deposited on flat and porous Ti, IEEE Xplore, 10.1109/ICBBE.2010.5515482 (2010) 1-4, ISBN 978-1-4244-4712-1.
34. Jakubowicz, J., Adamek, G., and Jurczyk, M. (2010). Unpublished data.
35. Jakubowicz, J., Jurczyk, K., and Adamek, G. (2010). Unpublished data.
36. Jeong, Y.-H., Lee, K., Choe, H.-Ch., Ko, Y.-M., and Brantley, W.A. (2009). Nanotube formation and morphology change of Ti alloys containing Hf for dental materials use, *Thin Sol. Films*, **517**, pp. 5365–5369.
37. Jessensky, O., Muller, F., and Gosele, U. (1998). Self-organized formation of hexagonal pore arrays in anodic alumina, *Appl. Phys. Lett.*, **72**, pp. 1173–1175.
38. Jurczyk, K., Niepodziana, K., and Jurczyk, M. (2006). Preparation and characterization of nanocomposite Ti-hydroxyapatite materials, *Europ. J. Med. Res.*, **11**, (Suppl. II), pp. 133–133.
39. Kaplan, F.S., Hayes, W.C., Keaveny, T.M., Boskey, A., Einhorn, T.A., and Iannotti, J.P. (1994). Form and function of bone, in *Orthopedic basic science* (ed. Simon, S.P.), American Academy of Orthopedic Surgeons, Columbus, OH., pp. 127–185.
40. Kaplan, F.S., Hayes, W.C., Keaveny, T.M., Boskey, A., Einhorn, T.A., and Iannotti, J.P. (1994). Biomaterials, in *Orthopedic basic science* (ed. Simon, S.P.), American Academy of Orthopedic Surgeons, Columbus, OH., pp. 460–478.
41. Kar, A., Raja, K.S., and Misra, M. (2006). Electrodeposition of hydroxyapatite onto nanotubular TiO_2 for implant applications, *Surf. Coat. Technol.*, **201**, pp. 3723–3731.
42. Kay, S., Thapa, A., Haberstroh, K.M., and Webster, T.J. (2002). Nanostructured polymer/nanophase ceramic composites enhance osteoblast and chondrocyte adhesion, *Tissue Eng.*, **8**, pp. 753–761.
43. Kim, H.M., Miyaji, F., Kokubo, T., and Nakamura, T. (1996). Preparation of bioactive Ti and its alloys via simple chemical surface treatment, *J. Biomed. Mater. Res.*, **32**, pp. 409–417.
44. Kim, S.E., Lim, J.H., Lee, S.Ch., Nam, S.-Ch., Kang, H.-G., and Choi, J. (2008). Anodically nanostructured titanium oxides for implant applications, *Electrochim. Acta*, **53**, pp. 4846–4851.

45. Kuroda, D., Niinomi, M., Morinaga, M., Kato, Y., and Yashiro, T. (1989). Design, mechanical properties of new β-type titanium alloys for implant materials, *Mater. Sci. Eng. A*, **243**, pp. 244–249.

46. Lakshmi, B.B., Patrissi, C.J., and Martin, C.R. (1997). Sol–gel template synthesis of semiconductor oxide micro- and nanostructures, *Chem. Mater.*, **9**, pp. 2544–2550.

47. Lee, J.-H, Kim, S.-E., Kim, Y.-J., Chi, Ch.-S., and Oh, H.-J. (2006). Effects of microstructure of anodic titania on the formation of bioactive compounds, *Mat. Chem. Phys.*, **98**, pp. 39–43.

48. Lim, J.I., Yu, B., Woo, K.M., and Lee, Y.-K. (2008). Immobilization of TiO_2 nanofibers on titanium plates for implant applications, *Appl. Surf. Sci.*, **255**, pp. 2456–2460.

49. Luo, B., Yang, H., Liu, S., Fu, W., Sun, P., Yuan, M., Zhang, Y., and Liu, Z. (2008). Fabrication and characterization of self-organized mixed oxide nanotube arrays by electrochemical anodization of Ti–6Al–4V alloy, *Mater. Lett.*, **62**, pp. 4512–4515.

50. Macak, J.M., Sirotna, K., and Schmuki, P. (2005). Self-organized porous titanium oxide prepared in Na_2SO_4/NaF electrolytes, *Electrochem. Acta*, **50**, pp. 3679–3684.

51. Macak, J.M., Tsuchiya, H., and Schmuki, P. (2005). High-aspect-ratio TiO_2 nanotubes, *Angew. Chem. Int. Ed.*, **44**, pp. 2100–2102.

52. Macak, J.M., Tsuchiya, H., Taveira, L., Ghicov, A., and Schmuki, P. (2005). Self-organized nanotubular oxide layers on Ti–6Al–7Nb and Ti–6Al–4V formed by anodization in NH_4F solutions, *Biomed. J. Mater. Res. Part A*, **75**, pp. 928–933.

53. Macak, J.M., Tsuchiya, H., Berger, S., Bauer, S., Fujimoto, S., and Schmuki, P. (2006). On wafer TiO_2 nanotube-layer formation by anodization of Ti-films on Si, *Chem. Phys. Lett.*, **428**, pp. 421–425.

54. Macak, J.M., Tsuchiya, H., Ghicov, A., Yasuda, K., Hahn, R., Bauer, S., and Schmuki, P. (2007). TiO_2 nanotubes: Self-organized electrochemical formation, properties and applications, *Curr. Op. Sol. St. Mat. Sci.*, **11**, pp. 3–18.

55. Marino, C.E.B., Nascente, P.A.P., Biaggio, S.R., Rocha-Filho, R.C., and Bocchi, N. (2004). XPS characterization of anodic titanium oxide films grown in phosphate buffer solution, *Thin Sol. Films*, **468**, pp. 109–112.

56. Miao, Z., Xu, D., Ouyang, J., Guo, G., Zhao, X., and Tang, Y. (2002). Electrochemically induced sol–gel preparation of single-crystalline TiO_2 nanowires, *Nano Lett.*, **2**, pp. 717–720.

57. Mohapatra, S.K., Raja, K.S., Misra, M., Mahajan, V.K., and Ahmadian, M. (2007). Synthesis of self-organized mixed oxide nanotubes by sonoelectrochemical anodization of Ti-8Mn alloy, *Electrochim. Acta*, **53**, pp. 590–597.

58. Mor, G.K., Varghese, O.K., Paulose, M., Mukherjee, N., and Grimes, C.A. (2003). Fabrication of tapered, conical-shaped titania nanotubes, *J. Mater. Res.*, **18**, pp. 2588–2593.

59. Muñoz, A.G. (2007). Semiconducting properties of self-organized TiO_2 nanotubes, *Electrochim. Acta*, **52**, pp. 4167–4176.

60. Narayanan, R., Seshadri, S.K., Kwon, T.Y., and Kim, K.H. (2007). Electrochemical nano-grained calcium phosphate coatings on Ti–6Al–4V for biomaterial applications, *Scripta Mater.*, **56**, pp. 229–232.

61. Niespodziana, K., Jurczyk, K., and Jurczyk, M. (2006). The manufacturing of Ti-hydroxyapatite nanocomposites for bone implant applications, *Nanopages*, **1**, pp. 219–229.

62. Oh, H.-J., Lee, J.-H., Jeong, Y., Kim, Y.-J., and Chi, C.-S. (2005). Microstructural characterization of biomedical titanium oxide film fabricated by electrochemical method, *Surf. Coat. Technol.*, **198**, pp. 247–252.

63. Oh, H.-J., Lee, J.-H., Kim, Y.-J., Suh, S.-J., Lee, J.-H., and Chi, Ch.-S. (2008). Surface characteristics of porous anodic TiO_2 layer for biomedical applications, *Mat. Chem. Phys.*, **109**, pp. 10–14.

64. Oh, I.-H., Nomura, N., and Hanada, S. (2002). Microstructures and mechanical properties of porous titanium compacts prepared by powder sintering, *Mater. Trans.*, **43**, pp. 443–446.

65. Oh, I.-H., Nomura, N., Masahashi, N., and Hanada, S. (2003). Mechanical properties of porous titanium compacts prepared by powder sintering, *Scripta Mater.*, **49**, pp. 1197–1202.

66. Oh, S., and Jin, S. (2006). Titanium oxide nanotubes with controlled morphology for enhanced bone growth, *Mater. Sci. Eng. C*, **26**, pp. 1301–1306.

67. Oh, S.H., Finones, R.R., Daraio, C., and Chen, L.H. (2005). Growth of nano-scale hydroxyapatite using chemically treated titanium oxide nanotubes, *Biomaterials*, **26**, pp. 4938–4943.

68. Oh, S.H., Daraio, C., Chen, L.H., and Pisanic, T. (2006). Significantly accelerated osteoblast cell growth on aligned TiO_2 nanotubes, *J. Biomed. Mater. Res.*, **78A**, pp. 97–103.

69. Okazaki, Y., Nishimura, E., Nakada, H., and Kobayashi, K. (2001). Surface analysis of Ti–15Zr–4Nb–4Ta alloy after implantation in rat tibia, *Biomaterials*, **22**, pp. 599–607.

70. Otsuka, K., and Ren, X. (1998). Recent developments in the research of shape memory alloys, *Intermetallics*, **7**, pp. 511–528.

71. Park, J., Bauer, S., Von der Mark, K., and Schmuki, P. (2007). Nanosize and vitality: TiO_2 nanotube diameter directs cell fate, *Nano Lett.*, **7**, pp. 1686–1691.

72. Pelton, A.R., Russel, S.M., and DiCello, J. (2003). The physical metallurgy of nitinol for medical applications, *JOM*, **5**, pp. 33–37.

73. Price, R.L., Waid, M.C., Haberstroh, K.M., and Webster, T.J. (2003). Selective bone cell adhesion on formulations containing carbon nanofibers, *Biomaterials*, **24**, pp. 1877–1887.

74. Raja, K.S., Misra, M., and Paramguru, K. (2005). Formation of self-ordered nano-tubular structure of anodic oxide layer on titanium, *Electrochim. Acta*, **51**, pp. 154–165.

75. Raja, K.S., Misra, M., and Paramguru, K. (2005). Deposition of calcium phosphate coating on nanotubular anodized titanium, *Mater. Lett.*, **59**, pp. 2137–2141.

76. Rohanizadeh, R., Al-Sadeq, M., and LeGeros, R.Z. (2004). Preparation of different forms of titanium oxide on titanium surface: Effects on apatite deposition, *J. Biomed. Mater. Res.*, **71A**, pp. 343–352.

77. Sander, M.S., Cote, M.J., Gu, W., Kile, B.M., and Tripp, C.P. (2004). Template-assisted fabrication of dense, aligned arrays of titania nanotubes with well-controlled dimensions on substrates, *Adv. Mater.*, **16**, pp. 2052–2057.

78. Seah, K.H.W., Thampuran, R., and Teoh, S.H. (1998). The influence of pore morphology on corrosion, *Corros. Sci.*, **40**, pp. 547–556.

79. Shih, Y.-H., Lin, Ch.-T., Liu, Ch.-M., Chen, Ch.-Ch., Chen, Ch.-S., and Ou, K.-L. (2007). Effect of nano-titanium hydride on formation of multi-nanoporous TiO_2 film on Ti, *Appl. Surf. Sci.*, **253**, pp. 3678–3682.

80. Sul, Y.T., Johansson, C.B., Jeong, Y., and Albrektsson, T. (2001). The electrochemical oxide growth behaviour on titanium in acid and alkaline electrolytes, *Med. Eng. Phys.*, **23**, pp. 329–346.

81. Sul, Y.T. (2003). The significance of the surface properties of oxidized titanium to the bone response: special emphasis on potential biochemical bonding of oxidized titanium implant, *Biomaterials*, **24**, pp. 3893–3907.

82. Supronowicz, P.R., Ajayan, P.M., Ullmann, K., Arulanandam, B.P., Metzger, D.W., and Bizios, R. (2002). Novel current-conducting composite substrates for exposing osteoblasts to alternating current stimulation, *J. Biomed. Mater. Res.*, **59**, pp. 499–506.

83. Sykaras, N., Iacopino, A., Marker, V., Triplett, R., and Woody, R. (2000). Implant materials, designs, and surface topographies: their effect on osseointegration. A literature review. *Int. J. Oral Maxillofac Implants*, **15**, pp. 675–690.

84. Taveira, L.V., Macak, J.M., Sirotna, K., Dick, L.F.P., and Schmuki, P. (2006). Voltage oscillations and morphology during the galvanostatic formation of self-organized TiO_2 nanotubes, *J. Electrochem. Soc.*, **153**, pp. B137–B143.

85. Thompson, G.E. (1997). Porous anodic alumina: Fabrication, characterization and applications, *Thin Solid Films*, **297**, pp. 192–201.

86. Tian, Z.R.R., Voigt, J.A., Liu, J., McKenzie, B., and Xu, H.F. (2003). Large oriented arrays and continuous films of TiO_2 based nanotubes, *J. Am. Chem. Soc.*, **125**, pp. 12384–12385.

87. Tsai, C.C., and Teng, H.S. (2004). Regulation of the physical characteristics of titania nanotube aggregates synthesized from hydrothermal treatment, *Chem. Mater.*, **16**, pp. 4352–4358.

88. Tsuchiya, H., Macak, J.M., Taveira, L., Balaur, E., Ghicov, A., Sirotna, K., and Shmuki, P. (2005). Self-organized TiO_2 nanotubes prepared in ammonium fluoride containing acetic acid electrolytes, *Electrochem. Commun.*, **7**, pp. 576–580.

89. Tsuchiya, H., Macak, J.M., Ghicov, A., Tang, Y.Ch., Fujimoto, S., Niinomi, M., Noda, T., and Schmuki, P. (2006). Nanotube oxide coating on Ti–29Nb–13Ta–4.6Zr alloy prepared by self-organizing anodization, *Electrochim. Acta*, **52**, pp. 94–101.

90. Tsuchiya, H., Macak, J.M., Muller, L., Kunze, J., Muller, F., and Greil, P. (2006). Hydroxyapatite growth on anodic TiO_2 nanotubes, *J. Biomed. Mater. Res.*, **77**, pp. 534–541.

91. Uchida, M., Kim, H.M., Kokubo, T., Fujibayashi, S., and Nakamura, T. (2003). Structural dependence of apatite formation on titania gels in a simulated body fluid, *J. Biomed. Mater. Res.*, **64**, pp. 164–170.

92. Varghese, O.K., Paulose, M., Shankar, K., Mor, G.K., and Grimes, C.A. (2005). Water-photolysis properties of micron-length highly-ordered titania nanotube-arrays, *J. Nanosci. Nanotechnol.*, **5**, pp. 1158–1165.

93. Wang, X., Li, Y, Xiong, J., Hodgson, P.D., and Wen, C. (2009). Porous Ti–Nb–Zr alloy scaffolds for biomedical applications. *Acta Biomaterialia*, **9**, pp. 3616–3624.

94. Ward, B.C., and Webster, T.J. (2006). Increased functions of osteoblasts on nanophase metals, *Mater. Sci. Eng. C*, **27**, pp. 575–578.

95. Webster, T.J., Siegel, R.W., and Bizios, R. (1999). Osteoblast adhesion on nanophase ceramics, *Biomaterials*, **20**, pp. 1221–1227.

96. Webster, T.J., Ergun, C., Doremus, R.H., Siegel, R.W., and Bizios, R. (2000). Enhanced functions of osteoblasts on nanophase ceramics, *Biomaterials*, **21**, pp. 1803–1810.

97. Webster, T.J., Siegel, R.W., and Bizios, R. (2000). Enhanced functions of osteoblasts on nanophase ceramics, *Biomaterials*, **21**, pp. 1803–1810.

98. Webster, T.J., Schandler, L.S., Siegel, R.W., and Bizios, R. (2001). Mechanisms of enhanced osteoblast adhesion on nanophase alumina involve vitronectin, *Tissue Eng.*, **7**, pp. 291–301.

99. Webster, T.J., and Ejiofor, J.U. (2004). Increased osteoblast adhesion on nanophase metals: Ti, Ti–6Al–4V, and Co–Cr–Mo, *Biomaterials*, **25**, pp. 4731–4739.

100. Wei, M., Ruys, A.J., Milthorpe, B.K., Sorrell, C.C., and Evans, J.H. (2001). Electrophoretic deposition of hydroxyapatite coatings on metal substrates: A nanoparticulate dual-coating approach, *J. Sol–Gel Sci. Techn.*, **21**, pp. 39–48.

101. Wen, C.E., Mabuchi, M., Yamada, Y., Shimojima, K., Chino, Y., and Asahina, T. (2001). Processing of biocompatible porous Ti and Mg, *Scripta Mater.*, **45**, pp. 1147–1153.

102. Wen, C.E., Xu, W., Hu, W.Y., and Hodgson, P.D. (2007). Hydroxyapatite/titania sol-gel coatings on titanium-zirconium alloy for biomedical applications, *Acta Biomater.*, **3**, pp. 403–410.

103. Whitters, C.J., Strang, R., and Brown, D. (1999). Dental materials: 1997 literature review, *J. Dent.*, **27**, pp. 401–435.

104. Woo, K.M., Chen, V.J., and Ma, P.X. (2003). Nano-fibrous scaffolding architecture selectively enhances protein adsorption contributing to cell attachment, *J. Biomed. Mater. Res.*, **67A**, pp. 531–537.

105. Wood, G.C. (1973). *Oxides and Oxide Films Vol. 2*, ed. Diggle J.W., Porous anodic films on aluminium, Marcel Dekker Inc., New York, pp. 167–279.

106. Xiao, X. F., Liu, R. F., and Zheng, Y. Z. (2005). Hydoxyapatite/titanium composite coating prepared by hydrothermal–electrochemical technique, *Mater. Lett.*, **59**, pp. 1660–1664.

107. Xie, J., Luan, B. L., Wang, J., Liu, X. Y., Rorabeck, C., and Bourne, R. (2008). Novel hydroxyapatite coating on new porous titanium and titanium-HDPE composite for hip implant, *Surf. Coat. Techn.*, **202**, pp. 2960–2968.

108. Xie, Y.B., and Li, X.Z. (2006). Preparation and characterization of TiO_2/Ti film electrodes by anodization at low voltage for photoelectrcatalytic application, *J. Appl. Electrochem.*, **36**, pp. 663–668.

109. Yang, B., Uchida, M., Kim, H.-M., Zhang, X., and Kokubo, T. (2004). Preparation of bioactive metal via anodic oxidation treatment, *Biomaterials*, **25**, pp. 1003–1010.

110. Yang, Ch.-L., Chen, F.-L., and Chen, S.-W. (2006). Anodization of the dental arch wires, *Mater. Chem. Phys.*, **100**, pp. 268–274.

111. Yang, D.-J., Kim, H.-G., Cho, S.-J., and Choi, W.-Y. (2008). Thickness-conversion ratio from titanium to TiO_2 nanotube fabricated by anodization method, *Mater. Lett.*, **62**, pp. 775–779.

112. Yang, W.-E., Hsu, M.-L., Lin, M.-Ch., Chen, Z.-H., Chen, L.-K., and Huang, H.-H. (2009). Nano/submicron-scale TiO_2 network on titanium surface for dental implant application, *J. Alloys Comp.*, **479**, pp. 642–647.

113. Yang, Y., Wang, X., and Li, L. (2008). Synthesis and growth mechanism of graded TiO_2 nanotube arrays by two-step anodization, *Mat. Sci. Eng. B*, **149**, pp. 58–62.

114. Yao, C., and Webster, T.J. (2006). Anodization: a promising nano-modification technique of titanium implants for orthopedic applications, *J. NanoSci. NanoTech.*, **6**, pp. 2682–2692.

115. Yasuda, K., and Schmuki, P. (2007). Control of morphology and composition of self-organized zirconium titanate nanotubes formed in $(NH_4)_2SO_4/NH_4F$ electrolytes, *Electrochim. Acta*, **52**, pp. 4053–4061.

116. Yen, S.K., and Lin, C.M. (2002). Cathodic reactions of electrolytic hydroxyapatite coating on pure titanium, *Mat. Chem. Phys.*, **77**, pp. 70–76.

117. Yook, S.-W., Yoon, B.-H., Kim, H.-E., Koh, Y.-H., and Kim, Y.-S. (2008). Porous titanium (Ti) scaffolds by freezing TiH_2/camphene slurries, *Mater. Lett.*, **62**, pp. 4506–4508.

118. Yu, X., Li, Y., Wlodarski, W., Kandasamy, S., and Kalantar-Zadeh, K. (2008). Fabrication of nanostructured TiO_2 by anodization: a comparison between electrolytes and substrates, *Sens. Actuators B*, **130**, pp. 25–31.

119. Zhang, Q.Y., Leng, Y., and Xin, R.L. (2005). A comparative study of electrochemical deposition and biomimetic deposition of calcium phosphate on porous titanium, *Biomaterials*, **26**, pp. 2857–2865.

120. Zhao, J., Wang, X., Chen, R., and Li, L. (2005). Fabrication of titanium oxide nanotube arrays by anodic oxidation, *Solid State Commun.*, **134**, pp. 705–710.

121. Zhu, X., Son, D.W., Ong, J.L., and Kim, K. (2003). Characterization of hydrothermally treated anodic oxides containing Ca and P on titanium, *J. Mater. Sci.: Mater. Med.*, **14**, pp. 629–634.

122. Zwilling, V., Darque-Ceretti, E., and Boutry-Forveille, A. (1999). Anodic oxidation of titanium and Ta–6V alloy in chromic media. An electrochemical approach, *Electrochim. Acta*, **45**, pp. 921–929.

123. Zwilling, V., Darque-Ceretti, E., Boutry-Forveille, A., David, D., Perrin, M.Y., and Aueouturier, M. (1999). Structure and physicochemistry of anodic oxide films on titanium and Ta–6V alloy, *Surf. Interface Anal.*, **27**, pp. 629–637.

Chapter 10

Carbon Materials

Jaroslaw Jakubowicz
Institute of Materials Science and Engineering,
Poznan University of Technology, Sklodowska-Curie 5 Sq., 60-965 Poznan, Poland
jaroslaw.jakubowicz@put.poznan.pl

10.1 Carbon Thin Films

Carbon nanomaterials with different structures and forms are a perspective for implant applications. Conventional carbon fibers have been considered for hard [30] and soft [32] tissue implants due to the excellent mechanical properties, such as flexural and fatigue strength and high strength/weight ratio. According to Chlopek *et al.* [12], there are many examples of possible fields of carbon biomaterials applications (Table 10.1). Rough surface of implants for hard tissue engineering is attractive for tissue growth, but unfortunately supports more bacteria adhesion because the larger surface area increases the possibility of bacteria contact and more places suitable for bacteria to stay and colonize. The surface roughness has a significant effect on the bacteria adhesion when

Bionanomaterials for Dental Applications
Edited by Mieczyslaw Jurczyk

ISBN 978-981-4303-83-5 (Hardcover), 978-981-4303-84-2 (eBook)
www.panstanford.com

the surface roughness is closed to a cell size of about 1.0 μm, due to possible bacteria entrapment [16]. The carbon thin film, for example, has antibacterial characteristics [66], which is very attractive for biomedical applications. Recently it was reported that the antibacterial property of stainless steel coated nanocrystalline diamond film was better than titanium [26].

Table 10.1 Examples of applications of carbon biomaterials [12]

Type of material	Function	Type of implant	Area of medicine
Carbon–carbon composites	Bone fixation	Screws, plates, nails, stems of endoprosthesis	Bone surgery
Braided carbon fibers	Tissue knitting, reconstruction of joint ligaments and tendons	Surgical sutures, ligament and tendons prosthesis	Orthopedics
Unwoven carbon	Filling bone and cartilage losses	Disks and rings	Bone surgery
Coatings of diamond-like carbon (DLC)	Coating of metal implants — corrosion protection	Joint endoprosthesis, screws	Bone surgery
Glassy carbon	Blood flow regulation	Heart valves	Cardiology

To enhance the antibacterial performance of carbon films, the functional doping with fluorine, platinum, and silver is useful [25, 33, 45]. The carbon thin film is prepared by chemical vapor deposition using hydrocarbon gas as the carbon source. Zhou *et al.* [66] investigated the influence of hydrogen on the antibacterial performance of carbon films, amorphous hydrogenated carbon (a-C:H) film, and hydrogen-free carbon (H-free C) film, deposited on the stainless steel 316L. Figure 10.1 shows the bacterial 24 h incubation results. The colony numbers of *E. coli* on the hydrogen-free and a-C:H films were reduced to about 15% and 33% of those in the original 316L substrate, respectively [66]. The relatively poor antibacterial performance of a-C:H film compared with H-free carbon films is explained by the degradation of the chemical inertness of carbon film due to hydrogen inclusion. The existence of hydrogen in the carbon film will form the C–H polar bonds, which will promote the van der Waals force between the polarized surface and the bacteria. Additionally the low surface free energy support good antibacterial property [66].

Linares *et al.* [37] found that it is possible to modify activated carbon with calcium carbonate on microporous carbon. These modified carbons with calcium carbonate are able to maintaining a proper pH. Figure 10.2 shows calcium carbonate crystals on activated carbon surface. No calcium carbonate crystals were observed on the activated carbon without modification in sodium carbonate solution. Figure 10.3 shows the antacid activity of the modified carbons.

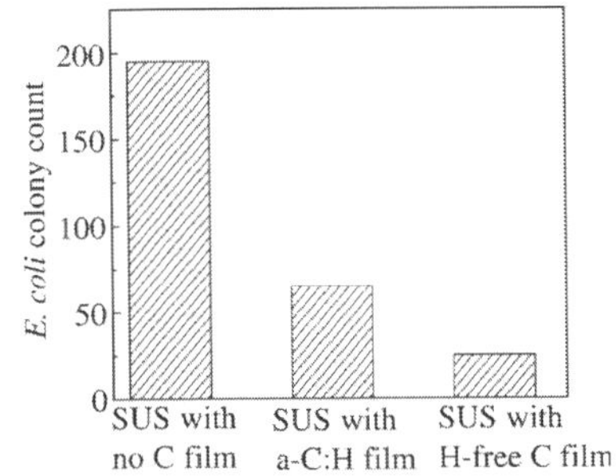

Figure 10.1 Antibacterial performance of SUS (316L) substrate without carbon film, with H-free carbon film, and with a-C:H film [66].

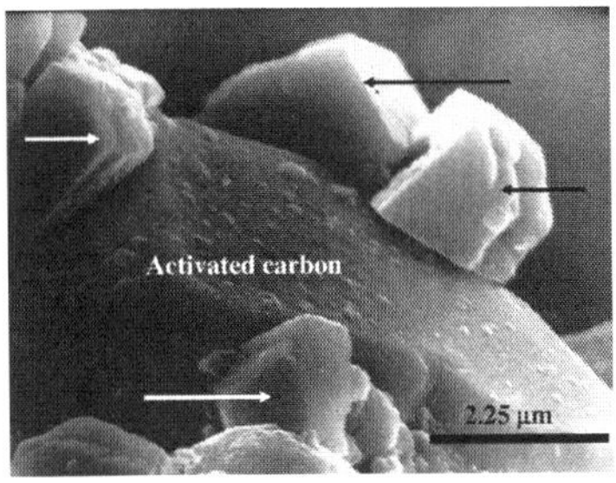

Figure 10.2 SEM image of $CaCO_3$ on activated carbon (3.5 M sample). Arrows indicate $CaCO_3$ crystals [37].

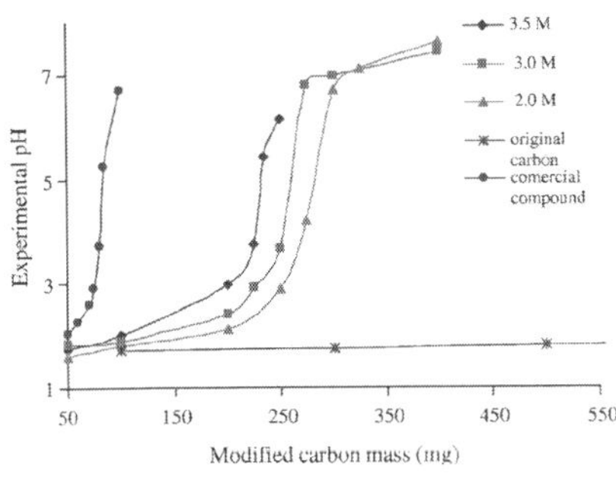

Figure 10.3 Neutralization capacity of original and modified activated carbons with calcium carbonate [17].

The ideal situation is to maintain a pH between 3 and 4, because it maintains an adequate pH to carry out the gastric function. The pH higher than 5 produces a gastric disturbance in the patient and the enzymatic activity of the pepsin is drastically diminished [17, 36]. Linares *et al.* did not found excess salts on the carbon surface.

A novel adherent, continuous, nanocrystalline hydroxyapatite/diamondlike carbon bilayer film was produced using pulsed laser deposition [46]. The diamondlike carbon (DLC) interlayer between the hydoxyapatite surface and the substrate improves hydroxyapatite film adhesion and prevents corrosion or third body wear in the bulk implant material [46]. The term diamondlike carbon refers to amorphous carbon thin films that contain some sp^3-hybridized atoms. These materials have atomic number densities greater than 3.19 g·atom/cm^3, and exhibit densities closer to that of diamond (3.51 g/cm^3) than that of graphite (2.26 g/cm^3) [47]. Amorphous hard carbon thin films demonstrate properties intermediate between those of graphite and those of diamond. These films contain a mixture of sp^3-hybridized carbon atoms and sp^2-hybridized carbon atoms. Within the amorphous carbon matrix micro- or nanocrystalline graphite domains may be observed [47].

Processing of diamondlike carbon thin films is followed by sputtering, bombardment with energetic species, electrostatic acceleration of carbon ions, arc discharge, or laser ablation [47]. Diamondlike carbon thin films have shown excellent compatibility to human cells [4, 38, 39, 47, 56]. The osteoblast cells exposed to DLC coatings did not demonstrate any change in generation of alkaline phosphatase, type I collagen, or osteocalcin [5] and neuronal growth occurs on DLC layer [23]. The TEM micrograph shown in Fig. 10.4 demonstrates several layers within the functionally gradient film structure [47]. These films exhibit significantly higher hardness values than the underlying Ti–6Al–4V alloy.

The titanium carbide/titanium carbonitride/titanium nitride region serves as a functionally gradient zone between the Ti–6Al–4V substrate and the diamondlike carbon layer. The silver regions within the functionally gradient diamondlike carbon–silver layer provide ductility to the film through cracking. Narayanan [47] envisaged applications for multilayer hydroxyapatite/diamondlike carbon films, in orthopedic and dental implants.

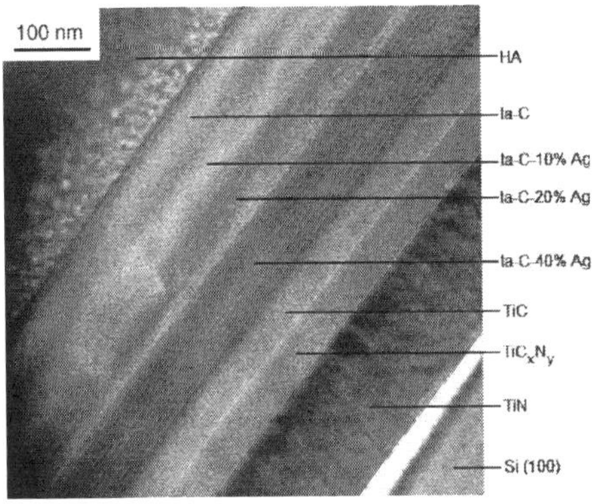

Figure 10.4 TEM micrograph of multilayered hydroxyapatite/diamondlike carbon nanocomposite on Si (100), including hydroxyapatite, diamondlike carbon, diamondlike carbon-silver, titanium carbide, titanium carbonitride, and titanium nitride layers [47].

Diamondlike carbon is an ideal surface coating mainly for prosthetic joints, because it is atomically smooth, low friction, wear resistant, inert, and immune to scratching by third body wear particles [34, 47], and it prevents the release of metal ions into the surrounding tissues. Guglielmotti *et al.* [20] found that DLC-coated zirconium and zirconium implants exhibited better tissue responses than titanium or aluminum implants. DLC-coated steel fracture fixation rods implanted in a human body for 7 months were found to prevent rod corrosion and metal ion release [67].

10.2 Carbon Nanotubes

The new carbon nanotubes are ultra lightweight and have excellent thermal and chemical stability. Compared with conventional carbon, carbon nanotubes are stronger and more flexible and have a higher tensile strength-to-weight ratio [1]. Medical applications of the carbon nanotubes in biosensors, drug delivery systems, scaffolds, and biomaterials are very promising. Carbon nanotubes are composed of the convoluted graphene lattice and can be in the form of a single-walled carbon nanotube (SWCNT) or multi-walled carbon nanotubes (MWCNTs), composed of single and multiplanes, respectively

(Fig. 10.5). Carbon nanotubes are prepared using arc-discharge [28], laser ablation [55], and chemical vapor deposition [11]. In the CVD process, the carbon nanotubes are grown using metal catalysts, such as nickel, so the growing nanotubes could be cytotoxic [21]. Thus, a purification step is usually required before carbon nanotubes can be used for biomedical applications. Refluxing carbon nanotubes in an oxidizing acid (for example nitric acid) is one of the most commonly used purification method [21]. This process oxidizes and removes the metal catalysts and carbonaceous deposits from the inside and outside of the tube.

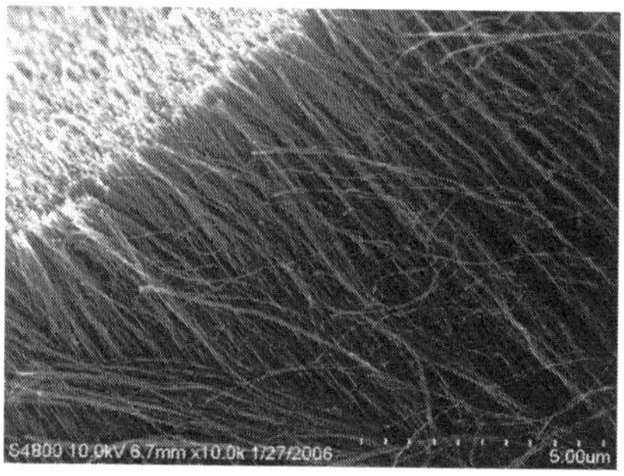

Figure 10.5 Scanning electron micrographs of multi-walled carbon nanotubes [21].

Several *in vitro* studies report that carbon nanotubes are cytotoxic [27, 44, 51]. However it was also reported that they improve neural signal transfer and support dendrite elongation and cell adhesion *in vitro* [42] and also support the smooth muscle [43], fibroblasts [13], and osteoblast as well [53]. The purity of nanotubes plays an important role in cytotoxicity, and functionalization of nanotubes with glycopolymers can lead to the diminishing of their cytotoxicity [48].

In recent years, the number of articles related to carbon nanotubes for use in biomedical applications, including tissue engineering, has doubled each year (Fig. 10.6) [21].

Carbon nanotubes can be used in bone tissue engineering, for example, for the enhancement of polymer and ceramic composites and for nanostructured coatings to improve the bioactivity of the implant surfaces [52]. The bone tissue is composed of nanoscale inorganic crystals, containing calcium, phosphate ions, and collagen fibrils with diameters ranging 20–40 nm [54]. The nanoscale materials improve growth and functioning of bone cells. A high surface area is useful for interactions with the cells and matrix of physiological environments. Carbon nanotubes have low density, high surface area and are strongest and stiffest materials available. The tensile strength of carbon nanotubes is in the range of 11–63 GPa [64]. The mechanical properties of SWCNTs have higher values than MWCNTs. The length-to-diameter ratio is greater than 1000.

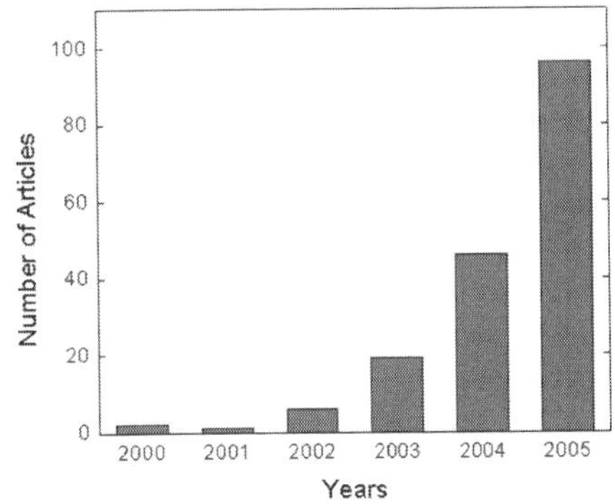

Figure 10.6 Number of articles per year published regarding carbon nanotubes for biomedical applications [21].

Carbon nanotubes have been introduced into composites with natural polymers, such as collagen and chitosan, as well as synthetic polymers or ceramics materials, improving their mechanical properties [7, 35, 59]. The amount of nanotubes required for mechanical properties improvement is in the range of tents of percent. Carbon nanotubes, with and without surface treatment fictionalization, have potential application in bone tissue applications [52].

MWCNTs with diameters of about 100 nm can be used to mimic neural fibers for neuronal growth. It has been shown that hippocampal neurons from 0- to 2-day-old Sprague-Dawley rats were able to grow on carbon nanotubes coated with 4-hydroxynonenal [22].

Carbon nanotubes can be applied in bone tissue engineering as nanoscale coatings for improved biological interactions [8, 10, 18]. These coatings have been formed by plasma spraying [8], electrophoretic deposition [10], film casting [18], and laser surface alloying [63]. Carbon nanotube-coated bioglass scaffolds form hydroxyapatite layers when soaked in simulated body fluid [10].

Two-week soaking of MWCNTs in simulated body fluid results in the formation of nanocrystalline hydroxyapatite [2]. Carbon nanotubes functionalized with phosphate groups enhance their biomineralization coatings and form layers of hydroxyapatite crystals, because of ionic interactions between the negatively charged functional groups and the positively charged calcium ions. Functionalized carbon nanotubes can provide additional capabilities for tissue engineering. Direct crystallization of hydroxyapatite on carbon nanotubes results in a thickness of 3 mm after 14 days of mineralization [65]. Carbon nanotubes are able to provide the initial structural reinforcement needed for newly created tissue scaffolds.

Carbon nanotube coatings and nanocomposites have been successfully applied as substrates for biomineralization, growth, proliferation, and normal functions of osteoblast-like and osteoblast cells [52]. These properties depend upon the type of carbon nanotubes, the chemical nature of treatment , and the composite composition [52]. Application of an electrical current to the conductive carbon nanotubes stimulates increases in the proliferation of the osteoblasts, the extracellular concentration of calcium, and the up-regulation of mRNA expression for collagen type-I [53].

Chlopek *et al.* [12] show good biocompatibility of the nanotubes, which is similar to that of polysulfone. Nanotubes show good cellular biocompatibility, because of high level of viability of the cells in contact with the nanotubes and unchanged level of osteocalcin released from osteoblasts. Chlopek *et al.* found a slight increase of collagen formation induced on nanotubes by both fibroblasts and osteoblasts, which may be significant for applications as substrates for the tissue regeneration [12].

Carbon nanotubes can be functionalized with drugs and biomolecules and proteins, for applications in drug delivery and antibiotics [14, 52]. Liu *et al.* investigated decoration of carbon nanotubes with chitosan [41]; the decoration of CNT with CHIT creates new CHIT–CNT nanomaterials. They have applied a non-destroyable surface decoration of carbon nanotubes with chitosan biopolymer via a controlled surface-deposition and cross-linking process. The method utilizes the emulsifying capacity of chitosan, a completely different water-solubility of chitosan in acidic and basic solutions, and the cross-linking reaction among chitosan polymers [41]. The method consists of the following steps:

(i) Dispersion of MWCNTs in chitosan acetic acid solution is carried out, and during the stirring, the CHIT–CNT blend the chitosan macromolecules are adsorbed on the surface of the CNTs, and acting as polymer cationic surfactants to stabilize the CNTs,

(ii) In the ammonia solution, the chitosan becomes non-dissolvable in aqueous media, and the chitosan deposits on the surface of the carbon nanotubes forming a layer of chitosan coating.

(iii) The blend is heated to 60°C and treated with glutaraldehyde for the cross-linking of the surface-deposited chitosan.

This treatment provides a strong fixing the chitosan coatings to the CNTs [41]. The images of carbon MWCNTs with chitosan are shown on Fig. 10.7. The chitosan surface-decoration results in the surfaces of carbon nanotubes fully covered with chitosan protuberances. The CNT bundles are slightly stretched after the surface decoration and CNTs-bundles are bound together by the chitosan coatings.

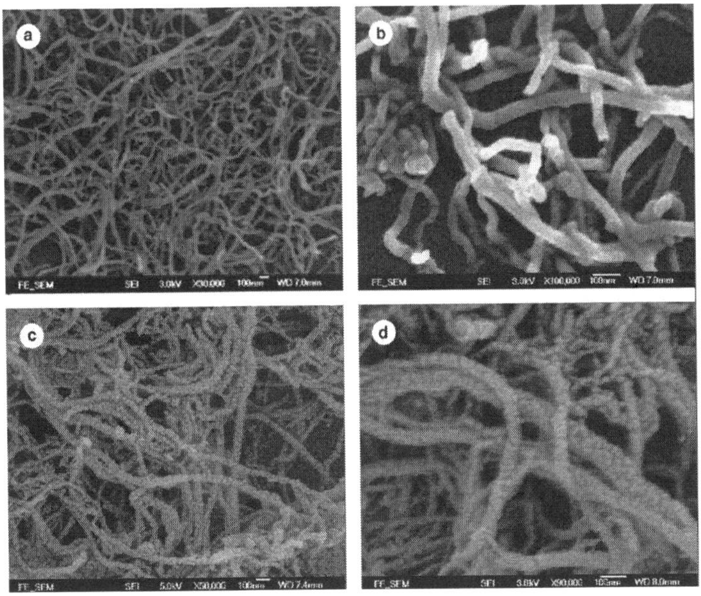

Figure 10.7 SEM image of pristine MWCNTs (a) and (b), and the corresponding chitosan surface-decorated MWCNTs (c) and (d) [41].

Akasaka and Watari investigated the oral bacterial adhesion to CNTs of different diameters and flexibility, and compared them with the widely used adsorbent, activated carbon (AC) particles [3]. Among the different carbon samples, the bacterial (*S. mutans*) precipitation efficiency of 30 nm diameter MWCNTs was the highest. SWCNTs were less effective because they were not easily dispersed before mixing with the bacteria. The 200 nm-MWCNTs are less effective because some of them did not precipitate but remained suspended in the solution. They found that 30 nm-MWCNTs and 200 nm-MWCNTs are highly adhesive to bacteria (Fig. 10.8). The MWCNTs with 30 nm diameter had the highest precipitation efficiency, which is attributed to their adequate dispersibility and aggregation activity.

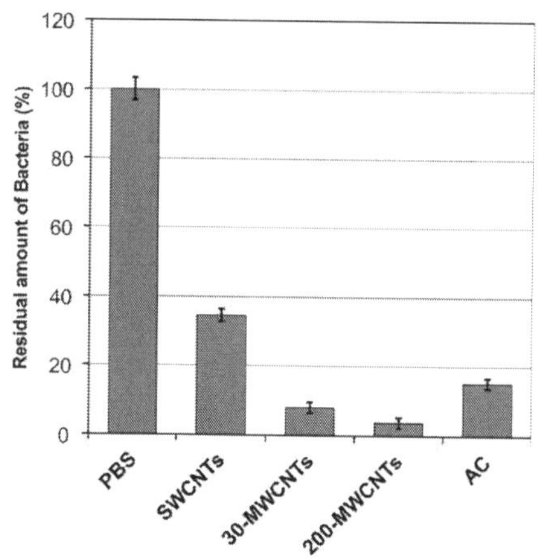

Figure 10.8 Residual amount of *S. mutans* in the supernatant after mixing with carbon samples at 0.66 mg ml^{-1} concentration [3].

Liu *et al.* investigated the silver coated carbon nanotubes, against bacterial attack [40]. The Ag coated CNTs improve the antibacterial property. The bactericidal rate *K* increases with the thickness of the Ag film (Fig. 10.9). For the same thickness, the bactericidal rate of Ag coated CNTs is much higher than that of Ag coated pyrolytic carbon. The carbon nanotubes enlarge the surface area of Ag coating, which provides high probability for Ag atoms to contact the bacteria leading to their death [29]. The antibacterial rate against *S. aureus* is lower than that against *E. coli* [40].

Wang *et al.* [60] used the polycarbosilane as a binder to consolidate CNTs to the form attractive for implant applications. The main disadvantage of that bionanocomposite is unsatisfactory biocompatibility due to the presence of the polycarbosilane [60].

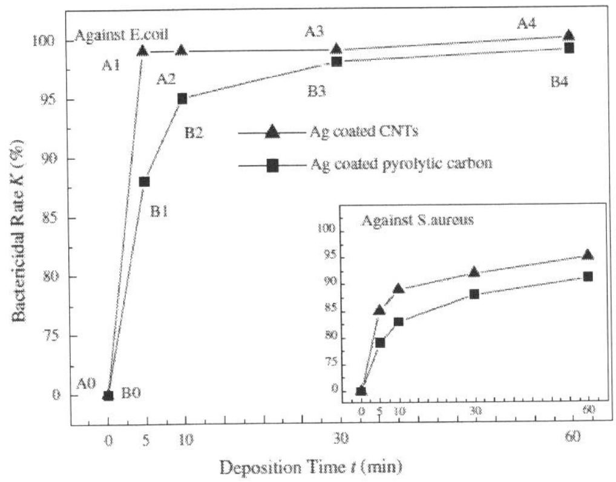

Figure 10.9 The bactericidal rate of Ag coated CNTs and Ag coated pyrolytic carbon. Growth rate ~3.33 nm/min [40].

A carbon nanotube monolith without any binders was obtained by Wang *et al.* using spark plasma sintering (SPS) [61]. *In vivo* testing reveals that pure bulk carbon nanotubes are not a strongly inflammatory substance and have no toxicity for bone regeneration. They suggest that a binderless carbon nanotube monolith with a strength matching that of bone could be a candidate bone substitute material and a bone tissue engineering scaffold material [61]. The SPS is a rapid sintering method, using self heating phenomena within the powder. The SPS method generates spark plasma between a powder particles, resulting in rapid temperature rise and leading to a fully dense CNT monolith obtain in a very short time (10 min), and at a relatively low temperate (1100°C) and pressure (40 MPa). Figure 10.10 shows a carbon nanotube powder and sintered carbon nanotube monolith [61]. The most important in monolith fabrication is initial fast heating rate (200°C/min) for activating the nanotubes surface, which are then sequentially decreases (up to 5°C/min) for stress releasing and keeping the nanotubes structure [61]. The density and mechanical properties of the sintered CNT monolith are at the same level to those of bone (Table 10.2) [6], which shows that the CNT monolith could be attractive for bone tissue repair. The strength of the CNT monolith was less than that of conventional implant materials, but very similar to those of bone, making it possible to achieve a sufficient bone regeneration at the tissue/material interface [61].

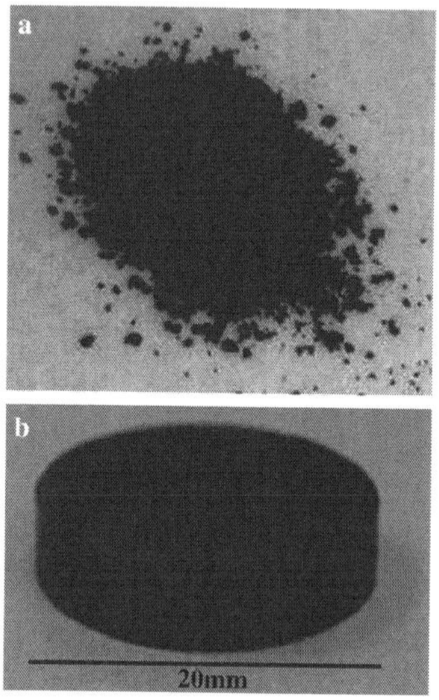

Figure 10.10 Carbon nanotube powders (a) and the sintered carbon nanotube monolith (b) [61].

Table 10.2 Comparison of the properties of CNTs monolith with bone [6, 61]

Material	Bulk density (g/cm^3)	Vickers hardness	Young modulus (GPa)	Compressive strength (MPa)	Flexural strength (MPa)
Bulk CNTs	1.95	44	20	249	172
Bone	1.9	<60	19	150	180

The *in vivo* reactions to the CNT monolith reveals that 1 week after implantation, the CNT monolith was surrounded by tissue with many cells like fibroblasts, fibroblasts with spindle-shaped cytoplasm, and some inflammatory round cells (Fig. 10.11a). 4 weeks after implantation, the CNT monolith was covered by loose fibrous connective tissue, and inflammation around materials was slight in comparison to that after 1 week (Fig. 10.11b). No severe inflammation such as necrosis, degeneration, or neutrophil infiltration was observed around the CNT materials [61]. At 1 week after implantation in the femur, active callus formation from the periosteum and immature newly formed bone were observed around the CNT monolith. The newly formed bone did not directly attach to the material (Fig. 10.11c). At 4 weeks, newly formed bone was remodeled to lamellar bone, and part of the lamellar bone attached to thc implant directly (Fig. 10.11d). Fibrous tissue was not observed by Wang *et al.* between bone tissue and the CNT implant [61].

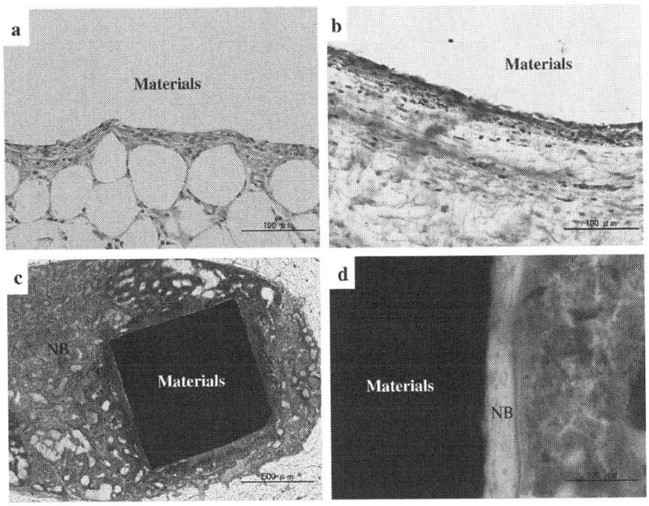

Figure 10.11 Tissue reactions in subcutaneous tissue at (a) 1 week and (b) 4 weeks; osteogenesis in the femur at (c) 1 week and (d) 4 weeks [61].

Binder-free MWCNT blocks cross-linked by de-fluorination were prepared by Sato *et al.* using thermal heating and a compression in vacuum [50]. The carbon nanotube blocks are lighter than graphite, and can be machined and polished. The binder-free MWCNT blocks possess good biocompatibility when tested in the subcutaneous tissue of rats [50]. The MWCNT blocks have a low apparent density 1.44 g/cm^3, average three times stronger in mechanical strength, bending strengths of 102.2 MPa and a bending modulus of 15.4 GPa, than that of commercial graphite.

10.3 Carbon Nanofibers

Carbon nanofibers are a promising new class of dental or orthopedic implant materials with improved osseointegrative properties [15]. Elias *et al.* show the possible use of carbon nanofiber compacts (Fig. 10.12) in implant applications.

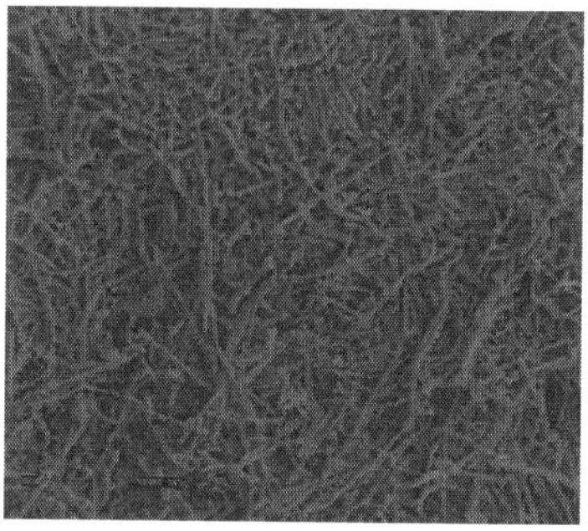

Figure 10.12 SEM image of carbon fiber (100 nm) compact (magnification 5000×) [15].

They found that the number of osteoblasts increases on nanophase carbon fiber compacts (Fig. 10.13) [15]. Compared to conventional compacts, alkaline phosphatase activity was significantly greater when osteoblasts were cultured on respective outer pyrolytic layer nanophase carbon fiber compacts (Fig. 10.14). For any time period tested, calcium content in the extracellular matrix increases with decreasing carbon fiber diameter (Fig. 10.15). Osteoblast proliferation increases with decreasing carbon fiber diameters after 3 and 7 days of culture [15]. Compared to larger-diameter carbon fibers, osteoblasts synthesized more alkaline phosphatase and deposited more extracellular calcium on nanometer-diameter carbon fibers after 7, 14, and 21 days of culture [15]. They suggested that protein-mediated enhanced osteoblast function may be occurring on the carbon nanofibers.

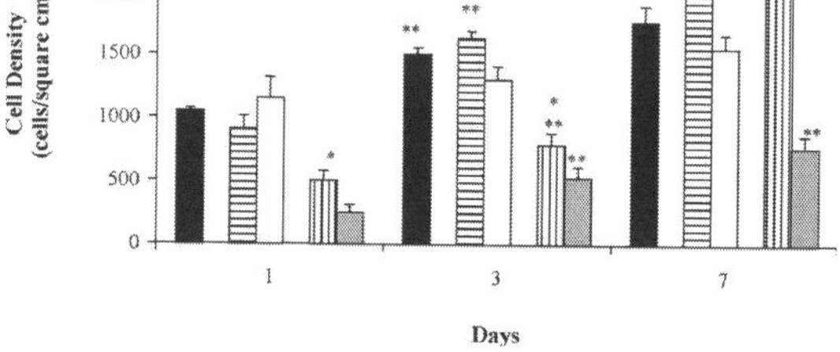

Figure 10.13 Enhanced osteoblast proliferations on nanophase carbon fiber compacts after 1, 3, and 7 days. Substrates: ■ borosilicate glass (reference substrate), ☰ 100 nm diameter with a pyrolytic outer core, □ conventional 200 nm diameter with a pyrolytic outer core, ▥ 60 nm diameter without a pyrolytic outer core, and ▒ conventional 125 nm diameter without a pyrolytic outer core [15].

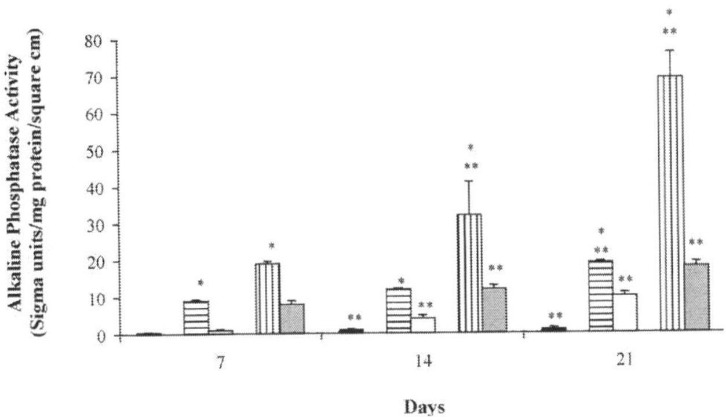

Figure 10.14 Enhanced alkaline phosphatase activity on nanophase carbon fiber compacts after 4, 14, and 21 days. Substrates: ■ borosilicate glass (reference substrate), ▤ 100 nm diameter with a pyrolytic outer core, □ conventional 200 nm diameter with a pyrolytic outer core, ▥ 60 nm diameter without a pyrolytic outer core, and ▩ conventional 125 nm diameter without a pyrolytic outer core [15].

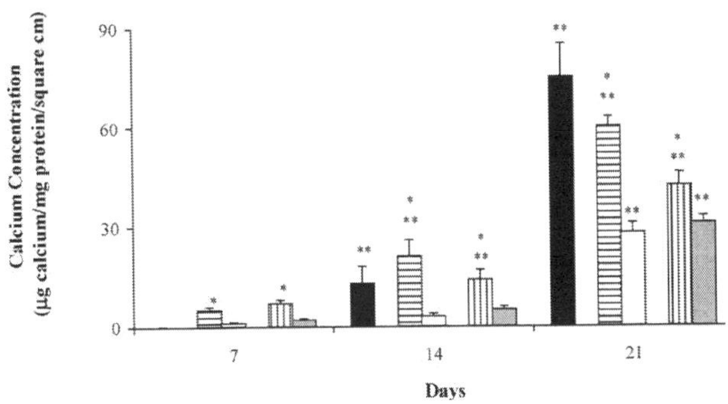

Figure 10.15 Enhanced deposition of calcium on nanophase carbon fiber compacts after 7, 14, and 21 days. Substrates: ■ borosilicate glass (reference substrate), ▤ 100 nm diameter with a pyrolytic outer core, □ conventional 200 nm diameter with a pyrolytic outer core, ▥ 60 nm diameter without a pyrolytic outer core, and ▩ conventional 125 nm diameter without a pyrolytic outer core [15].

Ishida *et al.* developed nanofibrillar forms of carbon using native cellulose rapid freeze drying of suspension and solvent exchange drying processes [24]. The pyrolysis in gaseous HCl acts on cellulose without wetting. For example, the cellulose materials gave a carbon yield of 20–25% in HCl gas, against 10–15% in nitrogen [24]. This method results in the straight microfibrillar morphology (Fig. 10.16). The ex-cellulose carbon obtained by 600°C treatment is nonocrystalline, but treatment at above 2000°C results in graphene planes and graphitic carbon crystallites while the apparent fibrillar morphology was not changed.

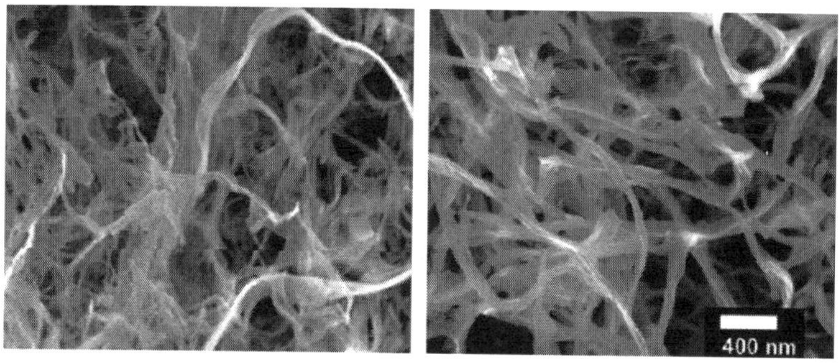

Figure 10.16 SEM image of 600°C carbon from spray-dried cotton microcrystal pyrolyzed in nitrogen (a) and HCl (b) [24].

Table 10.3 shows properties of the cellulose-treated carbon materials. The 600°C carbons have BET surface areas several times greater than those of original cellulose aerogels because of the formation of micropore. The 2200°C treatment caused significant losses in surface area because of carbon crystallite formation. Table 10.4 shows the properties of the different carbon material [24].

Table 10.3 Surface area and pore volume of carbon from tunicate cellulose [24]

Drying method, atmosphere[a]	600°C treated			2000°C treated[b]		
	Surface area (m^2/g)	Micropore[c] (cm^3/g)	Mesopore[c] (cm^3/g)	Surface area (m^2/g)	Micropore (cm^3/g)	Mesopore (cm^3/g)
FD, HCl	533	0.255	0.332	84	0.031	0.247
t-BuOH, N_2	667	0.309	0.468	137	0.051	0.364
t-BuOH, HCl	549	0.258	0.626	239	0.090	0.644

[a] Atmosphere for pyrolysis.

[b] In argon atmosphere.

[c] Micropore: 0.5 to 2 nm, mesopore: 2–50 nm.

Table 10.4 Comparison between carbon materials [24]

Material	Morphology	Surface area	Edge face	Adhesibility
Natural graphite	Flake	Small	Small	Weak
Conventional pyrolytic carbon	Aggregate	Small	Small	Weak
Fullerene	Sphere	Large	None	Weak
Carbon nanotubes	Tube	Large	None	Weak
Fibrous/rod-like carbon	Fibril	Large	(Large)*	(Strong)*

* Expected properties.

The regular arrays of carbon nanofibers could form a nanoporous matrix. The porous carbon with regular structures was obtained by Vinu *et al.* using the replica synthesis [58]. Figure 10.17A shows the process of mesoporous carbon materials. In this method, a carbon sources, such as sucrose, are first impregnated into silica template structures, followed by solidification and template removal, resulting in nanostructured porous carbon materials.

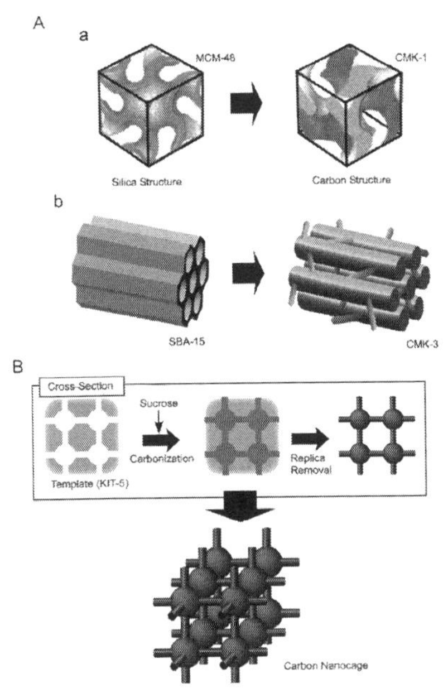

Figure 10.17 (A) Synthesis of mesoporous carbon materials: (a) CMK-1 from MCM-48; (b) CMK-3 from SBA-15. (B) Synthesis of carbon nanocage with cross-sectional view for synthesis and simplified three-dimensional illustration [58].

Mesoporous silica MCM-48 in cubic phase is one of the most appropriate templates because it has bicontinuous pore geometry, resulting in representative mesoporous carbon CMK-1. The SBA-15 template has hexagonal array of cylindrical pores. However, its interconnectivity of the silica channel with micropores leads to preservation of regular carbon structures (CMK-3) even after silica removal [58]. The synthesis of carbon nanocage is illustrated in Fig. 10.17B. The specific surface area and specific pore volume reaches up to 1600 $m^2 g^{-1}$ and 2.1 $cm^3 g^{-1}$, respectively, in the case of carbon nanocage at the lowest sucrose to silica ratio [58]. Because of large pore volume of the carbon nanocage, it would show superior capability in the biomaterials adsorption [57].

Grabinski *et al.* [49] investigated a risk of exposure to the carbon nanomaterials, included carbon fibers — CF (10 μm diameter), carbon nanofibers — CNF (100 nm diameter), MWCNT (10 nm diameter), and SWCNT (1 nm diameter). CF and CNF did not significantly affect cell viability (Fig. 10.18). MWCNT and SWCNT reduced cell viability in a time-dependent manner up to 48 h, with full recovery of mitochondrial function by the 72 h time point (Fig. 10.19) [49].

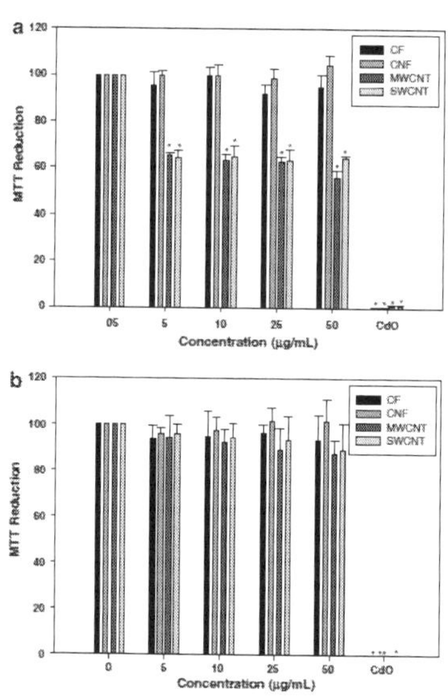

Figure 10.18 Cell viability at 24 and 72 h time points determined via MTT assay. Cell viability after exposure to CF, CNF, MWCNT, and SWCNT at four concentrations (5, 10, 25, and 50 lg/mL) after 24 h (a) and 72 h (b) [49].

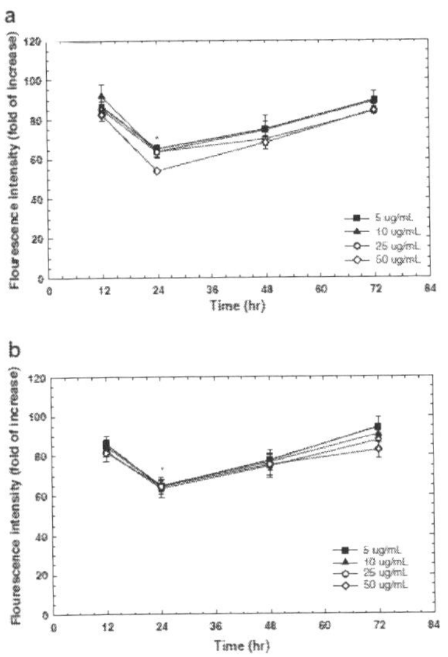

Figure 10.19 Cell viability over time determined via MTT assay. Cell viability after exposure to MWCNT (a), and SWCNT (b) at four concentrations (5, 10, 25, and 50 lg/mL) and four time points (12, 24, 48, and 72 h) [49].

After a 24 h exposure, cells exposed to MWCNT produced up to threefold higher increase in reactive oxygen species than those exposed to SWCNT. In their study, they suggest that high-aspect ratio carbon material toxicity is dependent on dimension and composition.

Surface of the carbon can be improved by HA layer deposition (Fig. 10.20, Fig. 10.21) from electrolyte consisting calcium nitrate, ammonium phosphate, and ammonium hydroxide [19].

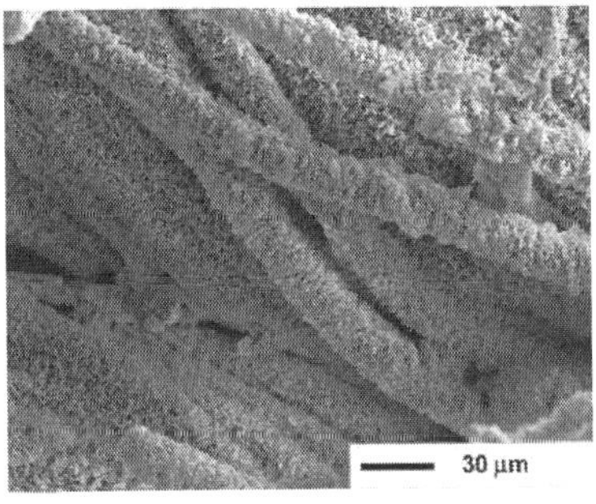

Figure 10.20 SEM image of hydroxyapatite (HA) deposited onto the carbon fibers [19].

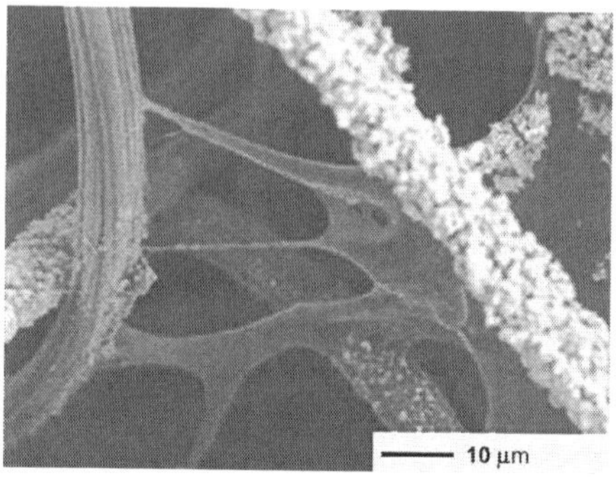

Figure 10.21 SEM of keratocytes adherent to HA coated M1 carbon matrix [19].

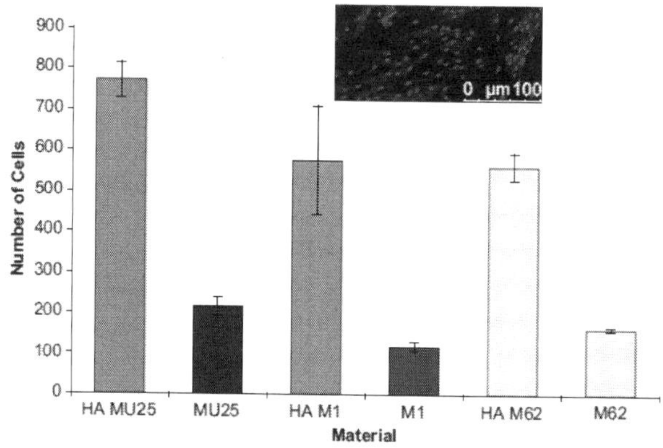

Figure 10.22 Keratocyte adhesion to the 3 types of HA coated and uncoated carbon meshes MU25, M1, and M62. Inside fluorescent microscopy image [19].

Live cell adhesion is possible on both coated and uncoated M1-type carbon meshes. The HA coating significantly increases cell adhesion to the carbon meshes (Fig. 10.22). The adherent cells show spanning of 10–40 μm across the coated M1 carbon fibers (Fig. 10.21) [19].

10.4 Carbon Nanoparticles

The biologically inert nature of the carbon itself limits the inflammatory response and material resorption. Providing a strong, long-lasting adhesive interface between a bone replacement implant and the surrounding tissue involves the use of bioactive materials, such as hydroxyapatite with properties similar to natural bone. A carbon-apatite materials, plays an essential role in forming the chemical bond of the bioactive material to the living bones. This carbon with apatite layer can be reproduced *in vitro* in SBF [9, 62]. Figure 10.23 shows the SEM images of the carbon microspheres before and after soaking in SBF for 5 and 10 days [62].

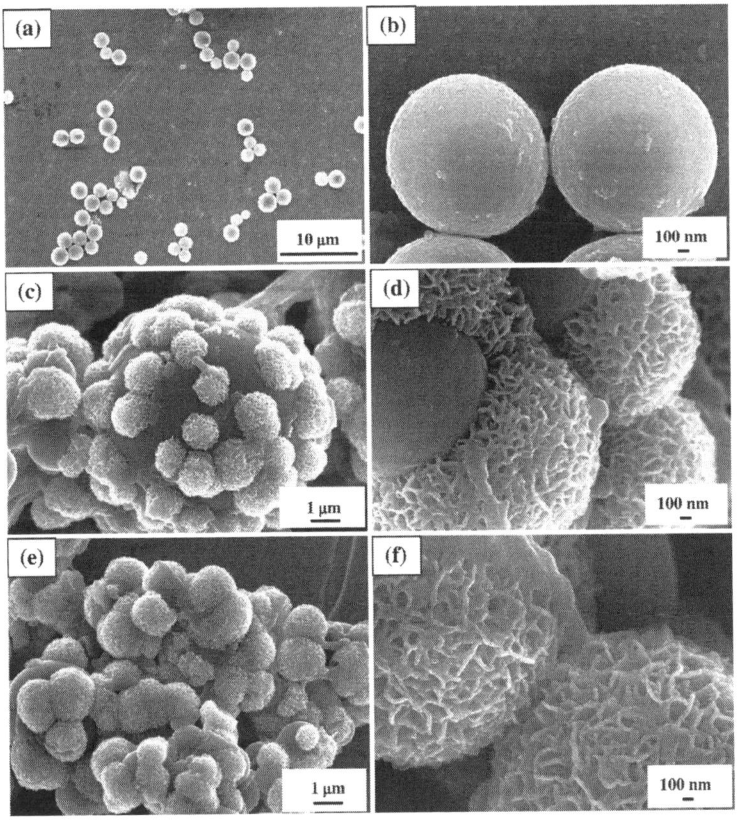

Figure 10.23 Carbon microspheres before and after soaking in SBF for 5 and 10 days (a) and (b) before soaking; (c) and (d) after soaking for 5 days; (e) and (f) after soaking for 10 days [62].

The carbon microspheres have size of about 2 μm (Fig. 10.23a). The surface of the carbon microspheres is smooth (Fig. 10.23b), and after 5 days of soaking in SBF, an apatite clusters with cotton morphology are formed on the surface (Fig. 10.23c) with size of the crystallites of about 200–300 nm (Fig. 10.23d). The apatite morphology is typical to the one shown in Chapter 9. Ten days of soaking results in more apatite forms on the surface compared with that after soaking for 5 days. Furthermore, the apatite forms a layer (Fig. 10.23e,f). The hydroxyl groups (OH^-) remained on the surface of the carbon microspheres after the process are important for the apatite formation in SBF, which provides favorable sites for calcium phosphate nucleation [31]. The potential mechanism of apatite formation on the surface was proposed by Wu and Chang (Fig. 10.24) [62]. Surface of the carbon microspheres is negative due to OH^- groups, which provide favorable sites for calcium phosphate nucleation. The Ca^{2+}, PO_4^{3-}, and CO_3^{2-} ions are assembled on the surface of carbon microspheres to form amorphous carbonate apatite. With the increase of soaking time, the amorphous carbonate apatite grown into crystalline carbonate apatite.

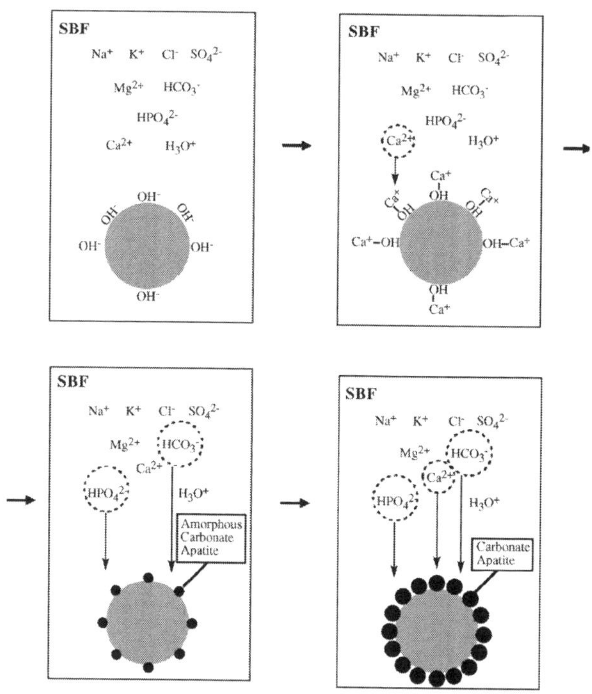

Figure 10.24 Schematic mechanism for the apatite formation on the surface of carbon microspheres in SBF [62].

References

1. Ajayan, P.M. (2000). Carbon nanotubes, in *Handbook of Nanostructured Materials and Nanotechnology, Volume 5: Organics, Polymers, and Biological Materials*, (ed. Nalwa, H.S.), Academic Press, San Diego, pp. 375–406.
2. Akasaka, T., Watari, F., and Sato, Y. (2006). Apatite formation on carbon nanotubes, *Mater. Sci. Eng. C*, **26**, pp. 675–678.
3. Akasaka, T., and Watari, F. (2009). Capture of bacteria by flexible carbon nanotubes, *Acta Biomater.*, **5**, pp. 607–612.
4. Allen, M., Law, F., and Rushton, N. (1994). The effects of DLC coatings on macrophages, fibroblasts and osteoblast-like cells *in vitro*, *Clin. Mater.*, **17**, pp. 1–10.
5. Allen, M., Myer, B., and Rushton, N. (2001). *In vitro* and *in vivo* investigations into the biocompatibility of diamondlike carbon (DLC) coatings for orthopaedic applications. *J. Biomed. Mater. Res.*, **58**, pp. 319–328.
6. Aoki, H. (1999). *Marvelous Biomaterials: Apatite*, Ishiyaku Publishers, Tokyo, pp. 54–58.
7. Ayutsede, J., Gandhi, M., and Sukigara, S. (2006). Carbon nanotube reinforced Bombyx mori silk nanofibers by the electrospinning process, *Biomacromolecules*, **7**, pp. 208–214.
8. Balani, K., Anderson, R., and Laha, T. (2007). Plasma-sprayed carbon nanotube reinforced hydroxyapatite coatings and their interaction with human osteoblasts *in vitro*, *Biomaterials*, **28**, pp. 618–624.
9. Barrias, C.C., Ribeiro, C.C., Lamghari, M., Miranda, C.S., and Barbosa, M.A. (2005). Proliferation, activity, and osteogenic differentiation of bone marrow stromal cells cultured on calcium titanium phosphate microspheres, *J. Biomed. Mater. Res.*, **72**, pp. 57–66.
10. Boccaccini, A.R., Chicatun, F., and Cho, J. (2007). Carbon Nanotube Coatings on Bioglass-Based Tissue Engineering Scaffolds, *Adv. Funct. Mater.*, **17**, pp. 2815–2822.
11. Cassell, A.M., Raymakers, J.A., Kong, J., and Dai, H.J. (1999). Large scale CVD synthesis of single-walled carbon nanotubes, *J. Phys. Chem. B*, **103**, pp. 6484–6492.
12. Chlopek, J., Czajkowska B., Szaraniec, B., Frackowiak, E., Szostak, K., and Be´guin, F. (2006). *In vitro* studies of carbon nanotubes biocompatibility, *Carbon*, **44**, pp. 1106–1111.
13. Correa-Duarte, M.A., Wagner, N., Rojas-Chapana, J., Morsczeck, C., Thie, M., and Giersig, M. (2004). Fabrication and biocompatibility of carbon

nanotube-based 3D networks as scaffolds for cell seeding and growth, *Nano Lett.*, **11**, pp. 2233–2236.

14. Daniel, S., Rao, T.P., and Rao, K.S. (2007). A review of DNA functionalized/ grafted carbon nanotubes and their characterization, *Sens. Actuators*, **122**, pp. 672–682.
15. Elias, K.L., Price, R.L., and Webster, T.J. (2002). Enhanced functions of osteoblasts on nanometer diameter carbon fibers, *Biomaterials*, **23**, pp. 3279–3287.
16. Flint, S.H., Brooks, J.D., and Bremer, P.J. (2000). Properties of the stainless steel substrate, influencing the adhesion of thermo-resistant *streptococci*, *J. Food Eng.*, **43**, pp. 235–242.
17. Fordtran, J.S., Morawski, S., and Richardson, Ch.T. (1973). *In vivo* and *in vitro* evaluation of liquid antacids, *New Engl. J. Med.*, **288**, pp. 923–928.
18. Gilmore, K.J., Moulton, S.E., and Wallace, G.G. (2007). Incorporation of Carbon Nanotubes into the Biomedical Polymer Poly(styrene-β-isobutylene-β-styrene), *Carbon*, **45**, pp. 402–410.
19. Grabinski, Ch., Hussain, S., Lafdi, K., Braydich-Stolle, L., and Schlager, J. (2007). Effect of particle dimension on biocompatibility of carbon nanomaterials, *Carbon*, **45**, pp. 2828–2835.
20. Guglielmotti, M.B., Renou, S., and Cabrini, R.L. (1999). A histomorphometric study of tissue interface by laminar implant test in rats, *Int. J. Oral Maxillofac. Implants*, **14**, pp. 565–570.
21. Harrison, B.S., and Atala, A. (2007). Carbon nanotube applications for tissue engineering, *Biomaterials*, **28**, pp. 344–353.
22. Hu, H., Ni, Y.C., Montana, V., Haddon, R.C., and Parpura, V. (2004). Chemically functionalized carbon nanotubes as substrates for neuronal growth, *Nano Lett.*, **4**, pp. 507–511.
23. Ignatius, M.J., Sawhney, N., Gupta, A., Thibadeau, B.M., Monteiro, O.R., and Brown, I.G. (1998). Bioactive surface coatings for nanoscale instruments: effects on CNS neurons, *J. Biomed. Mater. Res.*, **40**, pp. 264–274.
24. Ishida, O., Kim, D.-Y., Kuga, S., Nishiyama, Y., and Malcolm Brown, R. (2004). Microfibrillar carbon from native cellulose, *Cellulose*, **11**, pp. 475–480.
25. Ishihara, M., Kosaka, T., Nakamura, T., Tsugawa, K., Hasegawa, M., Kokai, F., and Koga, Y. (2006). Antibacterial activity of fluorine incorporated DLC films, *Diamond Relat. Mater.*, **15**, pp. 1011–1014.
26. Jakubowski, W., Bartosz, G., Niedzielski, P., Szymanski, W., and Walkowiak, B. (2004). Nanocrystalline diamond surface is resistant to bacterial colonization, *Diamond Relat. Mater.*, **13**, pp. 1761–1763.

27. Jia, G., Wang, H., Yan, L., Wang, X., Pei, R., and Yan, T. (2005). Cytotoxicity of carbon nanomaterials: single-wall nanotube, multi-wall nanotube, and fullerene, *Environ. Sci. Technol.*, **39**, pp. 1378–1383.
28. Journet, C., Maser, W.K., Bernier, P., Loiseau, A., de LaChapelle, M.L., and Lefrant, S. (1997). Large-scale production of single-walled carbon nanotubes by the electric-arc technique, *Nature*, **388**, pp. 756–758.
29. Kawashita, M., Tsuneyama, S., Miyaji, F., Kokubo, T., Kozuka, H., and Yamamoto, K. (2000). Antibacterial silver-containing silica glass prepared by sol-gel method, *Biomaterials*, **21**, pp. 393–398.
30. Kilfoil, B.M., Hesby, R.A., and Pelleu, G.B. (1983). The ten sile strength of a composite resin reinforced with carbon fibers, *J. Pros. Dent.*, **50**, pp. 40–43.
31. Kokubo, T., Kushitani, H., Saka, S., Kitsugi, T., and Yamamuro, T. (1990). Solutions able to reproduce *in vivo* surface-structure changes in bioactive glass-ceramic A-W, *J. Biomed. Mater. Res.*, **24**, pp. 721–734.
32. Kus, W.M., Gorecki, A., Strzelczyk, P., and Swiader, P. (1999). Carbon fiber scaffolds in the surgical treatment of cartilage lesions, *Ann. Trans.*, **4**, pp. 101–102.
33. Kwok, S.C.H., Zhang, W., Wan, G.J., McKenzie, D.R., Bilek, M.M.M., and Chu, P.K. (2007). Hemocompatibility and anti-bacterial properties of silver doped diamond-like carbon prepared by pulsed filtered cathodic vacuum arc deposition, *Diamond Relat. Mater.*, **16**, pp. 1353–1360.
34. Lappalainen, R., Heinonen, H., Anttila, A., and Santavirta, S. (1998). Some relevant issues related to the use of amorphous diamond coatings for medical applications, *Diamond Relat. Mater.*, **7**, pp. 482–485.
35. Li, G.Y., Wang, P.M., and Zhao, X. (2005). Mechanical behavior and microstructure of cement composites incorporating surface-treated multi-walled carbon nanotubes, *Carbon*, **43**, pp. 1239–1245.
36. Linares, C.F., Sánchez, S., Urbina de Navarro, C., Rodríguez, K., and Goldwasser, M.R. (2005). Study of cancrinite-type zeolites as possible antiacid agents, *Microporous Mesoporous Mater.*, **77**, pp. 215–221.
37. Linares, C.F., Quintero, J., Martínez, L., and González, G. (2007). A new antacid drug from activated carbon modified with calcium carbonate, *Mater. Lett.*, **61**, pp. 2362–2364.
38. Linder, S., Pinkowski, W., and Aepfelbacher, M. (2002). Adhesion, cytoskeletal architecture and activation status of primary human macrophages on a diamond-like carbon coated surface, *Biomaterials*, **23**, pp. 767–773.
39. Liu, E., Blanpain, B., Celis, J.P., Roos, J.R., Alvarezverven, G., and Priem, T. (1996). Tribological behaviour and internal stress of diamond coating

deposited with a stationary d.c. plasma jet, *Surf. Coat. Technol.*, **80**, pp. 264–270.

40. Liu, T., Tang, H.Q., Cai, X.M., Zhao, J., Li, D.J., Li, R., and Sun, X.L. (2007). A study on bactericidal properties of Ag coated carbon nanotubes, *Nuclear Instr. Met. Phys. Res. B*, **264**, pp. 282–286.
41. Liu, Y., Tang, J., Chen, X., and Xin, J.H. (2005). Decoration of carbon nanotubes with chitosan, *Carbon*, **43**, pp. 3178–3180.
42. Lovat, V., Pantarotto, D., Lagostena, L., Cacciari, B., Grandolfo, M., and Righi, M. (2005). Carbon Nanotube Substrates Boost Neuronal Electrical Signaling, *Nano Lett.*, **6**, pp. 1107–1110.
43. MacDonald, R.A., Laurenzi, B.F., Viswanathan, G., Ajayan, P.M., and Stegemann, J.P. (2005). Collagen-carbon nanotube composite materials as scaffolds in tissue engineering, *J. Biomed. Mater. Res. A*, **74**, pp. 489–496.
44. Maynard, A.D., Baron, P.A., Foley, M., Shvedova, A.A., Kisin, E.R., and Castranova, V. (2004). Exposure to carbon nanotube material: aerosol release during the handling of unrefined single-walled carbon nanotube material, *J. Toxicol. Environ. Health A*, **67**, pp. 87–107.
45. Morrison, M.L., Buchanan, R.A., Liaw, P.K., Berry, C.J., Brigmon, R.L., Riester, L., Abernathy, H., Jin, C., and Narayan, R.J. (2006). Electrochemical and antimicrobial properties of diamondlike carbon-metal composite films, *Diamond Relat. Mater.*, **15**, pp. 138–146.
46. Narayan, R.J. (2005). Hydroxyapatite–diamondlike carbon nanocomposite films, *Mat. Sci. Eng. C*, **25**, pp. 398–404.
47. Narayan, R.J. (2005). Nanostructured diamondlike carbon thin films for medical applications, *Mat. Sci. Eng. C*, **25**, pp. 405–416.
48. Reynolds, C.H., Annan, N., Beshah, K., Huber, J.H., Shaber, S.H., and Lenkinski, R.E. (2000). Gadolinium-Loaded Nanoparticles: New Contrast Agents for Magnetic Resonance Imaging, *J. Am. Chem. Soc.*, **122**, pp. 8940–8945.
49. Sandeman, S.R., Jeffery, H., Howell, C.A., Smith, M., Mikhalovsky, S.V., and Lloyd, A.W. (2009). The *in vitro* corneal biocompatibility of hydroxyapite coated carbon mesh. *Biomaterials*, **30**, pp. 3143–3149.
50. Sato, Y., Yokoyama, A., Kasai, T., Hashiguchi, S., Ootsubo, M., Ogino, S., Sashida, N., Namura, M., Motomiya, K., Jeyadevan, B., and Tohji, K. (2008). *In vivo* rat subcutaneous tissue response of binder-free multi-walled carbon nanotube blocks cross-linked by de-fluorination, *Carbon*, **46**, pp. 1927–1934.
51. Shvedova, A.A., Castranova, V., Kisin, E.R., Schwegler-Berry, D., Murray, A.R., and Gandelsman, V.Z. (2003). Exposure to carbon nanotubes

material: Assessment of nanotube cytotoxicity using human keratinocyte cells, *J. Toxicol. Environ. Health A*, **66**, pp. 1909–1926.

52. Spear, R.L., and Cameron, R.E. (2008). Carbon nanotubes for orthopaedic implants, *Int. J. Mater. Form.*, **1**, pp. 127–133.

53. Supronowicz, P.R., Ajayan, P.M., Ullmann, K.R., Arulanandam, B.P., Metzger, D.W., and Bizios, R. (2002). Novel current-conducting composite substrates for exposing osteoblasts to alternating current stimulation, *J. Biomed. Mater. Res.*, **59**, pp. 499–506.

54. Tencer, A.F., and Johnson, K.D. (1994). *Biomechanics in Orthopedic Trauma: Bone Fracture and Fixation*, Biomaterials used in fracture fixation, Martin Dunitz, London (1994) pp. 84–90.

55. Thess, A., Lee, R., Nikolaev, P., Dai, H.J., Petit, P., and Robert, J. (1996). Crystalline ropes of metallic carbon nanotubes, *Science*, **273**, pp. 483–487.

56. Thomson, L.A., Law, F.C., Rushton, N., and Franks, J. (1991). Biocompatibility of diamond-like carbon coating, *Biomaterials*, **12**, pp. 37–40.

57. Vinu, A., Miyahara, M., Sivamurugan, V., Mori, T., and Ariga, K. (2005). Large pore cage type mesoporous carbon, carbon nanocage: a superior adsorbent for biomaterials, *J. Mater. Chem.*, **15**, pp. 5122–5127.

58. Vinu, A., Mori, T., and Ariga, K. (2006). New families of mesoporous materials, *Sci. Technol. Adv. Mater.*, **7**, pp. 753–771.

59. Wang, S.F., Shen, L., and Zhang, W.D. (2005). Preparation and Mechanical Properties of Chitosan/Carbon Nanotubes Composites, *Biomacromolecules*, **6**, pp. 3067–3072.

60. Wang, W., Omori, M., Watari, F., and Yokoyama, A. (2005). Novel bulk carbon materials for implant by spark plasma sintering, *Dent. Mater. J.*, **24**, pp. 478–486.

61. Wang, W., Yokoyama, A., Liao, S., Omori, M., Zhu, Y., Uo, M., Akasaka, T., and Watari, F. (2008). Preparation and characteristics of a binderless carbon nanotube monolith and its biocompatibility, *Mat. Sci. Eng. C*, **28**, pp. 1082–1086.

62. Wu, C., and Chang, J. (2007). Bonelike apatite formation on carbon microspheres, *Mater. Lett.*, **61**, pp. 2502–2505.

63. Yao, C., Cuihua, G., and Tainua, Z. (2005). Laser-surface-alloyed carbon nanotubes reinforced hydroxyapatite composite coatings, *Appl. Phys. Lett.*, **86**, pp. 251905–251905.

64. Yu, M.-F., Lourie, O., and Dyer, M.J. (2000). Strength and Breaking Mechanism of Multiwalled Carbon Nanotubes Under Tensile Load, *Science*, **287**, pp. 637–640.

65. Zhao, B., Hu, H., Mandal, S.K., and Haddon, R.C. (2005). A bone mimic based on the self-assembly of hydroxyapatite on chemically functionalized single-walled carbon nanotubes, *Chem. Mater.*, **17**, pp. 3235–3241.

66. Zhou, H., Xu, L., Ogino, A., and Nagatsu, M. (2008). Investigation into the antibacterial property of carbon films, *Diamond Relat. Mater.*, **17**, pp. 1416–1419.

67. Zolynski, K., Witkowski, P., Kaluzny, A., Has, Z., Niedzielski, P., and Mitura, S. (1996). Implants with hard carbon layers for application in pseudoarthrosis femoris sin. ostitis post fracturam apertamolim factam, *J. Chem. Vapor Depos.*, **4**, pp. 232–239.

Chapter 11

Nanomaterials in Preventive Dentistry

Karolina Jurczyk[a,*] and Mieczyslawa U. Jurczyk[b]
[a]*Conservative Dentistry and Periodontology Department,*
University of Medical Sciences, Bukowska 70 Street, 60-812 Poznan, Poland
[b]*Division Mother's and Child's Health, University of Medical Sciences*
Polna 33 Street, 60-535 Poznan, Poland
*karolajur@gmail.com

Preventive dentistry is the modern way of reducing the amount of dental treatment necessary to maintain a healthy mouth. The two major causes of tooth loss are decay and gum disease. The better we prevent or deal with these two problems, the more chances people have in keeping their teeth for life. The combined efforts of the dentist, the hygienist and the patient, can help to prevent the need for treatment, and so to avoid the traditional pattern of fillings and extractions of teeth.

Tooth cleaning and polishing and fluoride application are all part of prevention program. Preventive dentistry has a wide range of opportunities for studies [1, 5, 15, 16]. There are interests in child dental health, geriatric dental health, fluoride and dental health, dental health services research, and dental labor force.

Nanotechnology will change preventive dentistry and healthcare [7, 9, 10, 29, 30]. Current work is focused on the recent develop-

Bionanomaterials for Dental Applications
Edited by Mieczyslaw Jurczyk

ISBN 978-981-4303-83-5 (Hardcover), 978-981-4303-84-2 (eBook)
www.panstanford.com

ments, particularly of nanoparticles and nanotubes for periodontal management, the materials developed such as the hollow nanospheres, core shell structures, nanocomposites, nanoporous materials, and nanomembranes will play a growing role in materials development for the dental industry.

11.1 Nanodentistry

Nanodentistry will make possible the maintenance of comprehensive oral health by nanomaterials, biotechnology including tissue engineering and nanorobotics. Freitas has described how medical nanorobots might utilize specific motility mechanisms to crawl or swim through human body tissues with navigational precision, acquire energy, sense and manipulate their surroundings, achieve safe cytopenetration (e.g., pass through plasma membranes such as the odontoblastic process without disrupting the cell), and employ any of a multitude of techniques to monitor, interrupt, or alter nerve impulse traffic in individual nerve cells, and in real time (Table 11.1) [7].

The visions described above by Freitas may sound unlikely, but recent advances in nanotechnological research and development have made such applications theoretically possible [7, 9, 29].

11.2 Biomimetic Nanomaterials

A new method for altered enamel surface remineralization has been proposed [8, 23, 25]. Dental caries is a widespread chronic disease caused by glucolytic biofilms. It is hypothesized that nano-sized hydroxylapatite crystallites occur in the oral cavity during extensive physiological wear of the hierarchical structured enamel surface due to dental abrasion and attrition. Modern bioinspired nanomaterials in preventive dentistry containing nano-sized hydroxylapatite particles have shown efficacy in reducing oral biofilm formation and yield re-mineralizing effects [8].

In the last years, several oral health care compounds, tooth pastes and mouth rinses were developed containing nano-sized bioinspired apatite particles in combination with or without proteinaceous additives like caseino-phosphopetides [23, 25]. The efficacy of these compounds in dental prophylaxis is attributed to

Table 11.1 New treatment opportunities in nanodentistry [7]

	Treatments	Comments
1	Tooth repair	Nanodental techniques for major tooth repair may evolve through several stages of technological development, first using genetic engineering, tissue engineering and tissue regeneration and later growing whole new teeth *in vitro* and installing them.
2	Tooth renaturalization	Dentition renaturalization procedures may become a popular addition to the typical dental practice, providing perfect methods for esthetic dentistry. This trend may begin with patients who desire to have their old dental amalgams excavated and their teeth remanufactured with native biological materials. But demand will grow for full coronal renaturalizations in which all fillings, crowns, and other necessary 20th century modifications to the visible dentition are removed, with the affected teeth remanufactured so as to be indistinguishable from the natural originals.
3	Hypersensitivity cure	Dentin hypersensitivity may be caused by changes in pressure transmitted hydrodynamically to the pulp. This etiology is suggested by the finding that hypersensitive teeth have 8 times higher surface density of dentinal tubules – and tubules with diameters twice as large – than nonsensitive teeth. There are many therapeutic agents for this common painful condition that provide temporary relief, but reconstructive dental nanorobots could selectively and precisely occlude selected tubules in minutes, using native biological materials, offering patients a quick and permanent cure.
4	Orthodontic nanorobots	Orthodontic nanorobots could directly manipulate the periodontal tissues including gingiva, periodontal ligament, cementum and alveolar bone, allowing rapid painless tooth straightening, rotating, and vertical repositioning in minutes to hours, in contrast to current molar uprighting techniques which require weeks or months to proceed to completion.

(*Contd*)

Table 11.1 (*Contd*)

	Treatments	Comments
5	Dental durability and cosmetics	Tooth durability and appearance may be improved by replacing upper enamel layers with covalently-bonded artificial materials such as sapphire or diamond which have 20–100 times the hardness and failure strength of natural enamel or contemporary ceramic veneers and good biocompatibility. Like enamel, sapphire is somewhat susceptible to acid corrosion, but sapphire can be manufactured in virtually any color of the rainbow, offering interesting cosmetic possibilities (e.g., iridescence) as alternatives to standard whitening and sealant procedures.
6	Nanorobotic dentifrice (dentifrobots)	Effective prevention has reduced caries in children and a caries vaccine may soon be available, but a subocclusal-dwelling nano-robotic dentifrice delivered by mouthwash or toothpaste could patrol all supragingival and subgingival surfaces at least once a day, metabolizing trapped organic matter into harmless and odorless vapors and performing continuous calculus debridement. These invisibly small (1–10 micron) dentifrobots, perhaps numbering 103–105 nanodevices per oral cavity and crawling at 1–10 microns/sec, might have the mobility of tooth amoebas but would be inexpensive purely mechanical devices that would safely deactivate themselves if swallowed and would be programmed with strict occlusal avoidance protocols. Properly configured dentifrobots could identify and destroy pathogenic bacteria residing in the plaque and elsewhere, while allowing the ~500 species of harmless oral microflora to flourish in a healthy ecosystem. Dentifrobots would also provide a continuous barrier to halitosis, since bacterial putrefaction is the central metabolic process involved in oral malodor. With this kind of daily dental care available from an early age, conventional tooth decay and gum disease will disappear into the annals of medical history.

size-specific effects of the apatite nano-particles corresponding to the ultra structure of the enamel. The nanostructured biomimetic materials expose a large surface to volume ratio and extraordinary physicochemical properties [18, 28]. An example is casein phosphopeptide (CPP) stabilized by amorphous calcium phosphate (ACP) [26]. These casein phosphopeptides have a high affinity to dental plaque and their incorporation into the oral biofilm provides a reservoir of calcium for re-mineralization [4]. CPP–ACP decreases the amount of calcium bridging between the pellicle and adhering bacteria as well as between the bacteria by interaction with calcium binding sites. Furthermore, specific receptor molecules are blocked. This reduces the general bacterial colonization as demonstrated *in situ* with germanium surfaces treated with CPP–ACP [23]. However, CPP–ACP does not mimic nano-sized enamel crystallites. Other biomimetic approaches are based on hydroxyapatite-nanocrystals resembling the nanostructure of abraded dental enamel crystallites. Non-aggregated as well as clustered hydroxylapatite nano-crystallite particles (100 × 10 × 5 nm) adsorb to bacterial surfaces *in vitro* [32]. The adsorbed nano-sized apatite interacts with the bacterial adhesins, and thus reduces bacterial adherence.

Different types and pharmaceutical forms of nano-hydroxylapatite promote re-mineralization and repair of de-mineralized enamel or micro-sized tooth surface defects [12, 27, 28]. Typically, these nano-crystallites mimic the size of natural dentinal hydroxylapatite (20 nm) or enamel apatite (100 nm), sometimes aggregated as clusters [28]. Different forms of the particles such as spheroidal or needle like crystallites have been tested and it turned out that they improve remineralization of artificial caries better than sodium fluoride typically found in toothpastes [17]. These artificial nano-apatites might be supplied during daily oral hygiene measures using toothpastes, mouth rinsing solutions, chewing gums, etc.

A new biomimetic mineral coating, which progressively fills and shadows surface scratches, covers, and safeguards the enamel structure by contrasting the acid and bacteria attacks [17, 27].

11.3 Antimicrobial Effect of Nanometric Bioactive Glass 45S5

The clinically interesting antimicrobial properties of commercially available, micron-sized bioactive glass 45S5 have been attributed to the continuous liberation of alkaline species during application [33, 34]. The release of Na^+ and Ca^{2+} ions from, and the incorporation of H_3O^+protons into, the corroding glass result in a high-pH environment in closed systems, which is not well-tolerated by microbiota [2, 31]. In addition, the release of silica has been linked to the antibacterial bioactive glass effect [34]. In contrast to commonly used disinfectants in dentistry, silica-containing bioactive glasses induce dentin mineralization, and thus are potentially interesting materials for the treatment of demineralized and infected dentin found in deep caries lesions and necrotic root canals [6].

Most recent advances in nanomaterials fabrication have given access to complex materials such as SiO_2–Na_2O–CaO–P_2O_5 bioactive glasses in the form of amorphous nanoparticles of 20 to 60 nm size [22, 33]. This substantially increases ionic release in suspension, and may thus result in enhanced antimicrobial efficacy [31]. As observed by scanning electron microscopy, the nanoparticulate bioglass was spherically shaped and highly agglomerated, while the conventional melt-derived glass was in the form of sharp-edged shards (Fig. 11.1). According to laser ablation inductively coupled plasma mass spectrometry, the composition of the nanometric bioactive glass was 44.7 wt% SiO_2, 4.9 wt% P_2O_5, 27.6 wt% CaO, and 22.8% Na_2O. Transmission electron microscopy confirmed particle size of the nanometric materials to be 30 ± 7.8 nm for bioactive glass (N = 200).

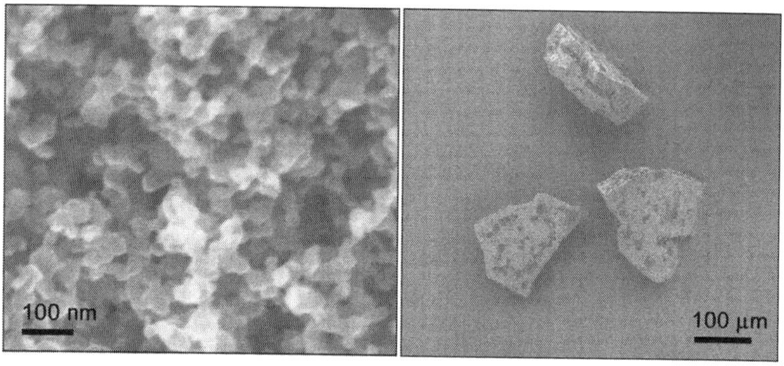

Figure 11.1 Scanning electron microscopic images of flame-derived, nanometric bioactive glass (left) and the micron-sized commercially available 45S5 glass (PerioglassTM, right) [33].

At the same solid-to-liquid ratio (wt/vol), the nanoparticulate 45S5 released approximately 10-fold more silica into the simulated body fluid than the conventional bioglass (Fig. 11.2). In contrast, 10-minute conventional bioglass supernatants contained substantially more calcium, 158.5 ± 8.4 ppm vs. 31.2 ± 5.8 ppm. Nanometric bioglass supernatants had a pH of 11.7, compared with 8.3 measured in counterparts obtained from conventional bioglass suspensions.

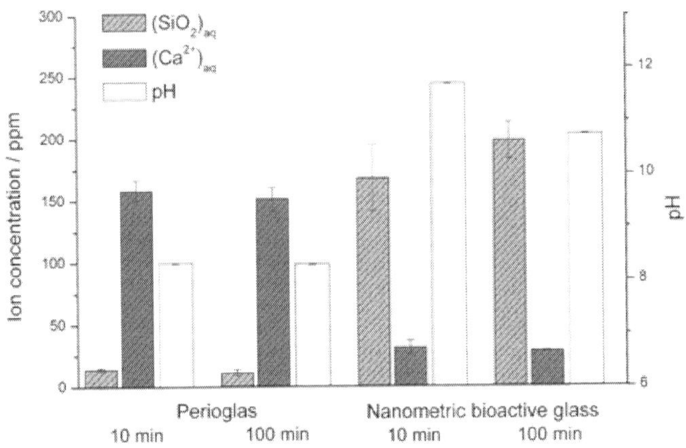

Figure 11.2 Silicon and calcium contents as well as pH levels of 1:10 (wt/vol) suspension supernatants of conventional and nanoparticulate bioglass in simulated body fluid. Error bars indicate standard deviations (N = 3) [33].

The current study showed a substantially higher release of alkaline species and better antibacterial efficacy of bioactive glass 45S5 in nanoparticulate form, compared with the conventional preparation [33]. The antibacterial effect of nanoparticulate bioactive glass appears to be directly linked to its high surface area and thus the resulting release of ionic components in solution. However, the low concentration of calcium in nanometric 45S5 supernatants suggests that calcium and phosphate released from this glass preparation precipitate immediately [3]. The release of silica should eventually result in the transformation of glass nanoparticles into inert Ca-P shells, which can have a substantial potential to (re)mineralize dentin [27, 33].

11.4 Nano-Sized Calcium-Deficient Apatite Crystals

Bioactive ceramic materials, such as hydroxyapatite [$Ca_{10}(PO_4)_6(OH)_2$, HA] were widely used as bone substitutes for several decades [14]. Stoichiometric hydroxyapatite has a hexagonal structure. On the other hand, the calcium-deficient hydroxyapatites [$Ca_{10-x}(PO_4)_{6-x}(HPO_4)_x(OH)_{2-x}$, $0 \leq x \leq 1$, CDHA] are of greater biological interest than stoichiometric HA since the Ca/P ratio in bone is nearer to 1.5 [14]. It has been suggested that calcium-deficient apatitic (CDHA) plays important roles in several processes such as bone remodeling and bone formation [13]. Furthermore, both compositions are chemically and structurally similar to the mineral constituent of human hard tissue. However, bone mineral (natural biocrystal) essentially has a CDHA structure with a Ca/P ratio of about 1.5 which, strictly speaking, is a Ca/P ratio similar to tricalcium phosphates (TCP), $Ca_3(PO_4)_2$, (Ca/P = 1.5) but structurally and chemically compositionally similar to stoichiometric hydroxyapatite, $Ca_{10}(PO_4)_6(OH)_2$, (Ca/P = 1.67). It was also reported that the chemical properties such as catalysis, ion-exchange, and degradation in solution are strongly dependent on Ca/P ratios [19].

Recently, various processes have been developed to prepare nano-sized apatite powders [13, 20, 21, 24]. The synthesis of single-phased needle-like CDHA nano-crystals with Ca/P ratios from 1.5 to 1.67 can be performed by a simple coprecipitation method with a good reproducibility [13].

Figure 11.3 shows the TEM bright-field (BF) image of the precipitated CDHA powders with different Ca/P ratios from 1.5 to 1.67. The powder morphology presents a needle-like shape and the needle-like particles are easily developed to form aggregates when higher concentration and fast mixing are applied during the solution preparation [11]. The dimension of these CDHA crystals in both radial (diameter) and axial (length) directions is 5–10 nm in diameter and 40–50 nm in length in terms of different Ca/P ratios. An axial zone in some needle-like CDHA particles can be clearly observed in the high-resolution TEM [13]. The lattice fringes were seen predominantly in long needle-like crystallites and parallel to (1 0 0) CDHA with a width of 0.82 nm. It was suggested that the HA crystals tend to grow along the [0 0 0 1] direction and the fastest growth direction results in needle-like morphology.

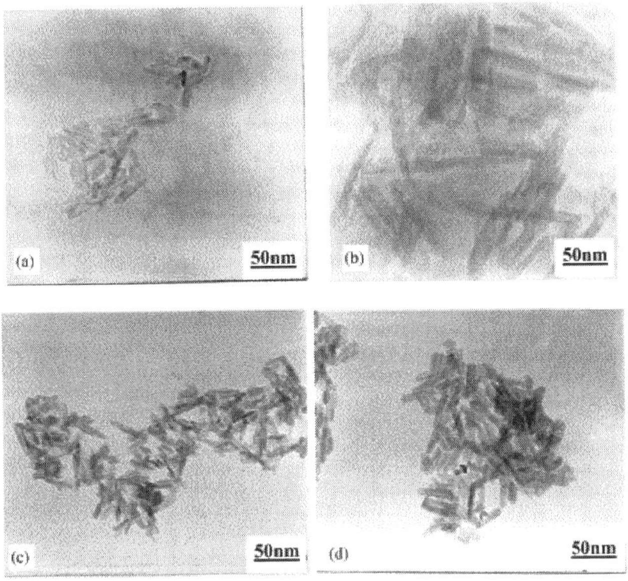

Figure 11.3 TEM micrographs of calcium-deficient apatites with different Ca/P ratios: (a) Ca/P = 1.5, (b) Ca/P = 1.55, (c) Ca/P = 1.6 and (d) Ca/P = 1.67 [13].

The amplitude of FT in the EXAFS spectra indicates that the structural disorder increases, following the sequence of 1.67 > 1.5 > 1.6 > 1.55, implying that stoichiometry and non-stoichiometry plays a more important role than the structural order–disorder. This may suggest that different biochemical properties between Ca/P = 1.5 and 1.67 is mostly due to the effect of stoichiometry and non-stoichiometry.

References

1. Al-Ahmad, A., Roth, D., Wolkewitz, M., Wiedmann-Al-Ahmad, M., Follo, M., Ratka-Kruger, P., Deimling, D., Hellwig, E., and Hannig, C. (2009). Change in diet and oral hygiene over an 8-week period: effects on oral health and oral biofilm, *Clin. Oral Investig.*, **14**, pp. 391–396.

2. Allan, I., Newman, H., and Wilson, M. (2001). Antibacterial activity of particulate bioglass against supra- and subgingival bacteria, *Biomaterials*, **22**, pp. 1683–1687.

3. Brunner, T.J., Grass, R.N., and Stark, W.J. (2006). Glass and bioglass nanopowders by flame synthesis, *Chem. Commun.*, **7**, pp. 1384–1386.

4. Cross, K.J., Huq, N.L., and Reynolds, E.C. (2007). Casein phosphopeptides in oral health–chemistry and clinical applications, *Curr. Pharm. Des.*, **13**, pp. 793–800.

5. Featherstone, J.D.B., Glena, R., Shariati, M., and Shields, C.P. (1990). Dependence of *in vitro* demineralization of apatite and remineralization of dental enamel on fluoride concentration, *J. Dental Res.*, **69**, pp. 620–625.

6. Forsback, A.P., Areva, S., and Salonen, J.I. (2004). Mineralization of dentin induced by treatment with bioactive glass S53P4 *in vitro*, *Acta Odontol. Scand.*, **62**, pp. 14–20.

7. Freitas, Jr. R.A. (2000). Nanodentistry, *J. Am. Dent. Assoc.*, **131**, pp. 1559–1566.

8. Hannig, C., and Hannig, M. (2010). Natural enamel wear — A physiological source of hydroxylapatite nanoparticles for biofilm management and tooth repair?, *Medical Hypotheses*, **74**, pp. 670-672.

9. Jhaveri, H.M., and Balaji, P.R. (2005). Nanotechnology. The future of dentistry a review. *Jr. I. Prosthetic.*, **5**, pp. 15–17.

10. Khang, D., Carpenter, J., Chun, Y.W., Pareta, R., and Webster, T.J. (2008). Nanotechnology for regenerative medicine, *Biomed. Microdevices*, **12**, pp. 575–587.

11. Layani, J.D., Cuisinier, F.J.G., Steuer, P., Cohen, H., Voegel, J.C., and Mayer, I. (2000). High-resolution electron microscopy study of synthetic carbonate and aluminum containing apatites, *J. Biomed. Mater. Res.*, **50**, pp. 199–207.

12. Li, L., Pan, H., Tao, J., Xu, X., Mao C., Gu, X., and Tang, R. (2008). Repair of enamel by using hydroxyapatite nanoparticles as the building blocks, *J. Mater. Chem.*, **18**, pp. 4079–4084.

13. Liou, S.C., Chen, S.Y., Lee, H.Y., and Bow, J.S. (2004). Structural characterization of nano-sized calcium deficient apatite powders, *Biomaterials*, **25**, pp. 189–196.

14. Liu, D.M., Troczynski, T., and Tseng, W.J. (2001). Water-based sol–gel synthesis of hydroxyapatite — process development, *Biomaterials*, **22**, pp. 1721–1730.

15. Lussi, A., Jaeggi, T., and Zero, D. (2004). The role of diet in the aetiology of dental erosion, *Caries Res.*, **38**, pp. 34–44.

16. Lussi, A., Hellwig, E., Zero, D., and Jaeggi, T. (2006). Erosive tooth wear: diagnosis, risk factors and prevention, *Am. J. Dentistry*, **19**, pp. 319–325.

17. Lv, K., Zhang, J., Meng X., and Li, X.F. (2009). Remineralization effect of the Nano-HA toothpaste on artificial caries, *Key Eng. Mater.*, **330–332**, pp. 267–270.

18. Mendonça, G., Mendonça, D.B.S., Aragao, F.J.L., and Cooper, L.F. (2008). Advancing dental implant surface technology — from micron to nanotopography — review, *Biomaterials*, **29**, pp. 3822–3835.

19. Meyer, J.L., and Fowler, B.O. (1982). Lattice defects in nonstoichiometric calcium hydroxyapatites. A chemical Approach, *Inorg. Chem.*, **21**, pp. 3029–3035.

20. Mortier, A., Lemaitre, J., Rodrique, L., and Rouxhet, P.G. (1989). Synthesis and thermal behavior of well-crystallized calcium-deficient phosphate apatite, *J. Solid State Chem.*, **78**, pp. 215–219

21. Narasaraju, T.S.B., and Phebe, D.E. (1996). Review, some physico-chemical aspects of hydroxylapatite, *J. Mater. Sci.*, **31**, pp. 1–21.

22. Radin, S., Ducheyne, P., Falaize, S., and Hammond, A, (2000). *In vitro* transformation of bioactive glass granules into Ca-P shells, *J. Biomed. Mater. Res.*, **49**, pp. 264–272.

23. Rahiotis, C., Vougiouklakis G., and Eliades, G. (2008). Characterization of oral films formed in the presence of a CPP–ACP agent: an *in situ* study, *J. Dent.*, **36**, pp. 272–280.

24. Raynaud, S., Champion, E., Bernache-Assollant, D., and Thoman, P. (2002). Calcium phosphate apatites with variable Ca/P atomic ratio I. Synthesis, characterization and thermal stability of powders, *Biomaterials*, **23**, pp. 1065–1072.

25. Reynolds, E.C., Cai, F., Shen P., and Walker, G.D. (2003). Retention in plaque and remineralization of enamel lesions by various forms of calcium in a mouthrinse or sugar-free chewing gum, *J. Dent. Res.*, **82**, pp. 206–211.

26. Reynolds, E.C. (2008). Calcium phosphate-based remineralization systems: scientific evidence?, *Aust. Dent. J.*, **53**, pp. 268–273.

27. Roveri, N., Battistello, E., Bianchi, C.L., Foltran, I., Foresti, E., Iafisco, M., Lelli, M., Naldoni, A., Palazzo, B., and Rimondini, L. (2009). Surface enamel remineralisation: biomimetic apatite nanocrystals and fluoride ions different effects, *J. Nanomater.*, article ID 746383, 9 pages, doi:10.1155/2009/746383.

28. Roveri, N., Palazzo B., and Iafisco, M. (2008). The role of biomimetism in developing nanostructured inorganic matrices for drug delivery, *Expert Opin. Drug Deliv.*, **5**, pp. 861–877.

29. Rybachuk, A.V., Chekman, I.S., and Nebesna, T.Y. (2009). Nanotechnology and nanoparticles in dentistry, *Pharmacol. Pharma.*, **1**, pp. 18–21.

30. Scheyler, T.L. (2000). Nanodentistry fact or fiction. *J. Am. Dent. Assoc.*, **131**, pp. 1567–1568.

31. Sepulveda, P., Jones, J.R., and Hench, L.L. (2002). *In vitro* dissolution of melt-derived 45S5 and sol–gel derived 58S bioactive glasses, *J. Biomed. Mater. Res.*, **61**, pp. 301–311.

32. Venegas, S.C., Palacios, J.M., Apella, M.C., Morando, P.J., and Blesa, M.A. (2006). Calcium modulates interactions between bacteria and hydroxyapatite, *J. Dent. Res.*, **85**, pp. 1124–1128.

33. Waltimo, T., Brunner, T. J., Vollenweider, M., Stark, W. J., and Zehnder, M. (2007). Antimicrobial Effect of Nanometric Bioactive Glass 45S5, *J. Dental Res.*, **86**, pp. 754–757.

34. Zehnder, M., Waltimo, T., Sener, B., and Söderling, E. (2006). Dentin enhances the effectiveness of bioactive glass S53P4 against a strain of Enterococcus faecalis, *Oral Surg. Oral. Med. Oral. Pathol. Oral. Radiol. Endod.*, **101**, pp. 530–535.

Chapter 12

Osteoblast Behavior on Nanostructured Implant Materials

Karolina Jurczyk[a,*] and Mieczyslawa U. Jurczyk[b]
[a]*Conservative Dentistry and Periodontology Department, University of Medical Sciences, Bukowska 70 Street, 60-812 Poznan, Poland*
[b]*Division Mother's and Child's Health, University of Medical Sciences, Polna 33 Street, 60-535 Poznan, Poland*
*karolajur@gmail.com

12.1 Introduction

The interactions between solid surfaces and cells are crucial to many biological phenomena for all biomaterials. This chapter provides an overview of metallic and ceramic biomaterials, along with a discussion of microstructure and surface changes that promote biocompatibility. For a material to be deemed biocompatible, any adverse reactions which may ensue at the blood/material or tissue/material interface must be minimal, while resistance to biodegeneration must be high. Implantable materials should not [27]: cause thrombus-formations, destroy, or sensitize the cellular elements of blood, alter plasma proteins (including enzymes) so as to trigger undesirable reactions, cause adverse immune responses, cause cancer and eratological effects, produce toxic and allergic responses, deplete electrolytes,

Bionanomaterials for Dental Applications
Edited by Mieczyslaw Jurczyk

ISBN 978-981-4303-83-5 (Hardcover), 978-981-4303-84-2 (eBook)
www.panstanford.com

and finally be affected by sterilization. Till now, there are no known materials which totally satisfy these criteria so when a foreign material is placed into a biological environment, inevitable reactions occurs which are detrimental to both the host and the material.

The surface properties of biomaterials are associated with cell adhesion and subsequent various cell behaviors, such as proliferation, migration, cytoskeletal arrangement, differentiation, and apoptosis [12, 99]. In particular, a large number of studies on cell adhesion to various substrate surfaces have been conducted. Cell adhesion and its performance have been reported to depend on the characteristics of substrates, including the chemical composition, surface charge, water wettability, roughness, and size of the cytophilic area [5, 23, 44, 62, 63, 67, 68, 90, 104, 106, 107, 133, 135, 138, 141–143]. Understanding the mechanisms whereby cells sense and respond to chemical, physical and biological signals from material surfaces will facilitate the development of novel biomaterials for the control of cell behavior.

All implantable materials possess inherent morphological, chemical, and electrical surface qualities which elicit reactionary responses from the surrounding biological environment. In fact, biocompatibility can be described as multifactorial in that simultaneous stimuli from any of these material properties can affect the host response.

Using nanotechnology for regenerative medicine becomes obvious when examining nature [153]. Bone is a nanocomposite that consists of a protein based soft hydrogel template (i.e., collagen, non-collagenous proteins (laminin, fibronectin, vitronectin), and water) and hard inorganic components (hydroxyapatite, HA, $Ca_{10}(PO_4)_6(OH)_2$) [139, 154]. Specifically, 70% of the bone matrix is composed of nanocrystalline HA [57].

In addition to the dimensional similarity to bone/cartilage tissue, nanomaterials also exhibit unique surface properties (such as surface topography, surface chemistry, surface wettability, and surface energy) due to their significantly increased surface area and roughness compared to conventional or micron structured materials. As known, material surface properties mediate specific protein (such as fibronectin, vitronectin, and laminin) adsorption and bioactivity before cells adhere on implants, further regulating cell behavior and dictating tissue regeneration [139]. Furthermore, an important criterion for designing medical implant materials is the formation of sufficient osseointegration between synthetic materials and bone tissue.

12.2 Biocompatibility

The biocompatibility of biomaterials is very closely related to cell behavior in contact with them, and particularly to cell adhesion to their surface [12]. Surface characteristics of materials, such as their topography, chemistry, or surface energy, play an essential part in osteoblast adhesion to biomaterials. The term "adhesion" in the biomaterial domain covers different phenomena [8, 12]: the attachment phase which occurs rapidly and involves short-term events like physicochemical linkages between cells and materials involving ionic forces, van der Walls forces, etc., and the adhesion phase occurring in the longer term and involving various biological molecules: extracellular matrix proteins, cell membrane proteins, and cytoskeleton proteins which interact together to induce signal transduction, promoting the action of transcription factors, and consequently regulating gene expression.

Studies have demonstrated that nanostructured materials with cell favorable surface properties may promote greater amounts of specific protein interactions to stimulate more efficiently new bone growth compared to conventional materials [29, 136, 140]. This may be one of the underlying mechanisms why nanomaterials are superior to conventional materials for tissue growth. Therefore, by controlling surface properties, various nanophase ceramic, polymer, metal, and composite scaffolds have been designed for bone/cartilage tissue engineering applications.

The biocompatibility of an artificial material in the body is complicated. The artificial implants, once implanted *in vivo*, induce a cascade of reactions in the biological micro-environment through interaction of the biomaterial with body fluid, proteins, and various cells [43, 45, 119, 145]. The sequence of local events often leads to the classic foreign body response and the formation of a fibrous tissue capsule around an implant. It is clear that a major factor influencing this unfavorable reaction of the body is the biomaterial surface. The specific interactions determine the path and speed of the healing process and the long-term integration of the biomaterial-body interface. Both the chemical composition of the surface and the surface topography are believed to be important in bone contacting implants. They regulate the type and the degree of the interactions that take place at the interface like adsorption of ions and biomolecules such as proteins, formation of calcium phosphate layers, and interaction with different types of cells (macrophages, bone marrow cells and osteoblasts). Thus, the nature of the initial interface that is developed between an artificial material and the attached tissue determines the ultimate success or failure of the materials. Tissue compatibility is the most important issue to be considered for the implant success.

Titanium is found to be well tolerated and nearly an inert material in the human body environment. In an optimal situation, titanium is capable of osseointegration with bone [43]. In addition, titanium forms a very stable passive layer of TiO_2 on its surface and provides superior biocompatibility. Even if the passive layer is damaged, the layer is immediately rebuilt. In the case of titanium, the nature of the oxide film that protects the metal substrate from corrosion is of particular importance, and its physicochemical properties such as crystallinity, impurity segregation etc, have been found to be quite relevant.

Titanium alloys show superior biocompatibility when compared to the stainless steel and Cr–Co alloys. The grain size of metal implant influences the osteoblast adhesion. *In vitro* studies carried out using ultra fine-grained Cp Ti (grade 2) and Ti64 alloy exhibited increased cell adhesion when compared to conventional materials. This increase in cell adhesion is attributed to the increase in surface energy at the grain boundaries.

Current trends in clinical dental implant therapy include use of endosseous dental implant surfaces embellished with nanoscale topographies. It has been shown, that implant surface character is one implant design factor affecting the rate and extent of osseointegration [21, 26, 89, 92]. According to Mendonça and coworkers, nanostructured surfaces possess unique properties that alter cell adhesion by direct (cell–surface interactions) and indirect (affecting protein–surface interactions) mechanisms [89]. Surface nanotopography appears to affect cell interactions at surfaces and alter cell behavior when compared to conventional sized topography [14, 16, 70, 89]. Nanotopography specific effects on cellular behavior have been demonstrated using a wide range of different cell types including epithelial cells, fibroblasts, myocytes, and osteoblasts.

Mendonça and coworkers recently reviewed the role of nanoscale topographic modification of titanium substrates for the purpose of improving osseointegration [89]. Depiction of broad range of nanoscale topography effects observed in cellular protein adsorption is altered by nanoscale modification of bulk material (Fig. 12.1). It is believed that, the changes in initial protein–surface interaction control osteoblast adhesion [14]. When implants come into contact with a biological environment, protein adsorption (e.g. plasma fibronectin) that occurs immediately will mediate subsequent cell attachment and proliferation. Changing the surface energy or wettability of a biomaterial represents a classical approach to altering cell interactions with the surface.

Webster and coworkers [136, 138, 142, 153] observed increased osteoblast adhesion on Ti, Ti6Al4V, and CoCrMo compacts with nanometer compared to conventionally sized particles. Respective metal formulations had similar chemistry and altered only in degree of nanometer roughness. Interestingly, osteoblasts were observed to adhere specifically at particle boundaries. Since nanophase metals have higher percentages of particle boundaries at the surface, this may explain the greater numbers of osteoblasts on nanophase compared to conventional metals.

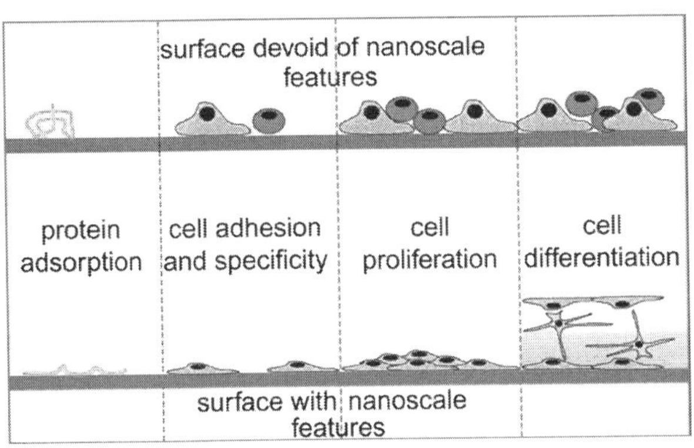

Figure 12.1 Depiction of broad range of nanoscale topography effects observed in cellular protein adsorption is altered by nanoscale modification of bulk material [89].

Previous studies have also demonstrated increased functions of osteoblasts (bone-forming cells) on nanophase compared to conventional ceramics (specifically, alumina, titania, and hydroxyapatite) [133–135], polymers (such as poly lactic-glycolic acid, and polyurethane) [58], carbon nanofibers/nanotubes and composites [58, 105].

Another study suggested higher adsorption of fibronectin on hydrophilic self-assembled monolayers (SAMs) surfaces with greater focal adhesion formation (integrin binding) evident in the osteoblast cells adhered to the hydrophilic SAM treated surfaces [113].

Both cell specificity and extent of cell adhesion are altered, too. Depending on the nano-architecture cell, spreading may be increased or decreased. Lim and coworkers [80] more directly related protein adsorption, cell adhesion and the active process of attachment by measurement of increased focal adhesion kinase (FAK) activity. Surface roughness at the nanoscale is an important determinant of protein interactions that ultimately direct cell activity in control of tissue formation at implant surfaces [101].

Nanofeatures of a surface affect both cell adhesion and cell motility. Andersson and coworkers [11] compared cell morphology and cytokine production on titanium substrates with 15 mm wide and 185 nm-deep grooves versus Ti substrates with 100 nm high, 168 nm diameter hemispherical nanopillars. The cells appeared partially aligned to the grooves and had a cytokine release similar to that found from cells on flat surfaces. Osteoprogenitor cell adhesion was enhanced on poly-L-lactide (PLLA) and polystyrene (PS) surface with nanoscale and micron-scale roughness compared to smooth surfaces. OCT-1 osteoblast-like cells grew along the surface with two different nanoscale surfaces (PLLA) and grew inside micron-scale pits of PS [127]. Similar conclusions were made when comparing nano- and micron-scale grain boundary effects on osteoblast cell adhesion and proliferation [142].

Cell proliferation appears to be enhanced by nanoscale topography, too. Webster and coworkers [135] observed increased osteoblast proliferation on the nanoscale materials tested. The mechanism(s) affecting this process is not defined. However, it can be speculated that many of the events associated with adhesion can affect signaling pathways that control proliferation.

Several investigators have demonstrated the relative diminution of fibroblast adhesion compared to osteoblast adhesion when nano- and micron-structured surfaces were evaluated [83, 104]. For example, on nano-sized materials, the affinity ratio between osteoblasts and fibroblasts was 3 to 1. In the conventional materials, the ratio was 1 to 1 [136]. Bacterial adhesion and proliferation is also diminished on nanophase materials [25]. Decreased bacterial colonization on nanostructured TiO_2 and ZnO is observed even though these surfaces promote osteoblast adhesion and differentiation.

For osteoblast, several investigators have shown nanoscale topography enhances osteoblast differentiation [133]. Webster revealed that alkaline phosphatase synthesis and calcium mineral content increased in cell layers formed on nano-sized materials after 21 and 28 days.

The chemical and topographical properties of the implant surface strongly influence the properties of the layer. Since proteins and cells range in size from nano- to micrometer, these are relevant length scales for the problem. Of equal importance are the properties of the cells, e.g. their ability to communicate through the extracellular matrix by signal molecules. During tissue healing, numerous bioactive signal molecules control the formation and some proteins have shown capability of stimulating healing near the implant.

The implant surface plays an important role in biological interactions [119, 145]. Various properties of the implant surface such as surface chemistry, energy, topography, and surface roughness are relevant factors for implant integration in bone tissue and, consequently, for osseointegration [1, 6, 7, 152]. The quality of titanium surfaces can be described in terms of surface chemistry, which refers to the critical surface tension (CST) or surface energy [66]. The CST is related to the contact angle of a liquid drop on the surface and, thus, provides an indicator of the potential of cell adhesion or surface wettability [33]. Complete moistening and distribution of a liquid on a surface indicates high surface energy, biocompatibility, and hydrophilicity of a material. It has been observed that chemically activated and hydrophilic sandblasted and acid-etched (SLA) surfaces resulted in a greater percentage of bone-implant contact in the first weeks of osseointegration [22].

The surface properties of the implants can be change by different methods of cleaning, sterilization, and storage [45, 61, 82]. For example, it has been observed that discs with an active SLA surface sterilized by gamma irradiation and continuously submersed in isotonic NaCl presented less contamination with hydrocarbons and carbonates from the atmosphere, producing a chemically clean and reactive surface [155].

Another aspect of surface chemistry determined by the oxide layer refers to the atomic and molecular structure of the biomaterial [59, 103]. Titanium forms various stable oxides such as TiO, TiO_2,

Ti_2O_3, and Ti_3O_4 on its surface [60]. The average thickness of the dense titanium oxide film is 10 nm [59]. This oxide thickness increases more rapidly when in contact with bone. Additionally, calcium and phosphorus ions increase and modify the oxide layer. It has been shown that titanium oxide acts as a nucleation substrate for calcium phosphate crystals [28, 109, 110]. The biological activity of TiO_2 also influences protein adsorption to the titanium surface [36]. The chemical composition and microstructure of a surface can regulate the adsorption of components present in extracellular fluid as a result of alterations in surface energy. *In vitro* studies have shown that rough and chemically activated surfaces provide the ideal conditions for direct protein adsorption and alter the adsorption of fibronectin and albumin due to modifications in their ionic state [119].

12.3 Nanostructured Biomaterials

Nanostructured biomaterials possess unique surface and mechanical properties similar to the bone and hence are considered to be the future generation biomaterials [39, 40, 132, 137, 144]. Owing to very high number of atoms on the surface, nanograined materials possess large surface energy. Thus, they exhibit entirely different behavior compared to the micron sized grains. The bone-forming cells generally attach themselves to the surface whose roughness is of nanometer range.

The nano roughness arises because of the fact that human bones consist of inorganic minerals of grain size varying from 20 to 80 nm long and 2 to 3 nm in diameter [34]. The variation in the surface energy due to the nanosurface roughness leads to desirable cellular responses on nanostructured titanium and other materials resulting in high osseointegration [65, 70, 71, 132, 137, 144]. The cell adhesion behavior on submicron, nanometer structured titanium surface was investigated and the obtained results were compared with a flat smooth titanium surface [65]. The study demonstrated that both nanometer and submicron surfaces have very high surface energy and adhesion of bone cells was very high. Additionally, nanograined alloys made of Cp Ti, Ti–6Al, 4V, and CoCr as well as nanoceramic biomaterials such as alumina, titania, and hydroxyapatite also exhibit increased cell adhesion [135, 142]. When the grain size was decreased from 167 to 24 nm, osteoblast adhesion got increased by 51% and fibroblast adhesion responsible for encapsulation was reduced by 235%.

The difference in the cell density between the conventional and nanomaterials is given in Table 12.1. It may be noted that, though different types of cells were utilized for cell culture studies on the alloys and ceramics, the cell density was observed to be relatively higher for the nanomaterials when compared to conventional counterparts.

Table 12.1 Cell density on nano-size (nanophase materials) and micron size (conventional materials) grains [43]

Material	Increase in surface area when compared to conventional materials	Roughness (nm)	Cell density[a] (cells/sq.cm.)
Ti (nano)	15%	11.9	2000[b]
Ti-6Al-4V (nano)	23%	15.2	1600[b]
Co-Cr-Mo (nano)	11%	35.6	1450[b]
Ti (conventional)			1400[b]
Ti-6Al-4V (conventional)			950[b]
Co-Cr-Mo (conventional)			600[b]
Alumina (24 nm) (nano)			6000[c]
Titania (39 nm) (nano)			8000[c]
Hydroxyapatite (67 nm), (nano)			9500[c]
Alumina (167 nm) (conventional)			5000[c]
Titania (4520 nm) (conventional)			7000[c]
Hydroxyapatite (179 nm) (conventional)			7000[c]

[a] Rounded values; [b] After 3 h; [c] After 5 days

Apart from the roughness, the pore size on the surface also has an influence on the protein adhesion. The protein, victronectin, is generally adsorbed on pores of smaller sizes (0.69, 0.95, and 0.66 nm of Al_2O_3, TiO_2, and HA), on the other hand, the protein that decreases cell adhesion such as laminin, generally adsorbs to pore size 2.54, 2.33, and 3.1 μm corresponding to Al_2O_3, TiO_2, and HA bioceramics [13]. Increased osteoblast adhesion was also observed on nano HA coated Ti-13Nb-11Zr alloy and further bone ingrowth toward implant was noted indicating ceramic surface coatings leading to high osseointegration [20].

12.3.1 Nanostructured Biointerfaces

In vivo, the interactions of a cell with its surroundings are mediated at the molecular and macromolecular level [3]. Specific interactions with, for example extracellular matrix components and soluble factors, or macromolecules in the outer membranes of adjacent cells provide necessary signaling and communication routes. Such interfaces have both topographic nanostructure and chemical/ biospecific interaction sites distributed at the nanoscale [2].

The use of diverse techniques to characterize material surfaces with great precision has led to a range of model experiments, studying in detail the influence of surface chemistry and surface topography *in vitro* and *in vivo* [3]. A constant desire has been to develop better surfaces of implant materials that are able to improve biocompatibility, hemocompatibility, or osseointegration. With the development and spread of nanofabrication approaches based on lithographic, chemical synthesis and self-assembly approaches have recently allowed material interfaces to be structured on the length scale of the macromolecular components of the extracellular matrix and cell membranes [126, 147].

A significant number of studies have focused on the effect of surface nanotopography on cell functions such as adhesion, motility, morphology, cytokine release, gene expression, and differentiation [3, 38]. The ability to define interfaces on a length scale which match that of the mediating macromolecules in cellular membranes and extracellular matrixes, has the potential to create artificial biointerfaces which are capable of communicating with/ signaling to adherent cells. Such artificial biointerfaces would be of immediate interest for application areas such as biomaterials, tissue engineering, substrates for generating cells for cell therapies, and cell-based electronics/sensors.

Additionally, the development of an implant/bone interface may be influenced by both nanoscale and micron-scale parameters of topography [89]. The role of surface parameters (both bulk chemistry and topography) requires consideration of molecular (ionic and biomolecular) interactions with the surface, cell adhesion phenomenon, and local biomechanical features of the established interface. It is clear that nanoscale modification will affect the chemical reactivity of an endosseous implant surface and alter the ionic and biomolecular interactions with the surface. Proposed changes include enhanced wettability, altered protein adsorption, and potential mineralization phenomenon [89]. Changes in wettability and altered protein adsorption lead to altered cell adhesion, likely involving both integrin and non-integrin receptors. The potential for mineralization and epitaxic crystal growth in support of early bone bonding could dramatically alter the biomechanical environment of the healing implant in favor of stability.

Various reports support the concept that nanotopography enhances osteoblastic differentiation which could also promote stability and favorably alter the biomechanical environment for healing [89]. However, initial clinical stability may require additional considerations of micron-scale topography and overall implant design. The pioneering investigations of Meirelles and coworkers suggest that nanometer-scale topography alone is not sufficient to assure robust osseointegration [87, 88]. Investigations which have isolated nanometer-scale topography as an experimental variable in osseointegration have required additional consideration of endosseous implant stability. It is possible that micron-level roughness is of additional value to the process of osseointegration [89].

12.3.2 Osteoblastic Cell Behavior on Nanostructured Surface of Metal Implants

The effect of nanotopography on osteoblastic cell behavior has been reported in the literature [19, 32, 77, 81, 85, 98, 100, 102, 114, 131]. Surface modifications at the nanometric scale may promote protein adsorption, cell adhesion and thus favor the osseointegration of metal implants. Chemical composition, surface energy, roughness, and topography are key factors for interaction with biological fluids, cells and tissues, and the modification of one of these parameters usually modifies the others.

In the field of dental implants, surface roughness has often been modified in order to control bone tissue apposition [9, 37, 76]. Surface roughness may be divided into three levels according to the scale of the features: macro-, micro- and nanometer-sized topologies. The macrostructure of implants has been adapted for primary anchorage of implants to bone in relation to the biomechanics of the skeleton. Most implant surfaces usually present a moderately rough surface in the micrometer range with R_a (Roughness average) values around

1–2 μm [9]. This microstructure favors mechanical anchorage to bone tissue. The third level of roughness, at the nanometer scale, plays an important role in the adsorption of proteins, adhesion of osteoblastic cells and thus the osseointegration rate. All the interactions between proteins and focal adhesion points of cells occur at the nanometer scale, activating intracellular molecular signaling pathways, and controlling cell fate [35].

Recent literature reports several methods for preparing surfaces structured at the nanometer scale and studying the behavior of cells on these surfaces. On metal surfaces, enhanced cell metabolic activity has been observed, such as the upregulation of bone sialoprotein and osteopontin [31], as well as a threefold increase in osteoblastic cell adhesion as compared with the surfaces without nanostructure. Furthermore, enhancement of calcium and phosphorus deposition has been observed on nanostructured titanium alloys and on CoCrMo surfaces but it was not observed on pure titanium [84, 131].

Recently, the behavior of osteoblastic cells was studied on mirror-polished (Smooth-SS) and nanostructured (Nano-SS) stainless steel surfaces [77]. Nanometer-scale features on stainless steel were prepared by anodization. The aim of this study was to compare osteoblastic cell behavior on smooth, plastic cultures, and these nanostructured stainless steel surfaces were made from an ordered array of pores. The Nano-SS exhibited a smooth surface with a regular array of nanometer-sized cavities visible at high magnification under SEM. The pores were organized in a hexagonal array like a honey-comb structure. The diameter and depth of the pores averaged 182 and 25 nm, respectively. Table 12.2 reports the surface roughness parameters of Smooth- and Nano-SS. Both the

Table 12.2 Surface roughness, contact angle and surface energy of mirror-polished (Smooth-SS) and nanostructured (Nano-SS) stainless steel discs compared with plastic culture [77]

Parameter	Plastic culture	Smooth-SS	Nano-SS
Roughness average Sa/Ra (nm)	0.65 ± 0.25	1.4 ± 0.3	6.2 ± 0.5
Roughness mean square Sq/Rq or Rms (nm)	0.86 ± 0.30	2.0 ± 0.3	7.6 ± 0.5
Peak-to-valley height Rt (nm)	3.2 ± 0.9	31 ± 3	41 ± 4
Skewness Rsk/Ssk (nm)	—	1.88	0.06
Dynamic contact angle (°)	61 ± 1.0	64.0 ± 0.5	89.5 ± 0.5 7
Surface energy (mJ/m^2)	48.7	43.5	34.1

mirror-polished and nanostructured surfaces had higher surface roughness than that of the plastic culture. The cell plastic culture and the smooth stainless steel discs presented an intermediate surface energy. The most hydrophobic material was the nanostructured stainless steel discs.

The surface composition of mirror-polished and nanostructured stainless steel surfaces was determined by X-ray photoelectron spectroscopy (Table 12.3). The stainless steel was covered by an oxide layer composed of iron and chromium oxides (Fe_2O_3, Cr_2O_3). The chromium content was higher on Nano-SS as compared with Smooth-SS while it was the opposite for iron. It may be due to a more rapid dissolution of iron during anodization. A higher concentration of carbon and a lower concentration of oxygen were observed on the nanostructured stainless steel discs as compared with the smooth stainless steel. This high carbon content on the Nano-SS in the iron oxide layer may be explained by the anodization process that was conducted in an organic solvent.

Table 12.3 Surface analyses determined by X-ray photoelectron spectroscopy of mirror-polished (Smooth-SS) and nano-structured stainless steel (Nano-SS) [77]

Element (atomic %)	Smooth-SS	Nano-SS
C	37.3 ± 0.7	53.6 ± 1.1
O	52.6 ± 1.0	40.6 ± 0.8
Cr	3.3 ± 0.7	3.6 ± 0.7
Fe	6.7 ± 0.1	2.2 ± 0.1

By comparing cell morphology on the two surfaces at high magnification, the osteoblastic cells exhibited a round shape on Smooth-SS while they appeared to have a "star-like" shape with more cytoplasmic extensions and filipodia on the Nano-SS after 4 h. The number of branched cells was relatively higher than the number of round cells on the Nano-SS as compared with Smooth-SS. Therefore, osteoblastic cells seemed to spread more rapidly on the nanostructured than on the smooth surface.

Cell proliferation of mouse pre-osteoblasts MC3T3-E1 was measured through their mitochondrial (MTS) activity at different time points on to the different substrates and controls. Compared with plastic culture, MTS cell viability was similar on Smooth-SS and Nano-SS at 8 days. At day 15, osteoblastic cells proliferated more rapidly on both smooth and nanostructured stainless steel substrates than on culture plastic.

The differentiation of osteoblastic cells was assessed using ALP (Alkaline phosphatase) activity normalized to protein content after culturing MC3T3-E1 on the different substrates for 8, 15, and 21 days. The differentiation increased with culture time on both Smooth-SS and Nano-SS substrates. ALP activity was better enhanced at 21 days on the nanostructured surface than on the Smooth-SS.

A positive influence for nanotopography on stainless steel with regard to the spreading of osteoblastic cells was observed [77]. Analysis of osteoblast morphology revealed that there were extensive interactions between osteoblasts and nanoscale features as they extended filipodia to a greater extent on Nano-SS than on Smooth-SS. Webster and Ejiofor [142] have reported that osteoblasts extended their filipodia and interacted with nanostructured surfaces faster than those on control surfaces.

The surface topography of bio-implant materials dramatically influences their cell response. Recently, a set of unique structures ranging from mesoporous nanoscaffolds, nanoflowers, nanoneedles, nanorods, and octahedral bipyramids were fabricated by systematically tuning the hydrothermal conditions such as reaction medium composition, concentration, temperature, and time duration [32]. The cytotoxicity of surface modified Ti was assessed using human primary osteoblastic cells, and more than 90% of the cells were found to be viable after 24 h of incubation. Protein adsorption studies revealed that the surface modified nanostructures on titanium adsorbed more proteins, suggesting that they are capable of promoting cell adhesion/attachment.

Recently, a monoclinic zirconia coating with a nanostructural surface was prepared on the Ti–6Al–4V substrate by an atmospheric plasma-spraying technique, and its microstructure and composition, as well as mechanical and biological properties, were investigated to explore potential application as a bioactive coating on bone implants [130]. X-ray diffraction, transmission electron microscopy, scanning electron microscopy, and Raman spectroscopy revealed that the zirconia coating was composed of monoclinic zirconia which was stable at low temperature, and its surface consists of nano-size grains 30–50 nm in size. The bond strength between the coating and the Ti–6Al–4V substrate was 48.4 ± 6.1 MPa, which is higher than that of plasma-sprayed HA coatings.

The hardness (H) and elastic modulus (E) of the plasma-sprayed monoclinic zirconia measured by nano-indentation are listed in Table 12.4. For comparison, the 3 mol.% Y_2O_3–ZrO_2 coating composed of the tetragonal phase [129] and 12.8 mol.% CaO–ZrO_2 coating composed of mainly the cubic phase [128] are summarized. The term H/E represents the degree of elastic response in elastic–plastic materials [72, 78]. It was taken as a useful indicator of a given incapacity of materials to absorb impact energy, that is to say, materials with lower H/E value possess stronger capacity to dissipate impact energy [73]. Table 12.4 shows that the elastic modulus and nano-hardness of the plasma-sprayed monoclinic zirconia coating are 139.72 ± 24.96, and 7.60 ± 1.77 GPa, respectively. In comparison with the 3 mol.% Y_2O_3–ZrO_2 and 12.8 mol.% CaO–ZrO_2 coatings, the plasma-sprayed monoclinic zirconia coating has the lowest H/E value, indicating that it was more resistant to exterior energy than the other two coatings. After Vickers indentation, no obvious cracks were formed around the impression on the plasma-sprayed monoclinic zirconia coating surface while spallation was observed on the 12.8 mol.% CaO–ZrO_2 coating, further indicating that the plasma sprayed monoclinic coating may have better toughness.

Table 12.4 Elastic modulus (E) and hardness (H) of the zirconia coating in comparison with 3 mol.% Y_2O_3–ZrO_2 [129] and 12.8 mol.% CaO–ZrO [128]

	Monoclinic ZrO_2 coating	**3 mol.% Y_2O_3–ZrO_2 coating**	**12.8 mol.% CaO–ZrO_2 coating**
Phase composition	M	T	C + 8.7% M
Elastic modulus E (GPa)	139.72 ± 24.96	136.41 ± 25.95	178.53 ± 27.31
Hardness H (GPa)	7.60 ± 1.77	9.20 ± 2.98	13.05 ± 2.92
H/E	0.055	0.067	0.073

Note: M, monoclinic phase; T, tetragonal phase; C, cubic phase.

The SEM micrographs of the osteoblast-like MG63 cells seeded on the plasma-sprayed monoclinic zirconia coating are displayed in Fig. 12.2. After seeding for 1 h, the cells on the zirconia coating exhibit a round morphology (Fig. 12.2a), which is consistent with the LSCM observation. A number of secretory vesicles are observed on the surface and many filopodia extend from the MG63 cells as shown in Fig. 12.2a. After seeding for 2 h, the cells become larger and flatter. The filopodia become more elongated and extend to all directions

(Fig. 12.2b). After seeding for 5 h, the cells enlarge and flatten obviously (Fig. 12.2c) and are in close contact with the coating via the extended filopodia. With further increase in the culturing time to 24 h, all the cells completely flatten (Fig. 12.2d). After culturing for 48 h, the cells appear to be more elongated and thicker (Fig. 12.2e). The results indicate that the MG63 cells can attach and adhere well to the plasma-sprayed monoclinic zirconia coating. Cell attachment and adhesion are the first phase of cell/material interaction and the efficacy and quality of this first phase will influence the capability of the cells to proliferate and differentiate upon contact with the implant.

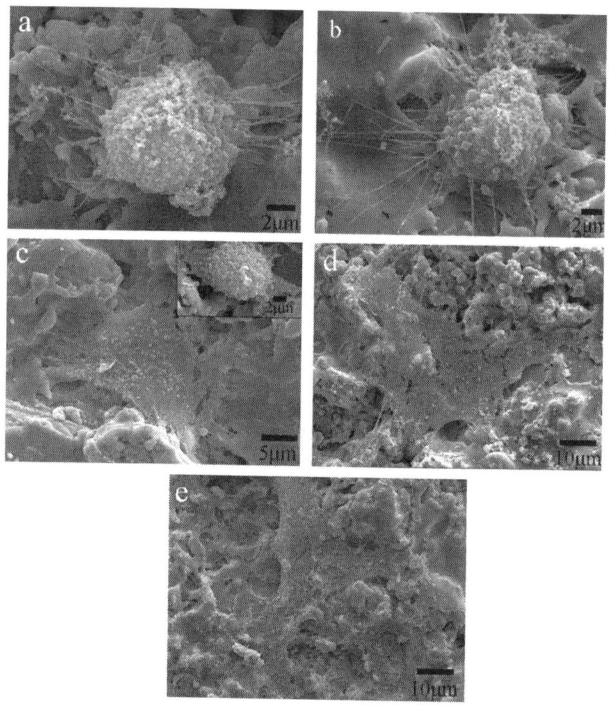

Figure 12.2 SEM pictures of MG63 cells cultured on the plasma-sprayed monoclinic zirconia coating for (a) 1 h, (b) 2 h, (c) 5 h, (d) 24 h, and (e) 48 h (the density of the cell suspension is 1×10^4 cell ml^{-1}) [130].

Figure 12.3 displays the morphology of the MG63 cells seeded on the coating surface for 3–11 days showing higher cell densities. After seeding for 3 days, the number and size of cells increase, indicating that cells grow and proliferate well on the surface (Fig. 12.3a). Cell confluence appears and a sheet-like layer is formed on the coating after the cells have been seeded for 7 days, as shown in Fig. 12.3b. The cells seeded on the coating for 11 days have a more elongated morphology and cover the whole surface (Fig. 12.3c).

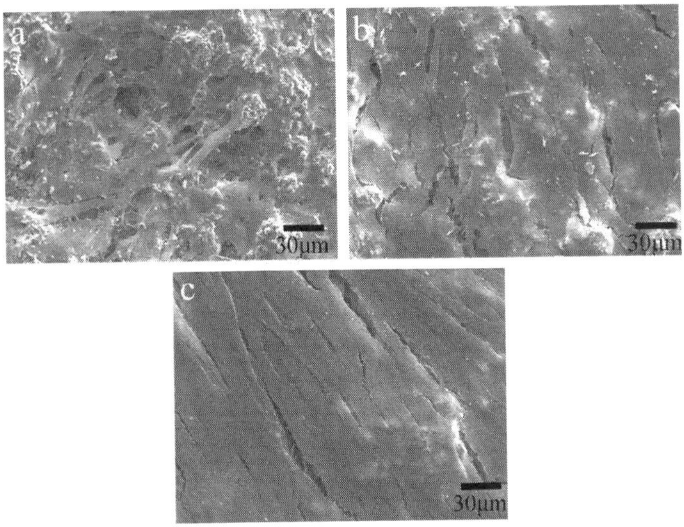

Figure 12.3 SEM photographs of MG63 cells cultured on the plasma-sprayed monoclinic zirconia coating for (a) 3 days, (b) 7 days, and (c) 11 days (the density of the cell suspension is 1×10^5 cell ml^{-1}) [130].

The good cytocompatibility of the plasma-sprayed monoclinic zirconia coating was possibly related to its surface nanostructure. It was reported that nano-sized topography improved cell attachment, adhesion, and proliferation on biomaterials [12, 127]. Webster has proposed that the critical grain size for osteoblast adhesion is 49–67 nm for Al_2O_3, and 32–56 nm for TiO_2 [133]. The surface of the plasma-sprayed monoclinic zirconia coating is constructed by grains with the size range of 30–50 nm, which appears suitable for cell attachment and viability.

The nanostructured surface that is composed of monoclinic zirconia is believed to be crucial to its bioactivity. Morphological observation and the cell proliferation test demonstrated that osteoblast-like MG63 cells could attach to, adhere to and proliferate well on the surface of the monoclinic zirconia coating, suggesting possible applications in hard tissue replacements [130].

These studies show that the nanostructuration of metal implants may improve the adhesion and differentiation of osteoblastic cells. However, further studies are necessary to corroborate these preliminary results *in vivo*.

Independently studies have shown that by varying the surface roughness of a biomaterial, bacteria adhesion decreases [25]. Simple means for the reduction of bacteria on and subsequent infection of titanium using nanometer sized Ti surface features was explored for medical applications by Webster and coworkers [108].

This study found that compared to conventional (nano-smooth) Ti, the nanorough Ti surfaces produced by electron beam evaporation decreased the adherence of all of the aforementioned bacteria the most. The conventional and nanorough Ti surfaces were found to have crystalline TiO_2 while the nanotubular and nanotextured Ti surfaces were found to be amorphous. The surface chemistries were similar for the conventional and nanorough Ti while the anodized Ti surfaces contained fluorine. Therefore, the results of this study *in vitro* demonstrated that certain nanometer sized Ti topographies may be useful for reducing bacteria adhesion while promoting bone tissue formation and, thus, should be further studied for improving the efficacy of Ti-based implants.

The unmodified titanium (Ti) possessed micron rough surface features as displayed under SEM (Fig. 12.4(a)). After electron beam evaporation, the Ti substrates possessed a high degree of nanometer surface features, thus, creating a more nanometer rough surface topography (Fig. 12.4(b)). Completion of anodization for 1 min in 0.5% hydrofluoric acid (HF) at 20 V resulted in a Ti substrate containing nanotextured surface features (Fig. 12.4(c)). Increasing the anodization time (10 min) and concentration of HF (1.5%) resulted in a Ti surface that contained nanotubular like structures with an inner diameter from 60 to 70 nm, as estimated from the SEM images (Fig. 12.4(d)).

Results from this study indicated that the prepared nanorough Ti surfaces are the best surfaces for inhibiting bacterial adhesion. Compared to conventional surfaces, nanostructured materials have excellent biocompatibility properties due to enhanced protein interaction (including adsorption and conformation) resulting in improved cellular adhesion and tissue growth [64, 136, 146]. It has been demonstrated that there is a linear relationship between nano-roughness, surface energy, and protein adsorption. More specifically, a surface that has more nanorough features possesses increased surface energy which leads to greater protein adsorption [64, 136, 146]. This study also confirmed the same correlation as it revealed that nanorough, nanotubular, and nanotextured Ti possessed higher degrees of nanometer features, higher surface energy, and increased fibronectin adsorption compared to conventional Ti. Furthermore, research has also shown that increased protein adsorption, such as fibronectin, results in decreased bacteria attachment [10, 69, 70].

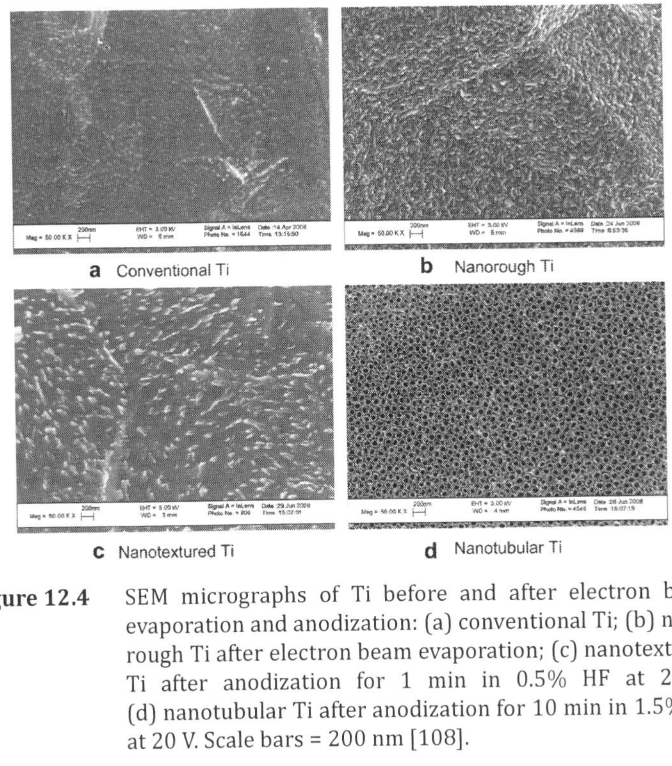

Figure 12.4 SEM micrographs of Ti before and after electron beam evaporation and anodization: (a) conventional Ti; (b) nanorough Ti after electron beam evaporation; (c) nanotextured Ti after anodization for 1 min in 0.5% HF at 20 V; (d) nanotubular Ti after anodization for 10 min in 1.5% HF at 20 V. Scale bars = 200 nm [108].

Compared to conventional Ti, nanorough Ti possessed no chemical difference, and, thus, the presence of nanometer features alone (higher surface energy) increased fibronectin adsorption, which decreased bacterial attachment.

Although total bacteria adhered the most to the anodized nanotubular surfaces, this study also revealed that the anodized surfaces (nanotubular and nanotextured Ti) decreased the percentage of living cells compared to the non-fluorinated surfaces (nanorough and conventional Ti). This could be a result of the antibacterial effects caused by the presence of fluorine, as shown by other studies [48, 111, 151]. Additionally, there was also a difference in the crystallinity between the anodized Ti surfaces, nanorough, and conventional Ti that can be linked to bacteria adherence. Nanotextured and nanotubular Ti contained amorphous TiO_2 while the nanorough and conventional surfaces contained crystalline TiO_2 (anatase and rutile phase). Research has shown that amorphous TiO_2 promoted bacteria attachment compared to anatase TiO_2 (which is known to possess antibacterial properties) [30,150].

Formation of porous TiO_x layers on Ti during electrochemical etching in H_3PO_4, CH_3COOH electrolytes modified by HF and NH_4F was described [50, 51]. The anodization resulted in porous TiO_x formation, useful in tissue growth and bone bonding. The pore dimensions increased due to the increase of HF or NH_4F content in H_3PO_4 electrolyte. During anodization at 10 V for 30 min, when the HF content increased from 0.5 to 10%, the pore diameter increased from 30 nm up to 8 μm, respectively. Anodization in CH_3COOH electrolyte resulted in non-uniform etching with flat hexagonal islands with nanopores inside surrounded by micropores.

The biocompatibility of the porous surface was investigated after one and five days on osteoblasts culture (Fig. 12.5). After the first day, we have observed relatively large amount of short filopodia (a,b). After 5 days, the cells showed adhesion and proliferation (c,d) and the osteoblasts covered most of the surface. On the porous sample, after one day, the cells strongly fixed to the insert, growing inside the pores. The ability of adhesion and growth on porous material is a specific feature of the osteoblasts.

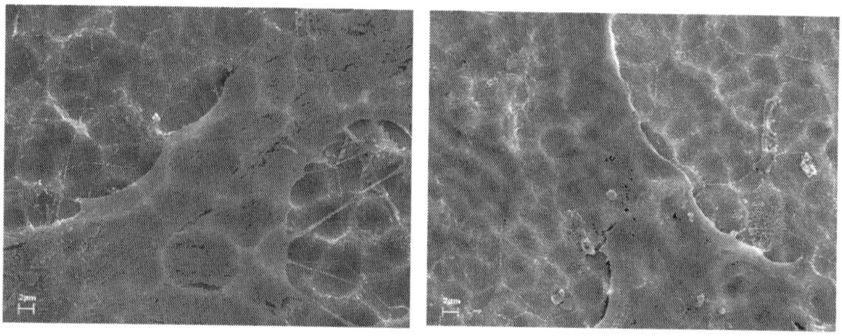

Figure 12.5 Human osteoblasts after 1st (a) and 5th (b) day of the culture on sample etched at 10 V/30 min in 1M H_3PO_4 + 10% HF electrolyte [51].

12.3.3 Adhesion of Osteoblast-Like Cells on Nanostructured Hydroxyapatite

In the field of biomaterials for substitution and reorganization of hard tissue, calcium phosphate ceramics are quite important because natural hard tissues are primarily composed of hydroxyapatite (HA) [15, 18, 32, 99, 135]. In consequence, the nanoscale topography of calcium phosphate ceramics determines the cellular performance of mesenchymal stem cells and osteoblast cells. Osteoblast proliferation was reported to be enhanced on nanophase HA in comparison with borosilicate glass, nanophase alumina, and nanophase titania [135].

Recently, the response of osteoblast-like cells seeded on hydroxyapatite substrates consisting of nano-sized crystals was investigated [99]. Various types of HA nanocrystals, such as nanofibers, nanoneedles, and nanosheets were selectively prepared as substrate through the hydrolysis of a solid precursor crystal of $CaHPO_4$ in alkaline solutions by varying the pH and ion con-centrations. Although all the substrates were macroscopically flat and smooth, the nanoscale topography influenced cell activity including the adhesion, proliferation, elongation, and formation of actin stress fibers. The presence of fine nanoneedles and nanofibers on the surface restricted the cellular activities, while the cells steadily proliferated on a nanoscopically smooth surface of large grains and on a substrate consisting of wide nanosheets. These results suggest that the adhesion and subsequent responses of osteoblast-like cells were affected by the contact domain size between the cell and the substrate. Isolated small domains of the nanostructured HA limited focal adhesion formation in the cells associated with the formation of stress fibers. Stable adhesion with contact domains larger than 100 nm in width was suggested to be required for cell survival. On the other hand, insufficient adhesion on the fine nanoneedles was found to lead to apoptosis.

Recently, research has been directed toward the development of electrostatic spray deposition technique, allowing the fabrication of dense, porous or nanostructured CaP coatings [4, 46, 47, 75, 79]. Figure 12.6 shows a typical osteoblast response to nanoscale hydroxyapatite particles deposited on to a glass surface using electrohydrodynamic atomization. Reports in the literature suggest that the special properties of these coatings are likely to influence protein interactions and subsequently, control *in vitro* cell proliferation and differentiation [116, 117]. Research has shown that the incorporation of transforming growth factor-beta 1 results in enhanced cellular function [118]. Thus, this route appears to offer the potential for chemical and topographical control of cell behavior and also drug delivery.

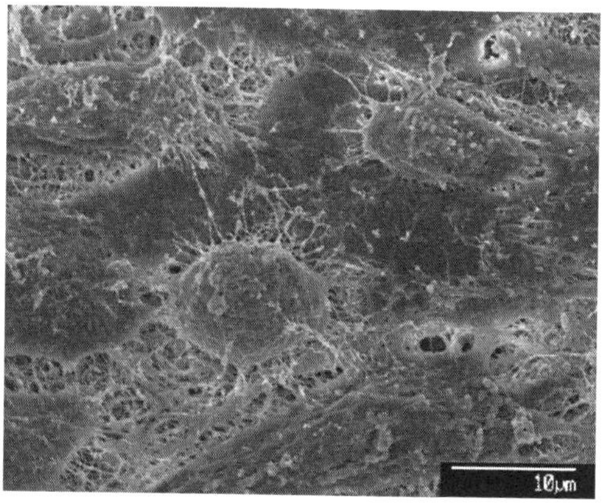

Figure 12.6 Osteoblast cells on a glass substrate coating with nanoscale HA deposited via electrohydrodynamic atomization [18].

12.3.4 Size Effect of Hydroxyapatite Nanoparticles on Proliferation and Apoptosis of Osteoblast Cells

Nano-hydroxyapatite (nano-HA), as a better candidate for an apatite substitute of bone in biomedical applications than microsized hydroxyapatite (m-HA), is suggested by many researches [115]. Nano-HA particles with diameters of ~20 nm (np20) and ~80 nm (np80) were synthesized and characterized. The size effects of these nano-HA's and m-HA were studied on human osteoblast-like MG-63 cells *in vitro*. These results demonstrate that both cell proliferation and cell apoptosis are related to the size of the HA particles. Np20 has the best effect on promotion of cell growth and inhibition of cell apoptosis. This work provides an interesting view of the role of nano-HA's as ideal biomedical materials in future clinical applications.

As shown in Fig. 12.7A, the MG-63 cells on np20, np80, and m-HA films proliferated with increasing culturing time (up to 5 days). Furthermore, cell number was significantly higher on np20 compared to m-HA after 3 and 5 days of incubation, though there was no significant difference between both surfaces after 1 day of incubation. The number of MG-63 cells on glass, np20, np80 and m-HA was 357 ± 34, 299 ± 24, 215 ± 21, and 135 ± 13 mm^{-2} after 5 day. These are consistent with the results of AO staining, which showed that the cell coverage of np20 film was higher than that of np80 and m-HA films after 5 days of culture (Fig. 12.7B).

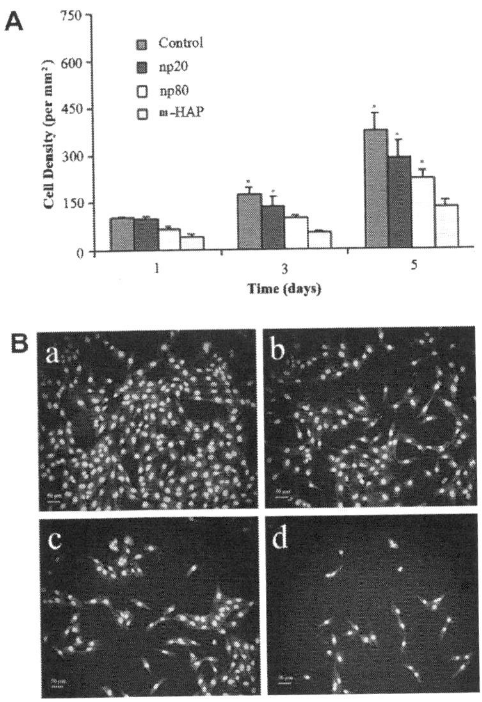

Figure 12.7 (A) Cell density of MG-63 on different substrates for 1, 3, and 5 days. $n = 6$; *significantly different from m-HA, $p < 0.05$. (B) Fluorescence microscopic photographs of MG-63 cells (cultured for 5 day) on (a) glass, (b) np20, (c) np80, and (d) m-HA [115]. See also Color Insert.

This study suggested that nano-HA particles could stimulate osteoblastic proliferation as compared with m-HA. In addition, this study further revealed that the cell proliferation was inversely related to the HA particle sizes. Np20 was the most effective of the three HA samples. This may be ascribed to the enhanced interfacial adhesion of HA nanoparticles to cells, as well as the improved penetration abilities of smaller HA's. Nano-HA's can provide a higher surface area per HA volume for cell growth than m-HA, which might result in increased cellular adherence and proliferation [49, 74, 133].

The origin of the effects of np20 may derive from the fact that the internalized HA nanoparticles were partially dissolved during lysosomal digesting and the obtained solutes such as Ca^{2+} ions diffused into the cytoplasm [17]. The proper Ca^{2+} concentrations favored osteoblast proliferation and differentiation [84].

Besides particle size, the crystallinity and the morphology of HA particles also affect cell behavior. The studies have provided evidence that the crystallinity of calcium phosphate affects the behavior of cultured osteogenic cells [42, 91]. The osteogenic differentiation is higher in cultures on amorphous calcium phosphate substrates compared to crystalline HA substrates. It is demonstrated that surface chemistry and topography of lower crystallinity is favorable for cell attachment and differentiation [14, 24, 41].

The morphology should be another important parameter for estimating the biological effect of calcium phosphate particles [115, 156]. The cell experiment showed that HA with spherical nanocrystals showed more favorable properties than rod-like HA for osteoblasts. The explanation for the favorable effect of sphere-like nano-HA on osteoblasts might be the well-organized surface which seems beneficial for filopodia protrusion.

Cytotoxicity was another important parameter to estimate the biological effect of HA nanoparticles. The cells in np80 and m-HA groups exhibited marked changes, including loss of contact with neighboring cells, contraction of the cells, swollen mitochondria, deformed nuclei, and condensed chromatin [115]. This means that larger sizes of HA nanoparticles are toxic to cells. Saunders and coworkers [112] pointed out that increased calcium phosphate in the matrix might trigger apoptosis in osteoblasts during bone resorption, and that bone cell apoptosis induced by high calcium and phosphate exhibited a marked change in mitochondrial function.

Substrate micro- and nano-topography, independently of substrate biochemistry, seem to have significant effects on cell behavior [86]. As reported till now, substrate topography (with micro- or nano-features) has direct effects on cell orientation, morphology, and cytoskeleton arrangements. As a general rule, the effects are more pronounced for smaller, nano-sized surface features, although there is a lack of knowledge on the basic mechanisms the cell uses to detect and respond to this nanotopography.

12.3.5 Biocompatibility of Ti-Bioceramic Nanocomposites

The application of Ti-bioceramic nanocomposites also focused our attention on the biocompatibility of synthesized bulk materials. In our work, hybrid Ti-x vol% 45S5 Bioglass, Ti-x vol% SiO_2, and Ti-x vol% HA bionanocomposites ($0 \leq x \leq 20$) were produced by the combination of mechanical alloying (MA) and powder metallurgical process [52–56, 93–97].

For example, the SEM images of the Normal Human Osteoblast (NHOst) cells cultured on Ti-10 vol% 45S5 Bioglass disks sterilized by autoclaving after 1 and 5 days are shown in Fig. 12.8. After the first day of incubation, cells show good adhesion to the surface of studied samples in the form of filopodia. After 5 days of incubation, the typical monolayer was observed. The same results were obtained in the case of other studied nanocomposites as well as microcrystalline titanium. Earlier, it has been demonstrated that metal (Ti, Ti6Al4V, and CoCrMo) surfaces utilizing submicron to nanometer particles, due to higher amounts of particle boundaries at their surfaces, promoted the adhesion of osteoblasts as compared to metals composed of respective micron particles [142].

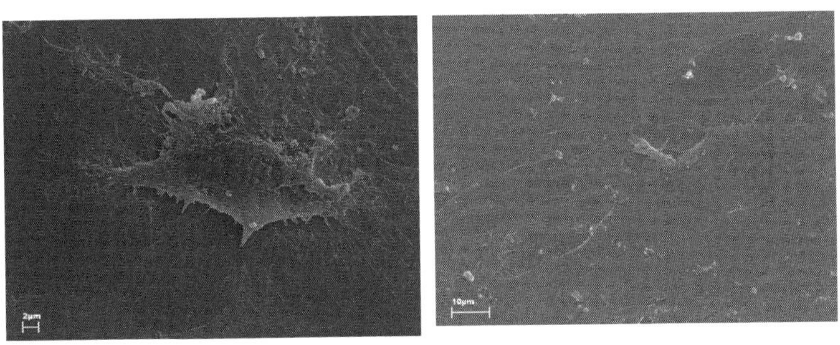

Figure 12.8 SEM images of the cells cultured on nanocomposite Ti-10 vol% 45S5 Bioglass after 1 (left, magnification 7000×) and 5 (right, magnification 3500×) days.

Cytotoxicity tests of the extracts of studied Ti-45S5 Bioglass materials under wear conditions are shown in Fig. 12.9. The relative viability of the cells (RVC) decreases when fraction increases. It is important to note that the RVC of nanoscale Ti-45S5 Bioglass is higher in comparison with microcrystalline titanium. The wear and fretting accelerates the corrosion of the studied samples in a biological environment such as cell culture medium. Two factors may influence cell growth on the disks: adsorbing protein onto the disks and released metal ions from the disks (see Table 12.5).

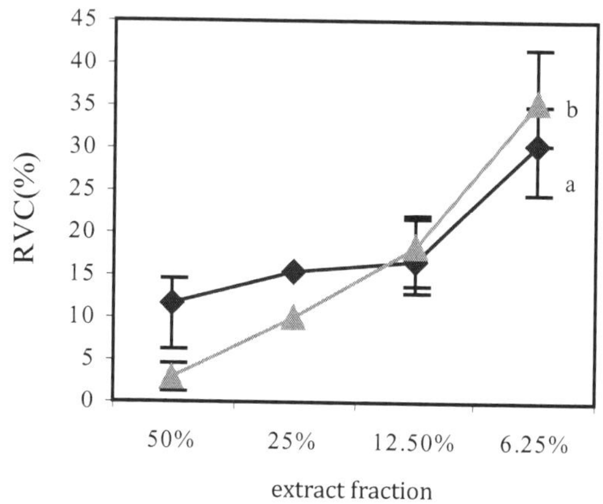

Figure 12.9 Cytotoxicity tests of the extracts of studied samples: (a) microcrystalline titanium, (b) Ti-10 vol% 45S5 Bioglass in dynamic conditions (see text for details).

Table 12.5 Quantifications of metallic elements in the extracts in microcrystalline titanium and nanocomposite Ti-10 vol% 45S5 Bioglass (<DL – concentration below detection level)

	Sample	
Element	**Microcrystalline Ti [mg/L]**	**Ti-10 vol% 45S5 Bioglass [mg/L]**
Ca	64 ± 8	82 ± 11
Cr	4.4 ± 0.7	4.1 ± 0.6
P	0	0
Ti	<DL	<DL

The quantification of metallic elements in each extract of studied nanocomposite Ti-45S5 Bioglass samples was performed for the elements such as Ca, Cr, P, and Ti under the optimum condition for each element and the results are shown in Table 12.5. In all extracts, chromium was detected at the concentration of 4.4 ± 0.7 and

4.1 ± 0.6 mg/L, respectively. Chromium is one of the essential elements for human, so slight amount of this element may contribute to cell proliferation, and resulting in higher cell growth. Additionally, in all studied extracts calcium was present at the concentration of 64 ± 8 and 82 ± 11 mg/L, respectively. The existence of Ca could promote the formation of apatite. Titanium element cannot be detected.

This study provided evidence of significant differences in the amount of calcium and phosphorus deposition by osteoblasts as well as their precipitation from culture media between common orthopedic (Ti6Al4V, CoCrMo) alloys due to nanometer particle sizes [142].

12.3.6 Biocompatibility of Ni-Free Stainless Steel

Ni-free austenitic stainless steels with nanostructure were synthesized by mechanical alloying (MA), heat treatment, and nitrogenation of elemental Fe, Cr, Mn, and Mo microcrystalline powders [120–125]. The phase transformation from ferritic to austenitic was confirmed by XRD analysis. The mechanical and corrosion properties of the produced biomaterials were investigated. Additionally, the biocompatibility of nickel-free austenitic stainless steels with nanostructure and microcrystalline 316L steel, were analyzed studying the behavior of Normal Human Osteoblast (NHOst) cells from Cambrex (CC-2538). An enhancement of the properties due to the nanoscale structures in the bulk consolidated materials was observed.

Biocompatibility tests were performed for synthesized bulk nanomaterials and 316L stainless steel. The AFM images of the cells cultured on $Fe_{54}Cr_{24}Mn_{21}Mo_1N$ disks sterilized by autoclaving for 1, 7, and 14 days are shown in Fig. 12.10. After incubation for 1 day, the cells were observed as scattered dots but after 7 days incubation the cells had grown to cover almost entire surface of the disks. There was a large difference in number of cells in the studied samples. Two factors may influence cell growth on the disks: adsorbing protein onto the disks and released metal ions from the disks.

Cytotoxicity tests of the extracts of the studied materials under wear conditions are shown in Fig. 12.11.

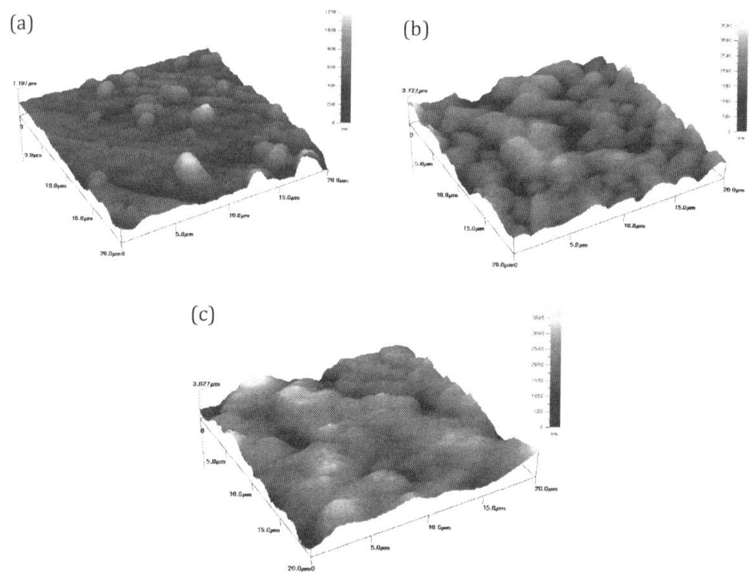

Figure 12.10 AFM images of the cells cultured on $Fe_{54}Cr_{24}Mn_{21}Mo_1N$ disks sterilized by autoclaving for 1 (a), 7 (b), and 14 days (c).

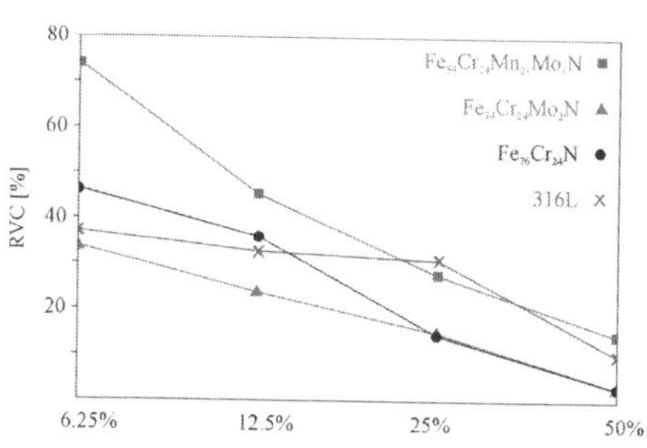

Figure 12.11 Cytotoxicity tests of the extracts of studied samples.

The relative viability of the cells (RVC) decreased when the concentration of cells in the medium increased. It is important to note that the RVC of nanoscale nickel-free austenitic stainless $Fe_{54}Cr_{24}Mn_{21}Mo_1N$ steels was higher than that of $Fe_{74}Cr_{24}Mo_2N$ and microcrystalline 316L steel. Wear and fretting accelerated the corrosion of the studied samples in a biological environment such as a cell culture medium.

Table 12.6 Quantifications of metallic elements in the extracts

	Sample		
Element	**316L [mg/L]**	**$Fe_{74}Cr_{24}Mo_2N$ [mg/L]**	**$Fe_{54}Cr_{24}Mn_{21}Mo_1N$ [mg/L]**
	4.6 ± 0.6	4.2 ± 0.5	4.3 ± 0.5
Fe	0.00	0.00	0.00
Mn	0.00	—	0.00
Mo	0.00	0.00	0.00
Ni	0.21 ± 0.03	0.10 ± 0.02	0.10 ± 0.02

The quantification of the metallic elements in each extract of the samples was undertaken for Cr, Fe, Mn, Mo and, Ni under the optimum condition for each element and the results are shown in Table 12.6.

In the extract of 316L, Ni was detected at a concentration of 0.21 ± 0.03 mg/L. Under dynamic conditions in the cell culture medium, the presence Ni was also observed with $Fe_{74}Cr_{24}Mo_2N$ and $Fe_{54}Cr_{24}Mn_{21}Mo_1N$, although nickel is not present as a constituent element. It should be noted that the presence of Ni was also reported from Ti–6Al–4V as an impurity at a concentration less than 0.01 mass% [150]. Iron and chromium oxides are the main component of the surface oxide on 316L and $Fe_{74}Cr_{24}Mo_2N$ and $Fe_{54}Cr_{24}Mn_{21}Mo_1N$, but small amounts of nickel, molybdenum, and manganese oxides were also detected in the surface oxide layer. However, the stability of nickel oxide is relatively low compared to chromium and other oxides, thus the preferential release of nickel could occur.

Additionally, in all extracts of 316L, $Fe_{74}Cr_{24}Mo_2N$, and $Fe_{54}Cr_{24}Mn_{21}Mo_1N$ chromium was detected at the concentration of 4.6 ± 0.6, 4.2 ± 0.5, and 4.3±0.5 mg/L, respectively. Chromium is one of the essential elements for humans, so a small amount of this element may contribute to cell proliferation, resulting in higher cell growth [148].

These results indicate that $Fe_{74}Cr_{24}Mo_2N$ and $Fe_{54}Cr_{24}Mn_{21}Mo_1N$ have superior cytocompatibility than the conventional metallic biomaterial, 316L. Additionally, the nitrogen adsorption treatment increases the corrosion resistance of $Fe_{74}Cr_{24}Mo_2N$ and $Fe_{54}Cr_{24}Mn_{21}Mo_1N$. The addition of nitrogen to stainless steel influences the corrosion resistance of the steel. Recently, microcrystalline Ni-free stainless steel was manufactured by the nitrogen adsorption treatment [148]. $Fe_{54}Cr_{24}Mn_{21}Mo_1N$ sample had a higher cell growth than 316L in static and dynamic conditions. Based on the results of Yamamoto and coworkers [148, 149], $Fe_{54}Cr_{24}Mn_{21}Mo_1N$ has a major possibility for application in the biomedical field.

References

1. Adell. R., Lekholm, U., Rockler, B., and Branemark, P.I. (1981). A 15-year study of osseointegrated implants in the treatment of the edentulous jaw, *Int. J. Oral Surg.*, **10**, pp. 387–416.
2. Abrahams, G.A., Goodman, S.L., Nealey, P.F., Franco, M., and Murphy, C.J. (2000). Nanoscale topography of the basement membrane underlying the corneal epithelium of the rhesus macaque, *Cell Tissue Res.*, **299**, pp. 39–46.
3. Agheli, H., Malmström, J., Hanarp, P., and Sutherland, D.S. (2006). Nanostructured biointerfaces, *Mater. Sci. Eng., C*, **26**, pp. 911–917.
4. Ahmad, Z., Huang, J., Edirisinghe, M.J., Jayasinghe, S.N., Best, S.M., Bonfield, W., Brooks, R.A., and Rushton, N. (2006). Electrohydrodynamic print-patterning of nanohydroxyapatite, *J. Biomed. Nanotechnol.*, **2**, pp. 201–207.
5. Akiyama, Y., Kikuchi, A., Yamato, M., and Okano, T. (2004). Ultrathin poly(Nisopropylacrylamide) grafted layer on polystyrene surfaces for cell adhesion/ detachment control, *Langmuir*, **20**, pp. 5506–5511.
6. Albrektsson, T., Branemark, P.I., Hansson, H.A., and Lindström, J. (1981). Osseointegrated titanium implants: requirements for ensuring a long-lasting direct bone-to-implant anchorage in man, *Acta Orthop. Scand.*, **52**, pp. 155–170.
7. Albrektsson, T., and Sennerby, L. (1990). Direct bone anchorage of oral implants: clinical and experimental considerations of the concept of osseointegration, *Int. J. Prosthodont.*, **3**, pp. 30–41.
8. Albrektsson, T., Berglundh, T., and Lindhe, J. (2003). Osseointegration: Historic background and current concepts, Chapter 3, in *Clinical Periodontology and Implant Dentistry*, (ed. Lindhe, J., Lang, N.P., and Karring., T.), Blackwell Murksgaard, Copenhagen, pp. 809–828.
9. Albrektsson, T., and Wennerberg, A. (2004). Oral implant sulfaces: part 2 - review focusing on clinical knowledge of different surfaces, *Int. J. Prosthodont.*, **17**, pp. 544–564.
10. Anagnostou, F., Debet, A., Pavon-Djavid, G., Goudaby, Z., Helary, G., and Migonney, V. (2006). Osteoblast functions on functionalized PMMA-

based polymers exhibiting Staphylococcus aureus adhesion inhibition, *Biomaterials*, **27**, pp. 3912–3919.

11. Andersson, A.S., Backhed, F., von Euler, A., Richter-Dahlfors, A., Sutherland, D., and Kasemo, B. (2003). Nanoscale features influence epithelial cell morphology and cytokine production, *Biomaterials*, **24**, pp. 3427–3436.
12. Anselme, K. (2000). Osteoblast adhesion on biomaterials — review, *Biomaterials*, **21**, pp. 667–681.
13. Ayad, S., Boot-Handford, R., Humphries, M.J., Kadler, K.E., and Shutleworth, A. (1994). *The Extracellular Matrix Facts Book*, Academic Press, San Diego.
14. Balasundaram, G., Sato, M., and Webster, T.J. (2006). Using hydroxyapatite nanoparticles and decreased crystallinity to promote osteoblast adhesion similar to functionalizing with RGD, *Biomaterials*, **27**, pp. 2798–2805.
15. Banerjee, A., Bandyopadhyay, A., and Bose, S. (2007). Hydroxyapatite nanopowders: Synthesis, densification and cell–materials interaction, *Mater. Sci. Eng., C*, **27**, pp. 729–735.
16. Baraton, M.I., Chen, X., and Gonsalves, K.E. (1997). FTIR study of nanostructured alumina nitride powder surface: determination of the acidic/basic sites by CO, CO_2, and acetic acid adsorptions, *Nanostruct. Mater.*, **8**, pp. 435–445.
17. Bauer, I.W., Li, S.P., Han, Y.C., Yuan, L., and Yin, Z. (2008). Internalization of hydroxyapatite nanoparticles in liver cancer cells, *J. Mater. Sci. Mater. Med.*, **19**, pp. 1091–1095.
18. Best, S.M., Porter, A.E., Thian, E.S., and Huang, J. (2008). Bioceramics: Past, present and for the future, *J. Eur. Cer. Soc.*, **28**, pp. 1319–1327.
19. Biggs, M.J., Richards, R.G., Gadegaard, N., Wilkinson, C.D.W., and Dalby, M.J. (2007). Regulation of implant surface cell adhesion: characterization and quantification of S-phase primary osteoblast adhesions on biomimetic nanoscale substrates. *J. Orthop. Res.*, **25**, pp. 273–282.
20. Bigi, A., Fini, M., Bracci, B., Boanini, E., Torricelli, P., Giavaresi, G. Aldini, N.N., Facchini, A., Sbaiz, F. And Giardino, R. (2008). The response of bone to nanocrystalline hydroxyapatite-coated $Ti_{13}Nb_{11}Zr$ alloy in an animal model, *Biomaterials*, **29**, pp. 1730–1736.
21. Boyan, B.D., Schwartz, Z., and Hambleton, J.C. (1993). Response of bone and cartilage cells to biomaterials *in vivo* and *in vitro*, *J. Oral Implantol.*, **19**, pp. 116–122.
22. Buser, D., Broggini, N., Wieland, M., Schenk, R.K., Denzer, A.J., Cochran, D.L., Hoffmann, B., Lussi, A., and Steinemann, S.G. (2004). Enhanced

bone apposition to a chemically modified SLA titanium surface, *J. Dent. Res.*, **83**, pp. 529–533.

23. Chen, S.C., Mrksich, M., Huang, S., Whitesides, M.G., and Ingber, E.D. (1997). Geometric control of cell life and death, *Science*, **276**, pp. 1425–1428.
24. Chou, L., Marek, B., and Wagner, W.R. (1999). Effects of hydroxylapatite coating crystallinity on biosolubility, cell attachment efficiency and proliferation *in vitro*, *Biomaterials*, **20**, pp. 977–985.
25. Colon, G., Ward, B.C., and Webster, T.J. (2006). Increased osteoblast and decreased Staphylococcus epidermidis functions on nanophase ZnO and TiO_2, *J. Biomed. Mater. Res., A*, **78**, pp. 595–604.
26. Cooper, L.F. (1998). Biologic determinants of bone formation for osseointegration: clues for future clinical improvements, *J. Prosthet. Dent.*, **80**, pp. 439–449.
27. Cooper, L.F., Masuda, T., Yliheikkila, P.K., and Felton, D.A. (1998). Generalizations regarding the process and phenomenon of osseointegration. Part II. *In vitro* studies, *Int. J. Oral Maxillofac. Implants*, **13**, pp. 163–174.
28. Damen, J.J.M., Ten Cate, J.M., and Ellingsen, J.E. (1991). Induction of calcium phosphate precipitation by titanium dioxide, *J. Dent. Res.*, **70**, pp. 1346–1349.
29. Degasne, I., Baslé, M.F., Demais, V., Huré, G., Lesourd, M., Grolleau, B., Mercier, L., and Chappard, D. (1999). Effects of roughness, fibronectin and vitronectin on attachment, spreading and proliferation of human osteoblast-like cells (Saos-2) on titanium surfaces, *Calcified Tissue Int.*, **64**, pp. 499–507.
30. Del Curto, B., Brunella, M.F., Giordano, C., Pedeferri, M.P., Valtulina, V., Visai, L., and Cigada, A. (2005). Decreased bacterial adhesion to surface-treated titanium, *Int. J. Artif. Organs*, **28**, pp. 718–730.
31. De Oliveira, P.T., and Nanci, A. (2004). Nanotexturing of titanium-based surfaces upregulates expression of bone sialoprotein and osteopontin by cultured osteogenic cells, *Biomaterials*, **25**, pp. 403–413.
32. Divya Rani, V.V., Manzoor, K., Menon, D., Selvamurugan, N., and Nair, S.V. (2009). The design of novel nanostructures on titanium by solution chemistry for an improved osteoblast response, *Nanotechnology*, **20**, 195101.
33. Doundoulakis, J.H. (1987). Surface analysis of titanium after sterilization: role in implant-tissue interface and bioadhesion, *J. Prosthet. Dent.*, **58**, pp. 471–478.

34. Ejiofor, J., and Webster, T.J. (2004). Biomedical implants from nanostructured materials, in *Dekker Encyclopedia of Nanoscience and Nanotechnology*, (ed. Schwarz, J.A., Contescu, C., and Putyera, K.), Marcel Dekker, Inc., New York, pp. 263–275.

35. El-Ali, J., Sorger, P.K., and Jensen, K.F. (2006). Cells on chips, *Nature*, **442**, pp. 403–411.

36. Ellingsen, J.E. (1998). Surface configurations of dental implants, *Periodontology 2000*, **17**, pp. 36–46.

37. Esposito, M., Coullhard, P., Thomsen, R., and Worthington, H.V. (2005). The role of implant surface modifications, shape and material on the success of osseointegrated dental implants. A Cochrane systematic review, *Eur. J. Prosthodont. Restor. Dent.*, **13**, pp. 15–31.

38. Flemming, R.G., Murphy, C.J., Abrahams, G.A., Goodman, S.L., and Nealey P.F. (1999). Effects of synthetic micro- and nano-structured surfaces on cell behavior, *Biomaterials*, **20**, pp. 573 –588.

39. Freitas Jr., R.A. (1999). *Nanomedicine, Volume I: Basic Capabilities*, Landes Bioscience, Georgetown, TX.

40. Freitas Jr., R.A. (2003). *Nanomedicine, Volume IIA: Biocompatibility*, Landes Bioscience, Georgetown, TX.

41. Frayssinet, P., Trouillet, J.L., Rouquet, N., Azimus, E., and Autefage, A. (1993). Osseointegration of macroporous calcium phosphate ceramics having a different chemical composition, *Biomaterials*, **14**, pp. 423–429.

42. Frayssinet, P., Tourenne, F., Rouquet, N., Conte, P., Delga, C., and Bonel, G. (1994). Comparative biological properties of HA plasma-sprayed coatings having different crystallinities, *J. Mater. Sci. Mater. Med.*, **5**, pp. 11–17.

43. Geetha, M., Singh, A.K., Asokamani, R., and Gogia, A.K. (2009). Ti based biomaterials, the ultimate choice for orthopaedic implants — a review, *Prog. Mater. Sci.*, **54**, pp. 397–425.

44. Gutwein, L.G., and Webster, T.J. (2004). Increased viable osteoblast density in the presence of nanophase compared to conventional alumina and titania particles, *Biomaterials*, **25**, pp. 4175–4183.

45. Hartman, L.C., Meenaghan, M.A., Schaaf, N.G., and Hawker, P.B. (1989). Effects of pretreatment sterilization and cleaning methods on materials properties and osseoinductivity of a threaded implant, *Int. J. Oral Maxillofac. Implants*, **4**, pp. 11–18.

46. Huang, J., Jayasinghe, S.N., Best, S.M., Edirisinghe, M.J., Brooks, R.A., and Bonfield, W. (2004). Electrospraying of a nano-hydroxyapatite suspension, *J. Mater. Sci.*, **39**, pp. 1029–1032.

47. Huang, J., Jayasinghe, S.N., Best, S.M., Edirisinghe, M.J., Brooks, R.A., Rushton, N., and Bonfield, W. (2005). Novel deposition of nano-sized siliconsubstituted hydroxyapatite by electrostatic spraying, *J. Mater. Sci. Mater. Med.*, **16**, pp. 1137–1142.

48. Hyde, F.W., Alberg, M., and Smith, K. (1997). Comparison of fluorinated polymers against stainless steel, glass and polypropylene in microbial biofilm adherence and removal, *J. Ind. Microbiol. Biotechnol.*, **19**, pp. 142–149.

49. Ito, H., Oaki, Y., and Imai, H. (2008). Selective synthesis of various nanoscale morphologies of hydroxyapatite via an intermediate phase, *Cryst. Growth Des.*, **8**, pp. 1055–1059.

50. Jakubowicz, J., Jurczyk, K., Niespodziana, K., and Jurczyk, M. (2009). Mechanoelectrochemical synthesis of porous Ti-based nanocomposite biomaterial, *Electrochem. Commun.*, **11**, pp. 461–465.

51. Jakubowicz, J., Jurczyk, K., and Jurczyk, M. (2010). Electrochemical formation and corrosion properties of porous TiO_x biomaterials, *Mater. Sci. Forum*, **636–637**, pp. 15–21.

52. Jurczyk, K., Niespodziana, K., and Jurczyk, M. (2006). Preparation and characterization of nanocomposite Ti-hydroxyapatite materials, *Eur. J. Med. Res.*, **11**, Suppl. II, p. 133.

53. Jurczyk, K., Niespodziana, K., Stopa, J., and Jurczyk M. (2007). Composite Ti-hydroxyapatite bionanomaterials for application in modern dentistry, *Polish J. Environ. Studies*, **16**, no 2CII, pp. 323–327.

54. Jurczyk, K., Niespodziana, K., Stopa, J., and Jurczyk, M. (2009). Titanium-ceramic dental bionanomaterials, *J. Clin. Periodontol.*, **36**, Suppl. 9, pp. 73–74.

55. Jurczyk, K., Niespodziana, K., Stopa, J., and Jurczyk, M. (2009). Nanotechnology in perspective of applications in modern dentistry, *Polish J. Environ. Stud.*, **18**, no 1A, pp. 21–27.

56. Jurczyk, M., Jurczyk, K., Niespodziana, K., and Miklaszewski, A. (2010). Hybrid Ti-ceramic bionanomaterials for dental engineering, *Eur. Cells Mater.*, **19**, Suppl. 1, p. 1.

57. Kaplan, F.S., Hayes, W.C., Keaveny, T.M., Boskey, A., Einhorn, T.A., and Iannotti, J.P. (1994). Form and function of bone, in *Orthopedic Basic Science* (ed. Simon, S.R.), American Academy of Orthopaedic Surgeons, Rosemont, pp. 127–185.

58. Kay, S., Thapa, A., Haberstroh, K.M., and Webster, T.J. (2002). Nanostructured polymer/nanophase ceramic composites enhance osteoblast and chondrocyte adhesion, *Tissue Eng.*, **8**, pp. 753–756.

59. Kasemo, B. (1983). Biocompatibility of titanium implants: surface science aspects, *J. Prosthet. Dent.*, **49**, pp. 832–837.

60. Kasemo, B., and Lausmaa, J. (1988). Biomaterial and implant surfaces: a surface science approach, *Int. J. Oral Maxillofac. Implants*, **3**, pp. 247–259.

61. Keller, J.C., Draughn, R.A., Wightman, J.P., Dougherty, W.J., and Meletiou, S.D. (1990). Characterization of sterilized cp titanium implants surface, *Int. J. Maxillofac. Implants*, **5**, pp. 360–367.

62. Keselowsky, B.G., Collard, D.M., and García, A.J. (2003). Surface chemistry modulates fibronectin conformation and directs integrin binding and specificity to control cell adhesion, *J. Biomed. Mater. Res., A*, **66**, pp. 247–259.

63. Keselowsky, B.G., Collard, D.M., and García, A.J. (2004). Surface chemistry modulates focal adhesion composition and signaling through changes in integrin binding, *Biomaterials*, **25**, pp. 5947–5954.

64. Khang, D., Kim, S.Y., Liu-Snyder, P., Palmore, G.T., Durbin, S.M., and Webster, T.J. (2007). Enhanced fibronectin adsorption on carbon nanotube/poly(carbonate) urethane: independent role of surface nano-roughness and associated surface energy, *Biomaterials*, **28**, pp. 4756–4768.

65. Khang, D., Lu, J., Yao, C., Haberstroh, K.M., and Webster, T.J. (2008). The role of nanometer and sub-micron surface features on vascular and bone cell adhesion on titanium, *Biomaterials*, **29**, pp. 970–983.

66. Kilpadi, D., and Lemons, J. (1994). Surface energy characterization of unalloyed titanium implants, *J. Biomed. Mater. Res.*, **28**, pp. 1419–1425.

67. Kilpadi, K.L., Chang, P.L., and Bellis, S.L. (2001). Hydroxylapatite binds more serum proteins, purified integrins, and osteoblast precursor cells than titanium or steel, *J. Biomed. Mater. Res.*, **57**, pp. 258–267.

68. Kim, P., Kim, D.H., Kim, B., Choi, S.K., Lee, S.H., Khademhosseini, A., Langer, R., and Suh, K.Y. (2005). Fabrication of nanostructures of polyethylene glycol for applications to protein adsorption and cell adhesion, *Nanotechnology*, **16**, pp. 2420–2426.

69. Kinnari, T.J., Peltonen, L.I., Kuusela, P., Kivilahti, J., Kononen, M., and Jero, J. (2005). Bacterial adherence to titanium surface coated with human serum albumin, *Otol. Neurotol.*, **26**, pp. 380–384.

70. Klabunde, K.J., Strak, J., Koper, O., Mohs, C., Park, D., Decker, S., Jiang, Y., Lagadic, I., and Zhang, D. (1996). Nanocrystals as stoichiometric reagents with unique surface chemistry, *J. Phys. Chem.*, **100**, pp. 12142–12153.

71. Kuroda, D., Hawana, T., Hiromoto, S., Katada, Y., and Asami, K. (2002). Surface characterization of nickel-free high nitrogen austenitic

stainless steel in simulated biological environment, *Mater. Trans.*, **43**, pp. 3093–3099.

72. Kweh, S.W.K., Khor, K.A., and Cheang, P. (2000). Plasma-sprayed hydroxyapatite (HA) coatings with flame-spheroidized feedstock: microstructure and mechanical properties, *Biomaterials*, **21**, pp. 1223–1234.
73. Lawn, B.R., and Howes, V.R. (1981). Elastic recovery at hardness indentations. *J. Mater. Sci.*, **16**, pp. 2745–2752.
74. Lee, H.J., Choi, H.W., Kim, K.J., and Lee, S.C. (2006) Modification of hydroxyapatite nanosurfaces for enhanced colloidal stability and improved interfacial adhesion in nanocompasites, *Chem. Mater.*, **18**, pp. 5111–5118.
75. Leeuwenburgh, S., Wolke, J., Schoonman, J., and Jansen, J.A. (2003) Electrostatic spray deposition (ESD) of calcium phosphate coatings, *J. Biomed. Mater. Res., A*, **66**, pp. 330–334.
76. Le Guehennec, L., Soueidan, A., Layrolle, P., and Amouriq, Y. (2007) Surface treatments of titanium dental implants for rapid osseointegration, *Dental Mater.*, **23**, pp. 844–854.
77. Le Guehennec, L., Martin, F., Lopez-Heredia, M.A., Louarn, G., Amouriq, Y., Cousty, J., and Layrolle, P. (2008) Osteoblastic cell behavior on nanostructured metal implants, *Nanomedicine* (Lond), **3**, pp. 61–71.
78. Leigh, S.H., Lin, C.K., and Berndt, C.C. (1997) Elastic response of thermal spray deposits under indentation tests, *J. Am. Ceram. Soc.*, **80**, pp. 2093–2099.
79. Lim, Y.M., Kim, B.H., Jeon, Y.S., Jeon, K.O., and Hwang, K.S. (2005) Calcium phosphate films deposited by electrostatic spray deposition and an evaluation of their bioactivity, *J. Ceram. Proc. Res.*, **6**, pp. 255–258.
80. Lim, J.Y., Hansen, J.C., Siedlecki, C.A., Runt, J., and Donahue, H.J. (2005 Human foetal osteoblastic cell response to polymer-demixed nanotopographic interfaces, *J. R. Soc. Interface*, **2**, pp. 97–108.
81. Lim, J.Y., Dreiss, A.D., Zhou, Z., Hansen, J.C., Siedlecki, C.A., Hengstebeck, R.W., Cheng, J., Winograd, N., and Donahue, H.J. (2007). The regulation of integrin-mediated osteoblast focal adhesion and focal adhesion kinase expression by nanoscale topography. *Biomaterials*, **28**, pp. 1787–1797.
82. Machnee, C.H., Wagner, W.C., Jaarda, M.J., and Lang, B.R. (1993). Identification of oxide layers of commercially pure titanium in response to cleaning procedures, *Int. J. Oral Maxillofac. Implants*, **8**, pp. 529–533.

83. McManus, A.J., Doremus, R.H., Siegel, R.W., and Bizios, R. (2005). Evaluation of cytocompatibility and bending modulus of nanoceramic/polymer composites, *J. Biomed. Mater. Res. A*, **72**, pp. 98–106.

84. Maeno, S., Niki, Y., Matsumoto, H., Morioka, H., Yatabe, T., Funayama, A.,Toyama, Y., Taguchi, T., and Tanaka, J. (2005). The effect of calcium ion concentration on osteoblast viability proliferation and differentiation in monolayer and 3D culture, *Biomaterials*, **26**, pp. 4847–4855.

85. Martínez, E., Lagunas, A., Mills, C.A., Rodríguez-Seguí, S., Estévez, M., Oberhansl, S., Comelles, J., and Samitier, J. (2009). Stem cell differentiation by functionalized micro- and nanostructured surfaces, *Nanomedicine (Lond).*, **4**, pp. 65–82.

86. Martinez, E., Engel, E., Planell, J.A., and Samitier, J. (2009). Effects of artificial micro-and nano-structured surfaces on cell behavior, *Ann. Anat.*, **191**, pp. 126–135.

87. Meirelles, L., Arvidsson, A., Albrektsson, T., and Wennerberg, A. (2007). Increased bone formation to unstable nano rough titanium implants, *Clin. Oral Implants Res.*, **18**, pp. 326–332.

88. Meirelles, L. (2007). On nano size structures for enhanced early bone formation (PhD, Gothenburg University).

89. Mendonça, G., Mendonça, D.B.S., Aragã˜o, F.J.L., and Cooper, L.F. (2008). Advancing dental implant surface technology — from micron to nanotopography, Review, *Biomaterials* **29**, pp. 3822–3835.

90. Miller, D.C., Haberstroh, K.M., and Webster, T.J. (2005). Mechanism(s) of increased vascular cell adhesion on nanostructured poly(lactic-co-glycolic acid) films, *J. Biomed. Mater. Res. A*, **73**, pp. 476–484.

91. Nagano, M., Nakamura, T., Kokubo, T., Tanahashi, M., and Ogawa, M. (1996). Differences of bone bonding ability and degradation behaviour *in vivo* between amorphous calcium phosphate and highly crystalline hydroxyapatite coating, *Biomaterials*, **17**, pp. 1771–1777.

92. Nanci, A., Wuest, J.D., Peru, L., Brunet, P., Sharma, V., Zalzal, S., and McKee, M.D. (1998). Chemical modification of titanium surfaces for covalent attachment of biological molecules, *J. Biomed. Mater. Res.*, **40**, pp. 324–335.

93. Niespodziana, K., Jurczyk, K., and Jurczyk M. (2006). The manufacturing of Ti-hydroxyapatite nanocomposites for bone implant applications, *Nanopages*, **1**, pp. 219–229.

94. Niespodziana, K., Jurczyk, K., and Jurczyk, M. (2007). Titanium-ceramic nanocomposite materials, *Adv. Mater. Sci.*, **7**, pp. 103–107.

95. Niespodziana, K., Jurczyk, K., and Jurczyk, M. (2009). Mechanical and corrosion properties of titanium hydroxyapatite nanocomposites, *Solid State Phenomena*, **151**, pp. 217–221.

96. Niespodziana, K., Jurczyk, K., Jakubowicz, J., and Jurczyk, M. (2010). Fabrication and properties of titanium — hydroxyapatite nanocomposites, *Mater. Chem. Phys.*, **123**, pp. 160–165.

97. Niespodziana, K., Jurczyk, K., Miklaszewski, A., and Jurczyk, M. (2010). Hybrid Ti-ceramic bionanomaterials for medical engineering, *Phys. Status Solidi C*, **7**, pp. 1363–1366.

98. Oh, S., Daraio, C., Chen, L.H. Pisanic, T.R., Finones, R.R., and Jin, S. (2006). Significantly accelerated osteoblast cell growth on aligned $Ti0_2$ nanotubes, *J. Biomed. Mater. Res. A*, **78**, pp. 97–103.

99. Okada, S., Ito, H., Nagai, A., Komotori, J., and Imai H. (2010). Adhesion of osteoblast-like cells on nanostructured hydroxyapatite, *Acta Biomater.*, **6**, pp. 591–597.

100. Oliveira, P.T., Zalzal, S.F., Beloti, M.M., Rosa, A.L., and Nanci, A. (2007). Enhancement of *in vitro* osteogenesis on titanium by chemically produced nanotopography, *J. Biomed. Mater. Res. A*, **80**, pp. 554–564.

101. Park, G.E., and Webster, T.J. (2005). A review of nanotechnology for the development of better orthopedic implants, *J. Biomed. Nanotechnol.*, **1**, pp. 18–29.

102. Popat, K.C., Chatvanichkul, K.I., Barnes, G.L., Latempa, T.J., Grimes, C.A., and Desai, T.A. (2007). Osteogenic differentiation of marrow stromal cells cultured on nanoporous alumina surfaces, *Biomed. Mater. Res.* A **80**, pp. 955–964.

103. Placko, H.E., Mishra, S., Weimer, J.J., and Lucas, L.C. (2000). Surface characterization of titanium-based implant materials, *Int. J. Oral Maxillofac. Implants*, **15**, pp. 355–363.

104. Price, R.L., Gutwein, L.G., Kaledin, L., Tepper, F., and Webster, T.J. (2003). Osteoblast function on nanophase alumina materials: influence of chemistry, phase, and topography, *J. Biomed. Mater. Res. A*, **67**, pp. 1284–1293.

105. Price, R.L., Waid, M.C., Haberstroh, K.M., and Webster, T.J. (2003). Increased, select bone cell adhesion on formulations containing carbon nanofibers, *Biomaterials*, **24**, pp. 1877–18.

106. Price, R.L., Haberstroh, K.M., and Webster, T.J. (2003). Enhanced functions of osteoblasts on nanostructured surfaces of carbon and alumina, *Med. Biol. Eng. Comput.*, **41**, pp. 372–375.

107. Price, R.L., Ellison, K., Haberstroh, K.M., and Webster, T.J. (2004). Nanometer surface roughness increases select osteoblast adhesion on carbon nanofiber compacts, *J. Biomed. Mater. Res. A*, **70**, pp. 129–138.

108. Puckett, S.D., Taylor, E., Raimondo, T., and Webster, T.J. (2010). The relationship between the nanostructure of titanium surfaces and bacterial attachment, *Biomaterials*, **31**, pp. 706–713.

109. Ramires, P.A., Romito, A., Cosentino, F., and Milella, E. (2001). The influence of titanium/hydroxyapatite composite coatings on *in vitro* osteoblasts behavior, *Biomaterials*, **22**, pp. 1467–1474.

110. Ratner, B.D., Johnston, A.B., and Lenk, T.J. (1987). Biomaterial surfaces, *J. Biomed. Mater. Res.*, **21**, pp. 59–90.

111. Raulio, M., Jarn, M., Ahola, J., Peltonen, J., Rosenholm, J.B., Tervakangas, S, Kolehmainen, J., Ruokolainen, T., Narko, P., and Salkinoja-Salonen, M. (2008). Microbe repelling coated stainless steel analysed by field emission scanning electron microscopy and physicochemical methods, *J. Ind. Microbiol. Biotechnol.*, **35**, pp. 751–760.

112. Saunders, R., Szymczyk, K.H., Shapiro, I.M., and Adams, C.S. (2007). Matrix regulation of skeletal cell apoptosis III: mechanism of ion pairinduced apoptosis, *J. Cell Biochem.*, **100**, pp. 703–715.

113. Scotchford, C.A., Gilmore, C.P., Cooper, E., Leggett, G.J., and Downes, S. (2002). Protein adsorption and human osteoblast-like cell attachment and growth on alkylthiol on gold self-assembled monolayers, *J. Biomed. Mater. Res.*, **59**, pp. 84–99.

114. Sgambato, A., Cittadini, A., Ardito, R., Dardeli, A., Facchini, A., Pria, P.D., and Colombo A. (2003). Osteoblast behavior on nanostructured titanium alloys, *Mater. Sci. Eng. C*, **23**, pp. 419–423.

115. Shi, Z., Huang, X., Cai, Y., Tang, R., Yang, D. (2009). Size effect of hydroxyapatite nanoparticles on proliferation and apoptosis of osteoblast-like cells, *Acta Biomater.*, **5**, pp. 338–345.

116. Siebers, M.C., Walboomers, X.F., Leeuwenburgh, S.C.G., Wolke, J.G.C., and Jansen, J.A. (2004). Electrostatic spray deposition (ESD) of calcium phosphate coatings, an *in vitro* study with osteoblast-like cells, *Biomaterials*, **25**, pp. 2019–2027.

117. Siebers, M.C., Walboomers, X.F., Leeuwenburgh, S.C.G., Wolke, J.G.C., and Jansen, J.A. (2006). The influence of the crystallinity of electrostatic spray deposition-derived coatings on osteoblast-liked cell behavior, *in vitro*, *J. Biomed. Mater. Res. A*, **78**, pp. 258–267.

118. Siebers, M.C., Walboomers, X.F., Leeuwenburgh, S.C.G., Wolke, J.C.G., Boerman, O.C., and Jansen, J.A. (2006). Transforming growth factor — beta 1 release from a porous electrostatic spray deposition-derived calcium phosphate coating, *Tissue Eng.*, **12**, pp. 2449–2456.

119. Togashi, A.Y., Cirano, F.R., and Lima, L.A.P.A. (2007). The role of implant surface chemistry in the biological bone response, *RPG Rev. Pós. Grad.*, **13**, pp. 340–344.

120. Tulinski, M., Jurczyk, K., and Jurczyk, M. (2007). Austenitic nanocrystalline nickel-free stainless steel for biomedical applications, *Acta Metall. Slov.*, **13**, pp. 223–228.

121. Tulinski, M., Jurczyk, K., and Jurczyk M. (2008). Nickel-free nanocrystalline austenitic stainless steels, *Materials Science Poland*, **26**, pp. 381–388.

122. Tulinski, M., Jurczyk, K., and Jurczyk M. (2008). Nanoscale nickel-free austenitic stainless steel, *Solid State Phenomena*, **140**, pp. 179–184.

123. Tulinski, M., and Jurczyk M. (2008). Mechanical and corrosion properties of Ni-free austenitic stainless steels, *Archiv. Metall. Mater.*, **53**, pp. 955–959.

124. Tulinski, M., and Jurczyk, M. (2009). Mechanical and corrosion properties of Ni-free austenitic stainless steel/hydroxyapatite nanocomposites, *Solid State Phenomena*, **151**, pp. 213–216.

125. Tulinski, M., and Jurczyk, M. (2010). Corrosion resistance of nickel-free austenitic stainless steels/hydroxyapatite composites, *Phys. Status Solidi C*, **7**, pp. 1359–1362.

126. Wallraff, G.M., and Hinsberg, W.D. (1999). Lithographic imaging techniques for the formation of nanoscopic features, *Chem. Rev.*, **99**, pp. 1801–1822.

127. Wan, Y., Wang, Y., Liu, Z., Qu, X., Han, B., Bei, J., and Wang, S. (2005). Adhesion and proliferation of OCT-1 osteoblast-like cells on micro- and nano-scale topography structured poly(L-lactide), *Biomaterials*, **26**, pp. 4453–4459.

128. Wang, G.C., Liu, X.Y., and Ding, C.X. (2008). Phase composition and *in vitro* bioactivity of plasma sprayed calcia stabilized zirconia coatings, *Surf. Coat. Technol.*, **202**, pp. 5824–5831.

129. Wang, G.C., Liu, X.Y., Gao, J.H., and Ding, C.X. (2009). *In vitro* bioactivity and phase stability of plasma-sprayed nanostructured 3Y-TZP coatings, *Acta Biomater.*, **5**, pp. 2270–2278.

130. Wang, G.C., Meng, F.H., Ding, C.X., Chuc, P.K., and Liu, X.O. (2010). Microstructure, bioactivity and osteoblast behavior of monoclinic zirconia coating with nanostructured surface, *Acta Biomater.*, **6**, pp. 990–1000.

131. Ward, B.C., and Webster, T.J. (2006). The effect of nanotopography on calcium and phosphorus deposition on metallic materials *in vitro*, *Biomaterials*, **27**, pp. 3064–3074.

132. Ward, B.C., and Webster, T.J. (2007). Increased functions of osteoblasts on nanophase metals, *Mater. Sci. Eng. C*, **27**, pp. 575–578.

133. Webster, T.J., Siegel, R.W., and Bizios, R. (1999). Osteoblast adhesion on nanophase ceramics, *Biomaterials*, **20**, pp. 1221–1227.

134. Webster, T.J., Siegel, R.W., and Bizios, R. (1999). Design and evaluation of nanophase alumina for orthopedic/dental applications, *Nanostruct. Mater.*, **12**, pp. 983–986.

135. Webster, T.J., Ergun, C., Doremus, R.H., Siegel, R.W., and Bizios, R. (2000). Enhanced functions of osteoblasts on nanophase ceramics, *Biomaterials*, **21**, pp. 1803–1810.

136. Webster, T.J., Ergun, C., Doremus, R.H., Siegel, R.W., and Bizios, R. (2000). Specific proteins mediate enhanced osteoblast adhesion on nanophase ceramics, *J. Biomed. Mater. Res.*, **51**, pp. 475–483.

137. Webster, T.J., (2001). Nanophase ceramics: The future orthopedic and dental implant material, in *Advances in Chemical Engineering. Nanostructured Materials*, (ed. Ying, J.Y.), Academic Press, San Diego, pp.125–166.

138. Webster, T.J., Siegel, R.W., and Bizios, R. (2001). Nanoceramic surface roughness enhances osteoblast and osteoclast functions for improved orthopaedic/dental implant efficacy, *Scripta Mater.*, **44**, pp. 1639–1642.

139. Webster, T.J., Schadler, L.S., Siegel, R.W., and Bizios, R. (2001). Mechanisms of enhanced osteoblast adhesion on nanophase alumina involve vitronectin, *Tissue Eng.*, **7**, pp. 291–301.

140. Webster, T.J., Ergun, C., Doremus, R.H., Siegel, R.W., and Bizios, R. (2001). Enhanced osteoclast like cell functions on nanophase ceramics, *Biomaterials*, **22**, pp. 1327–1333.

141. Webster, T.J. (2003). Nanophase ceramics as improved bone tissue engineering materials, *Am. Ceram. Soc. Bull.*, **82**, pp. 23–28.

142. Webster, T.J., and Ejiofor, J.U. (2004). Increased osteoblast adhesion on nanophase metals: Ti, Ti6Al4V, and CoCrMo, *Biomaterials*, **25**, pp. 4731–4739.

143. Webster, T.J., Hellenmeyer, E.L., and Price, R.L. (2005). Increased osteoblast functions on theta+delta nanofiber alumina, *Biomaterials*, **26**, pp. 953–960.

144. Webster, T.J. (2007). Nanotechnology for the improvement of all implants, *Mater. Sci. Forum*, **539–543**, pp. 511–516.

145. Wennerberg, A., and Albrektsson, T. (2009). Effects of titanium surface topography on bone integration: a systematic review, *Clin. Oral Implants Res.*, **20**, pp. 172–184.

146. Woo, K.M., Chen, V.J., and Ma, P.X. (2003). Nano-fibrous scaffolding architecture selectivelyenhances protein adsorption contributing to cell attachment, *J. Biomed. Mater. Res. A*, **67**, pp. 531–537.

147. Xia, Y., Rogers, J.A., Kateri, E.P., and Whitesides, G.M. (1999). Unconventional methods for fabricating and patterning nanostructures — review, *Chem. Rev.*, **99**, pp. 1823–1848.

148. Yamamoto, A., Kobayashi, T., Maruyama, N., Nakazawa, K., and Sumita, M. (1995). Fretting fatigue properties of Ti–6Al–4V alloy in pseudo-body fluid and evaluation of biocompatibility by cell culture method, *J. Jpn. Inst. Metals*, **59**, pp. 463–470.

149. Yamamoto, A., Kohyama, Y., Kuroda, D., and Hanawa, T. (2004). Cytocompatibility evaluation of Ni-free stainless steel manufactured by nitrogen adsorption treatment, *Mater. Sci. Eng. C*, **24**, pp. 737–743

150. Yang, B., Uchida, M., Kim, H.M., Zhang, X., and Kokubo, T. (2004). Preparation of bioactive titanium metal via anodic oxidation treatment, *Biomaterials*, **25**, pp. 1003–1010.

151. Yoshinari, M., Oda, Y., Kato, T., and Okuda, K. (2001). Influence of surface modifications to titanium on antibacterial activity *in vitro*, *Biomaterials*, **22**, pp. 2043–2048.

152. Zarb, G.A., and Schmitt, A. (1990). The longitudinal clinical effectiveness of osseointegrated dental implants: the Toronto study. Part III: problems and complications encountered. *J. Prosthet. Dent.*, **64**, pp. 185–194.

153. Zhang, L., and Webster, T.J. (2009). Nanotechnology and nanomaterials: Promises for improved tissue regeneration — review, *Nano Today*, **4**, pp. 66–80.

154. Zhang, L., Sirivisoot, S., Balasundaram, G., and Webster, T.J. (2009). Nanomaterials for Improved Orthopedic and Bone Tissue Engineering Applications, in *Advanced Biomaterials: Fundamentals, Processing and Applications* (ed. Basu, B., Katti, D., Kumar, A.), John Wiley & Sons, Inc., New Jersey, pp. 205–241.

155. Zhao, G., Schwartz, Z., Wieland, M., Rupp, F., Geis-Gerstorfer, J., Cochran, D.L., and Boyan, B. (2005). High surface energy enhances cell response to titanium substrate microstructure, *J. Biomed. Mater. Res.*, **74**, pp. 49–58.

156. Zhao, Y., Zhang, Y., Ning, F., Guo, D., and Xu, Z. (2007). Synthesis and cellular biocompatibility of two kinds of HAP with different nanocrystal morphology, *J. Biomed. Mater. Res., B Appl. Biomater*, **83**, pp. 121–126.

Chapter 13

Applications of Nanostructured Materials in Dentistry

Karolina Jurczyk[a] and Mieczyslaw Jurczyk[b,*]

[a]*Conservative Dentistry and Periodontology Department, University of Medical Sciences, Bukowska 70 Street, 60-812 Poznan, Poland*

[b]*Poznan University of Technology, Institute of Materials Science and Engineering, M. Sklodowska-Curie Sq. 5, 60-695 Poznan, Poland*

*mieczyslaw.jurczyk@put.poznan.pl

Current research focuses on improving the mechanical performance and biocompatibility of metal/alloy-based systems through variations in alloy composition, microstructure, and surface treatment [24, 27, 28, 34, 38, 47, 56]. In the case of titanium, significant efforts go into enhancing the strength characteristics of commercial purity grades in order to avoid potential biotoxicity of alloying elements, especially in dental implants [2, 4, 5, 17, 33].

To enhance the physicochemical and mechanical performance of implant materials through microstructure control, the top-down approaches known as mechanical alloying (see Chapters 6, 7, 8, and 9) and severe plastic deformation (SPD) techniques were applied. Recent studies clearly proved that nanostructuring of titanium can considerably improve not only the mechanical properties, but also the biocompatibility [17, 27, 28, 34, 38, 40, 48, 50–54]. On the other hand, this approach also has the benefit of enhancing the biological response of the cp titanium surface [27, 28, 40, 53].

Bionanomaterials for Dental Applications
Edited by Mieczyslaw Jurczyk

ISBN 978-981-4303-83-5 (Hardcover), 978-981-4303-84-2 (eBook)
www.panstanford.com

13.1 Bulk Nanostructured Titanium

Till now, a number of SPD methods for producing in bulk, ultra fine grain metals/alloys have been developed [48–54]. Valiev and co-workers apply a process known as equal channel angular pressing (ECAP), which is a viable processing route to grain refinement and property improvement [53]. Their study reports of nanostructured titanium (n-Ti), produced as long-sized rods with superior mechanical and biomedical properties and demonstrate its applicability for dental implants. It turns out that the extreme grain refinement of the bulk of the metal, down to nanoscale transpires to surface morphology that turns out to be conducive for enhanced adhesion and growth of living cells.

Commercially pure titanium (Grade 4) of the following composition was used: 0.052% C, 0.34% O_2, 0.3% Fe, 0.015% N, base material Ti (wt%). In the as-received condition, billets produced by hot rolling had a diameter of 40 mm with an average grain size of 25 μm. Nanostructuring was performed using SPD by equal-channel angular pressing (ECAP) with subsequent thermomechanical processing (TMP), which made it possible to manufacture rod semiproducts with a length of 3 m and a diameter of 7 mm [48–54].

This processing resulted in a large reduction in grain size, from the 25 μm equiaxed grain structure of the initial titanium rods to 150 nm after combined SPD and TMT processing, as shown in Fig. 13.1. The selected area electron diffraction pattern, Fig. 13.1c, further suggests that the ultra fine grains contained predominantly high-angle non-equilibrium grain boundaries with increased grain-to-grain internal stresses. It is important to note, that a similar structure for cp Ti can be produced in small discs using other SPD method such as high-pressure torsion (HPT) [49].

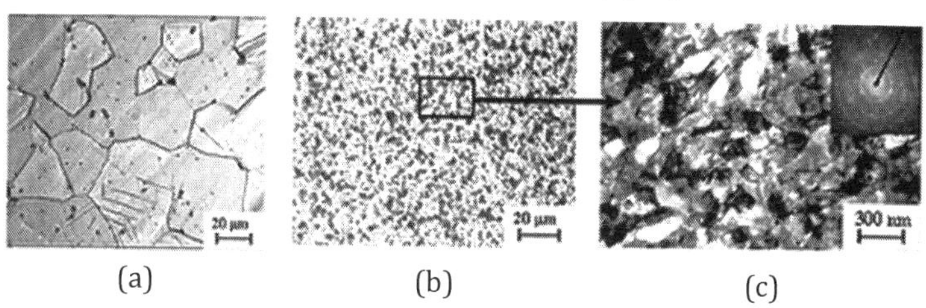

Figure 13.1 Microstructure of Grade 4 cp Ti: (a) the initial coarse grained rod; (b), (c) after ECAP + TMT (optical and electron photomicrographs) [53].

Mechanical properties of conventionally processed and nanostructured cp Grade 4 titanium is presented in Table 13.1. The strength of the nanostructured titanium is nearly twice that of conventional cp titanium. Additionally, it has been shown, that the fatigue strength of nanostructured cp titanium at 10^6 cycles is almost two times higher than for conventional cp titanium and exceeds that of the Ti–6Al–4V alloy.

Table 13.1 Mechanical properties of conventionally processed and nanostructured cp Grade 4 titanium [53]

Processing/ treatment conditions	UTS [MPa]	YS [MPa]	Elongation [%]	Reduction area [%]	Fatigue strength at 10^6 cycles
Conventional Ti (as received)	700	530	25	52	340
nTi ECAP + TMT	1240	1200	12	42	620
Ti–6Al–4V ELI annealed	940	840	16	45	530

Cytocompatibility tests utilizing fibroblast mice cells L929 were carried out. After nanostructuring, fibroblast colonization of the cp Grade 4 titanium surface dramatically increases (Fig. 13.2). For example, the surface cell occupation for conventional cp Ti was 53.0% after 72 hrs in contrast to 87.2% for nanostructured cp Grade 4 (Table 13.2). Compared to conventional titanium, high osteointegration rate should be expected with nanostructured cp Grade 4 titanium. Nanostructured (Nanoimplants®) implants have been successfully designed and fabricated (Fig. 13.3).

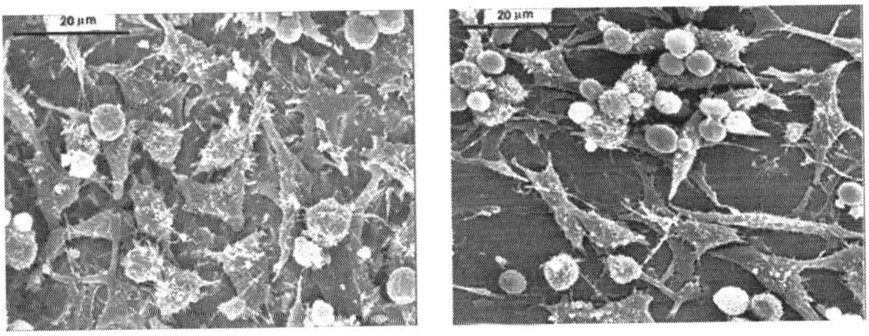

Figure 13.2 Occupation of the mice fibroblast cells L929 after 24 hours; Nanostructured (left) and conventional (right) cp Grade 4 titanium [53].

Table 13.2 Surface cell occupation for conventional and nanostructured cp Grade 4 titanium [53]

Material	Surface treatment	Occupied surface [pct.] after 72 hours
cp Gr. 4 Ti	Machining, followed by hydrofluoric acid etching	53.0
Nanostructured Gr. 4 Ti		87.2

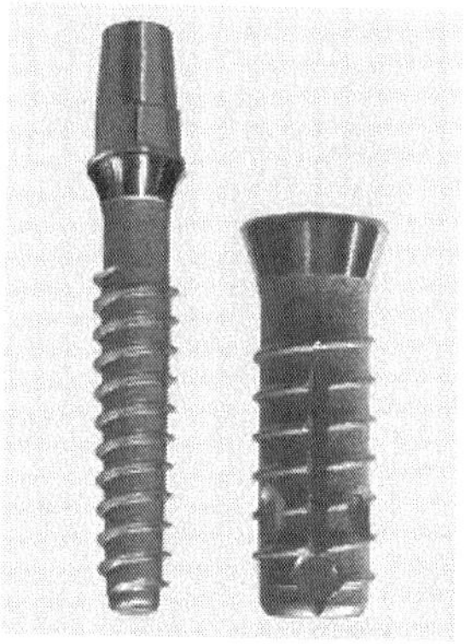

Figure 13.3 2.4 mm diameter Nanoimplant® (a) and 3.5 mm diameter Timplant® (b) [53].

The certified system of Timplant® manufactured according to standard EN ISO 13485:2003 was used during the development of the Nanoimplant® implant. The intraosseal nanoimplant of 2.4 mm diameter 2.4 mm has the strength equivalent to the conventional of 3.5 mm diameter implant. A number of over 250 Nanoimplants® have been implanted [41]. For example, a 55-year-old male with edentulous mandible and maxilla was treated by insertion of conical implants laterally and Nanoimplants® in the narrow anterior part (Fig. 13.4). Primary retention of all implants was very good. Nanostructuring of titanium by SPD processing has made the material with significantly superior mechanical performance when compared to conventional cp Grade 4 titanium and Ti–6Al–4V alloy (Figs. 13.5 and 13.6).

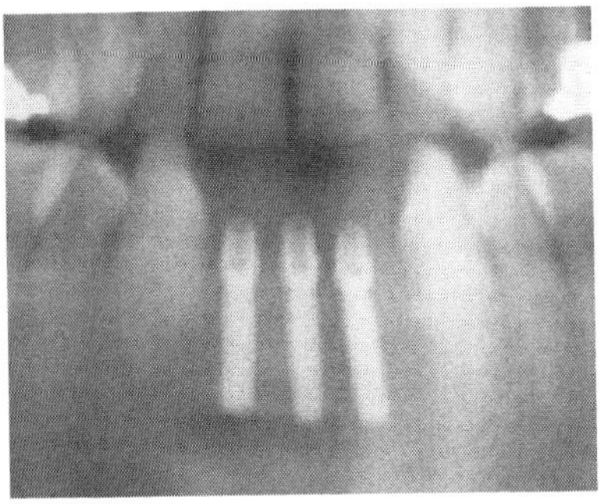

Figure 13.4 Implanted nanoimplants® [41].

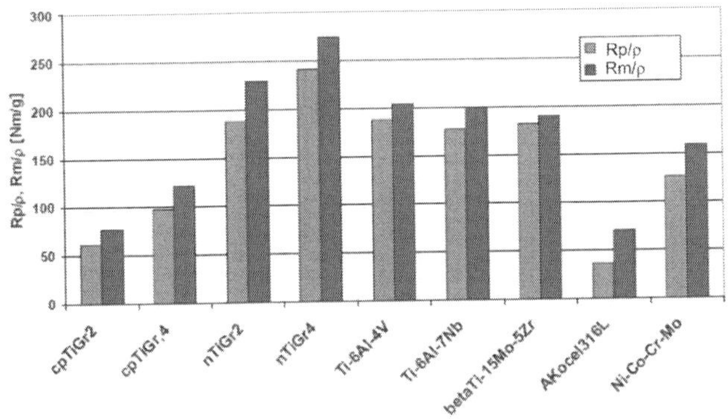

Figure 13.5 Selected mechanical properties of some implant alloys in comparison with nTiGr2 and nTiGr4 [41].

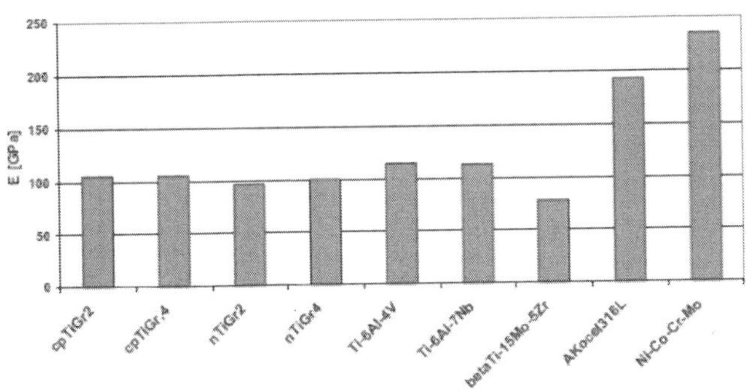

Figure 13.6 Young's modulus of different implant materials in comparison with nTiGr2 and nTiGr4 [41].

Currently, titanium and its alloys are used for dentistry devices such as implants, crowns, bridges, overdentures, and dental implant prosthesis components (screw and abutment). There are currently four cp Ti grades and one titanium alloy specially made for dental implant applications. These metals are specified according to ASTM as grades 1 to 5. Grades 1 to 4 are unalloyed while grade 5, with 6% aluminum and 4% vanadium, is the strongest. However, for permanent implant applications, the Ti–6Al–4V alloy has a possible toxic effect resulting from the released vanadium and aluminum [4, 5]. For this reason, vanadium- and aluminum-free alloys have been introduced for implant applications. These new alloys include Ti–6Al–7Nb (ASTM F1295), Ti–13Nb–13Zr (ASTM F1713), and Ti–12Mo–6Zr (ASTM F1813) [11].

13.2 Dental Implants with Nanosurface

Nanoscale modification can alter the chemistry and/or topography of the implant surface [36]. There are many different methods to impart nanoscale features to the implant surface (Table 13.3). Several of these methods have already been used to modify implants available commercially. Such changes alter the implant surface interaction with ions, biomolecules, and cells. These interactions can favorably influence molecular and cellular activities and alter the process of osseointegration.

Till now, a few nanoscale surface topography modifications have been used to enhance bone responses at clinical dental implants. For example, the Osseo-Speed surface (Astra Tech AB, Molndal, Sweden) possesses nanostructured features created by TiO_2 blasting followed by a proprietary hydrofluoric acid treatment [1, 7, 13]. Greater osteoblastic gene expression (Runx2, Osterix, Alkaline Ph-osphatase, and Bone Sialoprotein) was measured in cells adherent to the nanoscale HF-treated surface compared to the micron-scale surface [23]. This nanotopography is associated with the elevated levels of gene expression that indicate rapid osteoblastic differentiation. Other studies have demonstrated an increased bone formation, torque removal value [12]. In the rabbit tibia model of osseointegration, histomorphometric evaluations demonstrated higher bone-to-implant contact for the nanoscale OsseoSpeed implants compared to the micron-scale TiOblast™ implants (Astra Tech AB, Mölndal, Sweden) at 1 month (35 ± 14% vs 26 ± 8%) and 3 months (39 ± 11% vs 31 ± 6%) after placement [43].

Table 13.3 Methods for creating nanofeatures on cp titanium implants [36]

Methods	Characteristics
Self-assembly of monolayers	The exposed functional end group could be a molecule with different functions (an osteoinductive or cell adhesive molecule).
Physical approaches	
Compaction of nanoparticles	Conserves the chemistry of the surface among different topographies. Not readily applied over implant surfaces.
Ion beam deposition	Can impart nanofeatures to the surface based on the material used.
Chemical methods	
Acid etching	Combined with other methods (sand-blasting and/or peroxidation) can impart nanofeatures to the surface and remove contaminants.
Peroxidation	Produces a titania gel layer. Both chemical and topography changes are imparted.
Alkali treatment (NaOH)	Produces a sodium titanate gel layer allowing hydroxyapatite deposition. Both chemical and topography changes are imparted.
Anodization	Can impart nanofeatures to the surface creating a new oxide layer (based on the material used).
Nanoparticle deposition	
Sol–gel (colloidal particle adsorption)	Creates a thin-film of controlled chemical characteristics. Atomic-scale interactions display strong physical interactions.
Discrete crystalline deposition	Superimposes a nanoscale surface topographical complexity on the surface.
Lithography and contact printing technique	Many different shapes and materials can be applied over the surface. Approaches are labor intensive and require considerable development prior to clinical translation and application on implant surface.

The Astra Tech TiOblast™ is the precursor of the OsseoSpeed™ surface [25]. This surface was grit blasted with titanium dioxide particles to achieve an isotropic, moderately roughened surface. The TiOblast surface shows excellent outcomes when compared with machined titanium surfaces in theoretical models, in terms of *in vitro* cell biocompatibility evaluations and *in vivo* studies [8, 16, 21, 22, 37].

On the other hand, OsseoSpeed™ was launched in the fall 2004 and was a further development of the moderately roughened (grit blasted with titanium dioxide particles) titanium surface TiOblast™ [43]. OsseoSpeed gains its additional surface characteristics via a chemical (fluoride) treatment and a slight topographic modification of the TiOblast surface. Incorporation of small amounts of fluoride ions in the oxide layer results in a slight increase on the micrometer scale in surface roughness and the appearance of a nanoscale topography have been reported for the OsseoSpeed surface. *In vitro* and animal experiments indicate that the OsseoSpeed surface leads to increased bone formation and stronger bone-to-implant bonding at shorter healing times than TiOblast or machined titanium surfaces. Enhanced osteoblast differentiation, platelet activation, and thrombogenic properties of the fluoride-treated surface have been reported [3, 13, 14, 18, 20, 44, 57].

Published data shows that the OsseoSpeed implant can be safely used with a range reported for survival rate from 94.5% to 100%, including the use of immediate loading protocol [10] even in the atrophic edentulous maxilla [46], in sinus lifted maxillary posterior jaw sites [45], immediate installation in extraction sockets [35], and implants placed in atrophied mandibles close to the nerve [42].

Another nanoscale surface implant presently available in the clinical marketplace involves a CaP nanoparticle modification of a minimally rough titanium alloy implant (see Figs. 13.7 and 13.8) [9]. The NanoTite™ Implant starts with the industry-proven OSSEOTITE® Surface at the core. Next, discrete nanometer scale crystals of calcium phosphate (CaP) are deposited onto the OSSEOTITE Surface substrate. These crystals are bonded to the substrate, occupying approximately 50% of the surface, thereby being differentiated from the traditional plasma-sprayed CaP coatings that have been in commercial use for more than 20 years. The nanoscale topography and potential biologic benefits associated with the CaP crystals may play a key role in enhanced site response, potentially improving clinical predictability and outcomes. Preclinical studies demonstrate a substantial improvement on the rate and extent of osseointegration for the NanoTite Implant versus the Osseotite Implant.

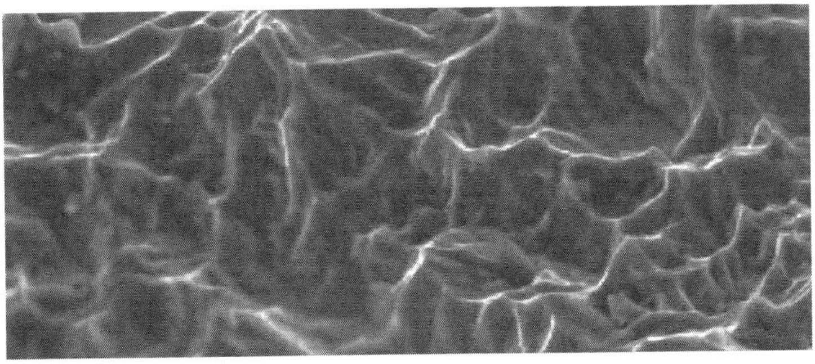

Figure 13.7 OSSEOTITE® Surface at 20,000× [http://biomet3i.com].

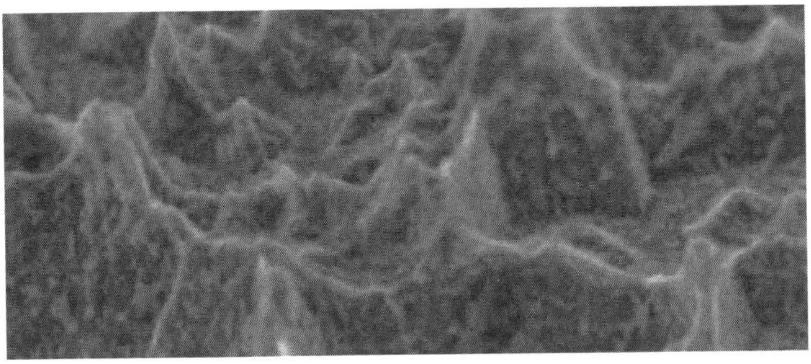

Figure 13.8 NanoTite™ Surface at 20,000× [http://biomet3i.com].

The nanoscale CaP surface created by DCD (Nanotite, 3i) was evaluated [39]. The histologic evaluation of clinical implants revealed bone-to-implant contact of 19 ± 14.2% and 32.2 ± 18.5% for the Osseotite (3i) control and the Nanotite (3i) experimental implants, respectively. Additionally, greater bone formation at 4 and 8 weeks was observed [19].

Recently, ion-beam assisted deposition (IBAD) process, which provides increased integration with the implant surface, known as high-energy sputter deposition, has been used to create a commercially available dental implant surface [6]. In the NanoTite™

process, a high-energy ion beam source aims a beam of ions at the surface of a target treated with HA. These high-energy ions eject the basic chemical elements of the HA from the target/substrate and create a molecular cloud whose molecules bond with the surface of the Bicon Integra-Ti™ implant. The bone formation was higher in the experimental group than in the control group (sandblasted/acid-etched) after 2 (13.56% vs 24.04%) and 4 weeks (14.22% vs 27.39%) [6].

13.3 Nanostructured Materials for Permanent and Bioresorbable Medical Implants

The development of metallic implant materials is driven by the biocompatibility requirements and also by the need for improved mechanical performance of biomedical implants. Additionally, the implants should strongly bond with the bone for the unfailing and long-term exploitation. For that reason, beside good mechanical and corrosion properties, the implant surface should be sufficiently rough for tissue growth and bone bonding [27–29]. Different paradigms govern this development for permanent and temporary implants. While materials for permanent implants, e.g., for bone or tooth replacement, obviously need to be as inert in bodily fluids as possible, those for temporary implants must degrade at a rate suitable for the targeted application.

Recently, two alloy systems based on titanium and magnesium are investigated [17, 26, 55]. Indeed, Ti forms a protective surface layer of titania and is considered to be bio-inert, while Mg is extremely reactive and biodegradable. It is known, that the mechanical performance of implant materials is possible to enhance through microstructure control. An equal channel angular pressing (ECAP) is a viable processing route to grain refinement and property improvement. It turns out that the extreme grain refinement of the bulk of the metal down to nanoscale transpires to surface morphology that turns out to be conducive for enhanced adhesion and growth of living cells (Fig. 13.9). Indeed, proliferation of the preosteoblast cells on the surface of nanostructured Ti processed by ECAP was shown to be hugely enhanced, by a factor of about 20. Improved adhesion and accelerated rate of proliferation following ECAP processing of titanium was recently reported for osteoblast cells [40].

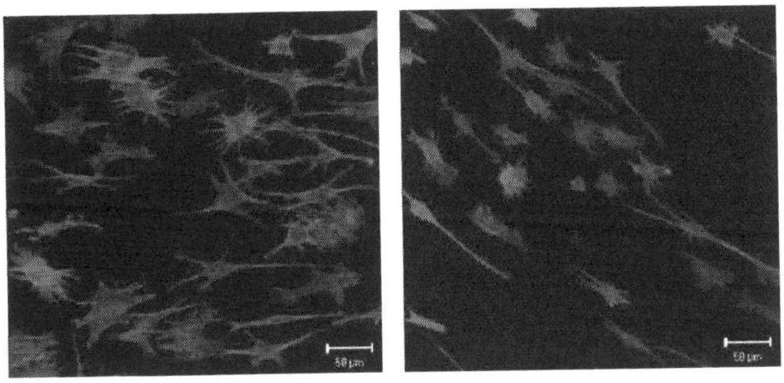

Figure 13.9 Enhanced growth of preosteoblast cells on nanostructured titanium (left) as compared to coarse-grained one (right) [17].

Bioactivity of titanium can be enhanced further by a combination of coating techniques, such as micro-arc oxidation (MAO), anodizing, and coating with hydroxyapatite (Fig. 13.10) [32, 33]. When HA was deposited on a MAO-treated surface, bioactivity was higher than in the case when only surface treatment was performed.

Figure 13.10 SEM micrographs of surface-modified titanium implants: Morphology of MAO-treated surface (left top), anodized surface (right top) and cross-sectional view of HA coating on Ti substrate (left bottom) [32, 33].

The potential for using Mg alloys in bioresorbable vascular stents or bone implants has recently attracted a huge interest of researchers [15, 26, 30, 31, 58]. Reducing the danger of inflammations and avoiding the need for repeat surgery by using temporary, biodegradable metallic implants and at the same time capitalizing on their good mechanical strength is, indeed, a very attractive possibility. Many groups worldwide have rapidly moved into this area. Despite some problems with biotoxicity of certain alloying elements, structural Mg alloys have been used in biocompatibility tests, both *in vitro* and *in vivo*. Clinical tests have demonstrated the viability of Mg alloys as stent implant materials. Recent results supported by the published work suggest, however, that *in vivo* tests are indispensable already at this stage, as biocompatibility of Mg alloys *in vitro* does not fully represent what happens *in vivo* [15, 58].

Recent work has demonstrated that bulk grain refinement techniques, such as equal-channel angular pressing, are potent tools to improve the fatigue strength and the bio-corrosion resistance of common structural Mg alloys [55]. We are exploring the ways to further improve the properties of Mg based alloys and make them fit for applications in bioresorbable implants.

However, bulk grain refinement may be insufficient to bring the bio-corrosion rate down to the levels required by the clinical needs. A natural way to contain corrosion of Mg and achieve controllable corrosion rates is by surface modification, particularly through smart coating design. For example, corrosion of Mg alloys is retarded markedly when a thin MgF_2 layer is formed on the surface in a fluoridation process. Furthermore, when a bioactive material, such as hydroxyapatite, is deposited on top of the MgF_2 layer, both the corrosion resistance and biocompatibility are enhanced significantly (Fig. 13.11).

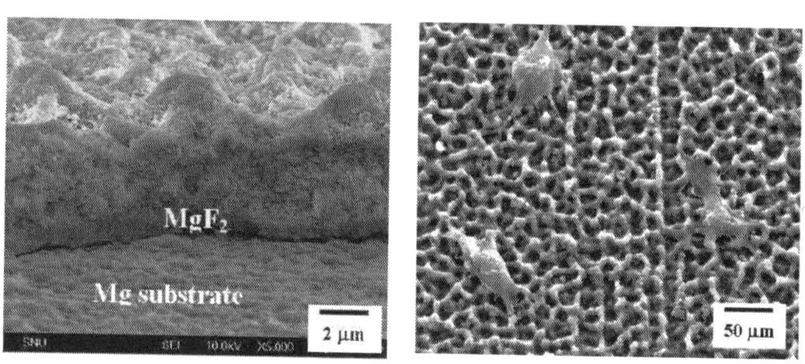

Figure 13.11 Coating layers of MgF_2 and hydrohyapatite (HA) on Mg (left) and osteoblast cells attached on the surface of Mg coated with MgF_2 and HA (right) [55].

13.4 Nanostructured Dental Composite Restorative Materials

In recent years, materials used for dental restorations have comprised principally acrylate or methacrylate polymers. However, acrylic materials exhibit high coefficients of thermal expansion relative to the coefficient of thermal expansion for the tooth structure, therefore these substances by themselves proved to be less than satisfactory. The disparity in thermal expansion, coupled with high shrinkage upon polymerization, resulted in poor marginal adaptability and ultimately led to secondary decay. Furthermore, the wear and abrasion characteristics and the overall physical, mechanical, and optical properties of these unfilled acrylic resinous materials were quite poor. Composite dental restorative materials containing methacrylate resins and fillers were thus developed.

The nanotechnology and nanoscience approach is to design and develop new dental composite materials with superior properties [29]. The filler composition comprises at least a bound, nanostructured, and silica. The silica is in the form of nano-sized particles, preferably spherical particles. The individual particles have the largest dimension or diameter in the range from about 10 to about 100 nm and preferably from about 10 to about 50 nm. The silica particles are furthermore bound to each other so as to result in chains having lengths in the range from about 50 nm to about 400 nm. In order to improve bonding with the resin matrix, the bound colloidal silica filler particles may optionally be treated with a silane. These filled compositions are useful for a variety of dental treatments and restorative functions including crown and bridge materials, fillings, adhesives, sealants, luting agents or cements, denture base materials, orthodontic materials and sealants, and other dental restorative materials.

In addition to the bound, nanostructured silica, the filler composition may further comprise one or more of the inorganic fillers currently used in dental restorative materials. Preferred additional fillers include those which are capable of being covalently bonded to the resin matrix itself or to a coupling agent which is covalently bonded to both. Examples of suitable filling materials include but are not limited to, silica, silicate glass, quartz, barium silicate, strontium silicate, barium borosilicate, strontium borosilicate, borosilicate, lithium silicate, lithium alumina silicate, amorphous silica, ammoniated or deammoniated calcium phosphate and alumina, zirconia, tin oxide, and titania. Suitable fillers have a particle size in the range from about 0.1–5.0 μm, and may further comprise unbound silicate colloids of about 0.001 to about 0.07 μm. These additional fillers may also be silanized.

The amount of total filler composition in the dental composite can vary widely, being in the range from about 1 to about 90% by weight (wt%) of the total composition. The amount used is determined by the requirements of the particular application. Thus, for example, crown and bridge materials generally comprise from about 60 to about 90 wt% filler, luting cements comprise from about 20 to about 80 wt% filler, sealants generally comprise from about 1 to about 20 wt% filler, adhesives generally comprise from about 1 to about 30 wt% filler and restorative materials comprise from about 50 to about 90% filler, with the remainder in all cases being the resin composition.

The expected outcomes are new formulations for dental composites with improved adhesive bonding to dentin and enamel surfaces, improved durability, esthetics, and biocompatibility. Nanostructured dental composites can have superior mechanical properties (e.g., increased elastic modulus, strength, or resistance to fatigue fracture) that can easily be tuned by small modifications of their building blocks.

References

1. Abron, A., Hopfensperger, M., Thompson, J., and Cooper L.F. (2001). Evaluation of a predictive model for implant surface topography effects on early osseointegration in the rat tibia model, *J. Prosthet. Dent.*, **85**, pp. 40–6.
2. Adamek, G., and Jakubowicz, J. (2010). Mechanoelectrochemical synthesis and properties of porous nano Ti–6Al–4V alloy with hydroxyapatite layer for biomedical applications, *Electrochem. Commun.*, **12**, pp. 653–656
3. Berglundh, T., Abrahamsson, I., Albouy, J.P., and Lindhe, J. (2007). Bone healing at implants with a fluoride-modified surface: an experimental study in dogs, *Clin. Oral Implants Res.*, **18**, pp. 147–152.
4. Boyer, R., Welsch, G., and Collings, E. (1994). *Materials Properties Handbook: Titanium Alloys*, ASM International, Materials Park, OH.

5. Brunette, D.M., Tengvall, P., Textor, M., and Thomsen, P. (2003). *Titanium in Medicine*, Springer-Verlag Berlin, Heidelberg.
6. Coelho, P.G., and Suzuki, M. (2005). Evaluation of an IBAD thin-film process as an alternative method for surface incorporation of bioceramics on dental implants, *J. Appl. Oral.*, **13**, pp. 87–92.
7. Cooper, L.F., Zhou, Y., Takebe, J., Guo, J., Abron, A., Holmen, A., and Ellingsen, J.E. (2006). Fluoride modification effects on osteoblast behavior and bone formation at TiO_2 grit blasted c.p. titanium endosseous implants, *Biomaterials*, **27**, pp. 926–936.
8. Cooper, L.F., Masuda, T., Whitson, S.W., Yliheikkila, P., and Felton, D.A. (1999). Formation of mineralizing osteoblast cultures on machined, titanium oxide grit-blasted, and plasma-sprayed titanium surfaces, *Int. J. Oral Maxillofac. Implants*, **14**, pp. 37–47.
9. Davarpanah, M., Martinez, H., Etienne, D., Zabalegui, I., Mattout, P., Chiche, F., and Michel, J. (2002). A prospective multicenter evaluation of 1,538 BIOMET 3i implants: one to five year data, *Int. J. Oral Maxillofac. Implants*, **17**, pp. 820–828.
10. Donati, M., La Scala, V., Billi, M., Di Dino, B, Torrisi, P., and Berglundh, T. (2008). Immediate functional loading of implants in single tooth replacement: a prospective clinical multicenter study, *Clin. Oral Implants Res.*, **19**, pp. 740–748.
11. Elias, C.N., Lima, J.H.C., Valiev, R., and Meyers, M.A. (2008). Biomedical applications of titanium and its alloys, *JOM, J. Miner. Metals Mater. Soc.*, **60**, pp. 46–49.
12. Ellingsen, J.E., Johansson, C.B., Wennerberg, A., and Holmen, A. (2004). Improved retention and bone-to-implant contact with fluoride-modified titanium implants, *Int. J. Oral Maxillofac. Implants*, **9**, pp. 659–566.
13. Ellingsen, J.E. (2005). Pre-treatment of titanium implants with fluoride improves their retention in bone, *J. Mater. Sci.: Mater. Med.*, **6**, pp. 749–53.
14. Ellingsen, J.E., Thomsen, P., and Lyngstadaas, S.P. (2006). Advances in dental implant materials and tissue regeneration, *Periodontol. 2000*, **41**, pp. 136–156.
15. Erbel, R., Di Mario, C., Bartunek, J., Bonnier, J., de Bruyne, B., Eberli, F.R., Erne, P., Haude, M., Heublein, B, Horrigan, M, Ilsley, C, Bose, D, Koolen, J, Luscher, TF, Weissman, N, Waksman, R. (2007). Temporary scaffolding of coronorary arteries with bioabsorbable magnesium stents: a prospective, non-randomized multo-centre trial, *Lancet*, **369**, pp. 1869–1875.

16. Ericsson, I., Johansson, C.B., Bystedt, H., and Norton M.R. (1994). A histomorphometric evaluation of bone-to-implant contact on machine-prepared and roughened titanium dental implants. A pilot study in the dog, *Clin. Oral Implants Res.*, **5**, pp. 202–206.

17. Estrin, Y., Kasper, C., Diederichs, S., and Lapovok, R. (2008). Accelerated growth of preosteoblastic cells on ultrafine grained titanium, *J. Biomed. Mater. Res. A*, **90**, pp. 1239–1242.

18. Fandridis, J., and Papadopoulos, T. (2008). Surface characterization of three titanium dental implants, *Implant Dent.*, **17**, pp. 91–99.

19. Goene, R.J., Testori, T., and Trisi, P. (2007). Influence of a nanometer-scale surface enhancement on de novo bone formation on titanium implants: a histomorphometric study in human maxillae, *Int. J. Periodontics Restorative Dent.*, **27**, pp. 211–219.

20. Godarzi, R., Rasmusson, L., Dasmah, A., and Albrektsson, T. (2008). Effects of implant design and surface on osseointegration. An experimental study in the dog mandible, *Appl. Osseointegration Res.*, **7**, pp. 58–60.

21. Gotfredsen, K., Nimb, L., Hjorting-Hansen, E., Jensen, J.S., and Holmen, A. (1992). Histomorphometric and removal torque analysis for TiO_2-blasted titanium implants. An experimental study on dogs. *Clin. Oral Implants Res.*, **3**, pp. 77–84.

22. Gotfredsen, K., Wennerberg, A., Johansson, C., Skovgaard, L.T., and Hjorting-Hansen, E. (1995). Anchorage of TiO_2-blasted, HA-coated, and machined implants: an experimental study with rabbits, *J. Biomed. Mater. Res.*, **29**, pp. 1223–1231.

23. Guo, J., Padilla, R.J., Ambrose, W., De Kok, I.J., and Cooper, L.F. (2007). Modification of TiO_2 grit blasted titanium implants by hydrofluoric acid treatment alters adherent osteoblast gene expression *in vitro* and *in vivo*, *Biomaterials*, **28**, pp. 5418–5425.

24. Han, C.H., Johansson, C.B., Wennerberg, A., and Albrektsson, T. (1998). Quantitative and qualitative investigations of surface enlarged titanium and titanium alloy implants, *Clinical Oral Implants Res.*, **9**, pp. 1–10.

25. Hansson, S. (2000). Surface roughness parameters as predictors of anchorage strength in bone: a critical analysis, *J. Biomech.*, **33**, pp. 1297–1303.

26. Hoog, C., Birbilis, N., Zhang, M.X., and Estrin, Y. (2008). Surface grain size effects on the corrosion of magnesium, *Key Eng. Mater.*, **384**, pp. 229–240.

27. Jakubowicz, J., Jurczyk, K., Niespodziana, K., and Jurczyk, M. (2009). Mechanoelectrochemical synthesis of porous Ti-based nanocomposite biomaterial, *Electrochem. Commun.*, **11**, pp. 461–465.

28. Jakubowicz, J., and Adamek, G. (2009). Preparation and properties of mechanically alloyed and electrochemically etched porous Ti–6Al–4V, *Electrochem. Commun.* **11**, pp. 1772–1775.

29. Jia, W.T., and Jin, S.H. (2002). Dental composite materials, US Patent 6417246.

30. Kaese, V., Pinkvos, A., Haferkamp, H., Niemeyer, M., and Bach, F.W. (2005). Process for producing bioresorbable implants, US Patent 6854172

31. Krivoruchko, M., Allen, J., and Birdsall, M. (2010). Bioresorbable stent, United States Patent 7651527.

32. Lee, S.H., Kim, H.E., and Kim, H.W. (2007). Nano-sized hydroxyapatite coatings on Ti substrate with TiO_2 buffer layer by e-beam deposition, *J. Am. Ceram. Soc.*, **90**, pp. 50–56.

33. Li, L.H., Kong, Y,M., Kim, H.W., Kim, Y.W., Kim, H.E., Heo, S.J. and Koak, J.Y. (2004). Improved biological performance of Ti implants due to surface modification by micro-arc oxidation, *Biomaterials*, **25**, pp. 2867–2875.

34. Lindhe, J., Karring, T., and Lang, N.P. (2003). *Clinical Periodontology and Implant Dentistry, 4th edition*, Blackwell Publishing Ltd, Oxford, UK.

35. Lops, D., Chiapasco, M., Rossi, A., Bressan, E., and Romeo, E. (2008). Incidence of inter-proximal papilla between a tooth and an adjacent immediate implant placed into a fresh extraction socket: 1-year prospective study, *Clin. Oral Implants Res.*, **19**, pp. 1135–1140.

36. Mendonça, G., Mendonça, D.B.S., Aragão, F.J.L., and Cooper L.F., (2008). Advancing dental implant surface technology — from micron to nanotopography — review, *Biomaterials*, **29**, pp. 3822–3835.

37. Mustafa, K., Silva Lopez, B., Hultenby, K., Wennerberg, A., and Arvidson, K. (1998). Attachment and proliferation of human oral fibroblasts to titanium surfaces blasted with TiO_2 particles. A scanning electron microscopic and histomorphometric analysis, *Clin. Oral Implants Res.*, **9**, pp. 195–207.

38. Niespodziana, K., Jurczyk, K., and Jurczyk, M. (2009). Mechanical and corrosion properties of titanium hydroxyapatite nanocomposites, *Solid State Phenomena*, **151**, pp. 217–221.

39. Orsini, G., Piattelli, M., Scarano, A., Petrone, G., Kenealy, J., Piattelli, A., and Caputi, S. (2007). Randomized, controlled histologic and histomorphometric evaluation of implants with nanometer-scale

calcium phosphate added to the dual acid-etched surface in the human posterior maxilla, *J. Periodontol.*, **78**, pp. 209–218.

40. Park, J.W., Kim, Y.J., Park, C.H., Lee, D.H., Ko, Y.G., Jang, J.H., and Lee, C.S. (2009). Enhanced osteoblast response to an equal channel angular pressing-processed pure titanium substrate with microrough surface topography, *Acta Biomater.*, **5**, pp. 3272–3280,

41. Petruzelka, J., Dluhos, L., Hrusak, D., and Sochova, J. (2006). Nanostructured titanium. Application in dental implants, *Sbornik vedeckych praci Vysoke skoly banske* (Technicke University Ostrava) **1**, pp. 177–185.

42. Pinholt, E.M. (2006). Surface engineered dental implant insertion in conjunction with bilateral inferior mandibular nerve transposition — a case report, *Appl. Osseointegration Res.*, **5**, pp. 59–61.

43. Rocci, M., Rocci, A., Martignoni, M., and Albrektsson, T. (2008). A comparative study of TiOblast and OsseoSpeed implants retrieved from humans, *Appl. Osseointegration Res.*, **7**, pp. 26–30.

44. Stanford, C.M. (2008). Surface modifications of dental implants, *Aust. Dent. J.*, **53**, pp. 26–33.

45. Thor, A. (2008). TiOblast and OsseoSpeed implant in sinus lift surgery, *Appl. Osseointegration Res*, **7**, pp. 17–25.

46. Toljanic, J., Thor, A., Baer, R., and Ekstrand, K. (2008). Immediate fixed restoration of implants in the atrophic edentulous maxilla, *Dent. Today*, **27**, pp. 56–63.

47. Tulinski, M., and Jurczyk, M. (2009). Mechanical and corrosion properties of Ni-free austenitic stainless steel/hydroxyapatite nanocomposities, *Solid State Phenomena*, **151**, pp. 213–216.

48. Valiev, R.Z., Islamgaliev, R.K. and Alexandrov, I.V. (2000). Bulk nanostructured materials from severe plastic deformation, *Prog. Mater. Sci.*, **45**, pp. 103–189.

49. Valiev, R.Z., Sergueeva, A.V., and Mukherjee, A.K. (2003). *Scr. Mater.*, **49**, pp. 669–673.

50. Valiev, R.Z., Estrin, Y., Horita, Z., Langdon, T.G., Zehetbauer, M.J., and Zhu, Y.T. (2006). Producing bulk ultrafine-grained materials by severe plastic deformation, *JOM, J. Miner. Metals Mater. Soc.*, **58**, pp. 33–39.

51. Valiev, R.Z., Zehetbauer, M.J., Estrin, Y., Höppel, H.W., Ivanisenko, Y., Hahn, H., Wilde, G., Roven, H.J., Sauvage, X., and Langdon, T.G. (2007). The innovation potential of bulk nanostructured materials, *Adv. Eng. Mater.*, **9**, pp. 527–533.

52. Valiev, R.Z., Semenova, I.P., Latysh, V.V., Shcherbakov, A.V., and Yakushina, E.B. (2008). Nanostructured titanium for biomedical applications: new developments and challenges for commercialization, *Nanotechnol. Russia*, **3**, pp. 593–601.

53. Valiev, R.Z., Semenova, I.P., Latysh, V.V., Rack, H., Lowe, T.C., Petruzelka, J., Dluhos, L., Hrusak, D., and Sochova, J. (2008). Nanostructured titanium for biomedical applications, *Adv. Eng. Mater.*, **10**, pp. 1–3.

54. Valiev, R.Z., Semenova, I.P., Latysh, V.V., Rack, H., Lowe, T.C., Petruzelka, J., Dluhos, L., Hrusak, D., and Sochova, J. (2009). Nanostructured titanium for biomedical applications, *Adv. Eng. Mater.*, **10**, B15 – B17.

55. Wang, H., Estrin, Y., and Zuberova, Z. (2008). Bio-corrosion of a magnesium alloy with different processing histories, *Mater. Lett.*, **62**, pp. 2476–2479.

56. Wang, K. (1996). The use of titanium for medical applications in the USA, *Mater. Sci. Eng. A*, **213**, pp. 134–137.

57. Wennerberg, A., and Albrektsson, T. (2006). Implant surfaces beyond micron roughness. Experimental and clinical knowledge of surface topography and surface chemistry, *Appl. Osseointegration Res.*, **5**, pp. 40–44.

58. Zhang, S.X., Zhang, X.N., Zhao, C.L., Li, J.A., Song, Y., Xie, C., Tao, H., Zhang, Y., He, Y., Jiang, Y, and Bian, Y. (2010). Research on an Mg–Zn alloy as a degradable biomaterial, *Acta Biomater.*, **6**, pp. 626–640.

Index

Colour Insert

Chapter 5

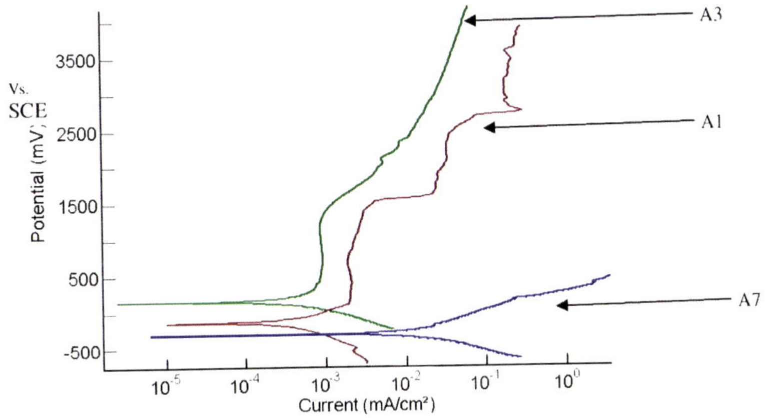

Figure 5.7

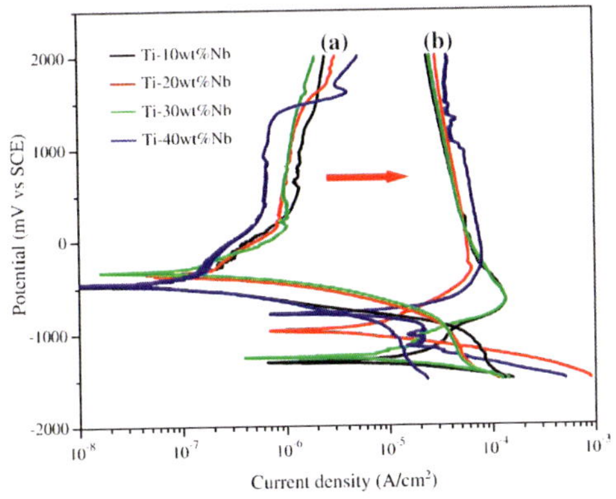

Figure 5.18

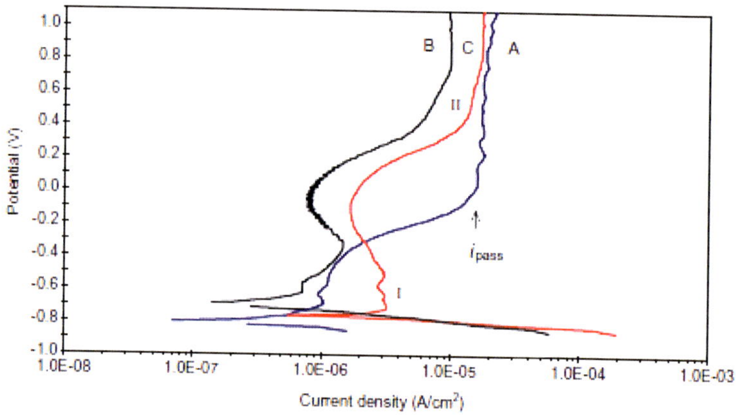

Figure 5.25

Chapter 6

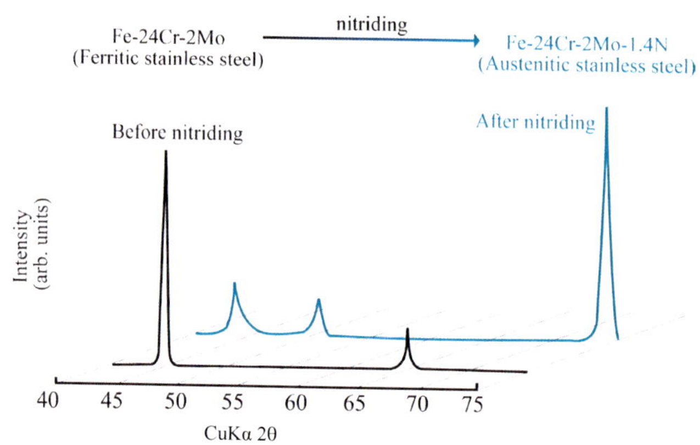

Figure 6.4

Chapter 8

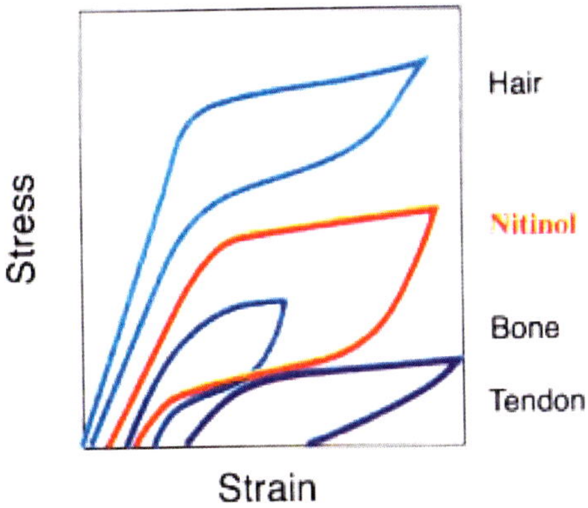

Figure 8.27

Chapter 9

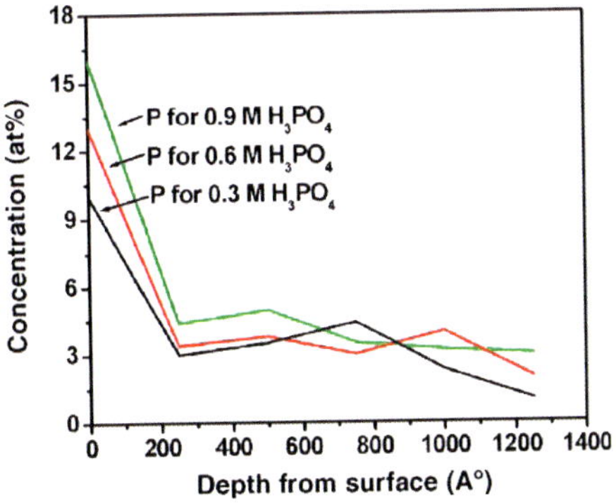

Figure 9.24

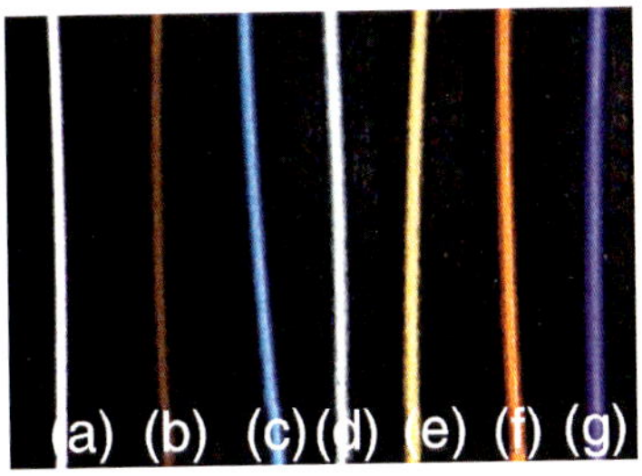

Figure 9.27

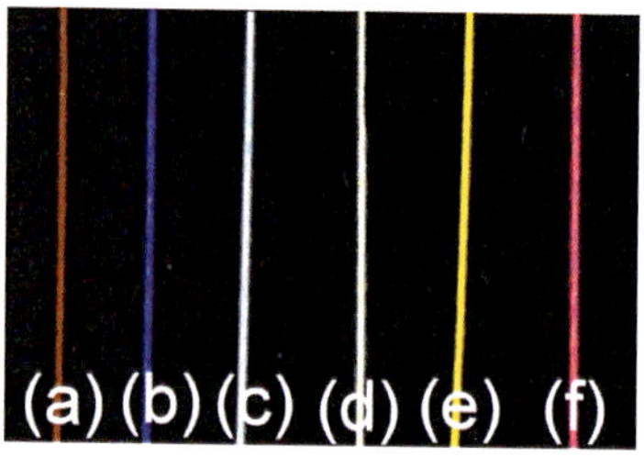

Figure 9.28

Chapter 12

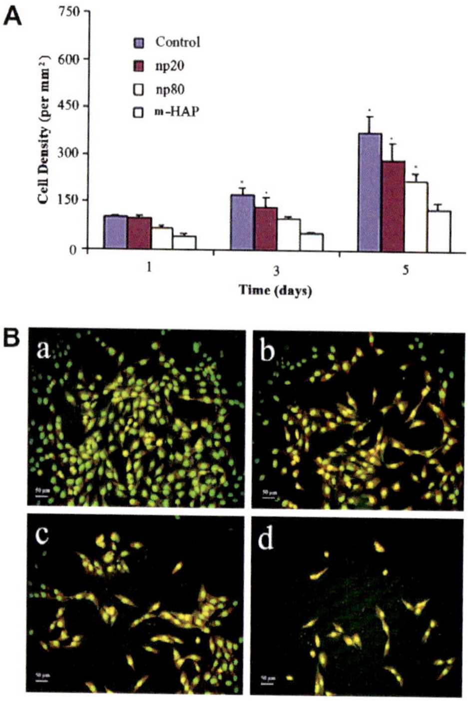

Figure 12.7